D0948211

Solving the

Naval Radar Crisis

Solving the
Naval Radar Crisis

The Eddy Test – Admission to the Most Challenging Training Program of World War II

by
Raymond C. Watson, Jr.

Foreword from
Louis Brown

Order this book online at www.trafford.com/07-2812
or email orders@trafford.com

Most Trafford titles are also available at major online book retailers.

Note for Librarians: A cataloguing record for this book is available from Library and Archives Canada at www.collectionscanada.ca/amicus/index-e.html

ISBN: 978-1-4251-6884-1

www.trafford.com

North America & international
toll-free: 1 888 232 4444 (USA & Canada)
phone: 250 383 6864 • fax: 250 383 6804
email: info@trafford.com

The United Kingdom & Europe
phone: +44 (0)1865 722 113 • local rate: 0845 230 9601
facsimile: +44 (0)1865 722 868 • email: info.uk@trafford.com

10 9 8 7 6 5 4 3 2 1

Contents

Dedicated to the memory of

William Crawford Eddy
Captain, USNR

> By his keen foresight, unwavering zeal, and meticulous attention to detail in discharging all duties pertaining to the recruiting, selection, and training of radio technicians, Captain Eddy contributed materially to the successful prosecution of the war against the enemy.
>
> *From his Legion of Merit Citation, December 1945*

Foreword

A Radar History of World War II (Institute of Physics, 1999) is widely regarded as the most definitive book available on the origins and early military applications of this electronic marvel. It was written by Dr. Louis Brown, a renowned scientist and historian at the Department of Terrestrial Magnetism of the Carnegie Institution of Washington. Dr. Brown was well known in the physics/geochemistry community for building instruments to explore nuclear phenomena, but his passion was radar history. His scholarly books and articles in this area give an unbiased insight of the worldwide developmen of this technology.

When the major sections of *Naval Radar Crisis* were drafted, Dr. Brown consented to read and criticize this work. His detailed comments gave guidance to significant rewrites, but, of equal importance, he encouraged expanding the coverage. Unfortunately, his sudden and unexpected death occurred before being invited to write a Foreword for this book. However, his wife, Lore Elizabeth Brown, has graciously consented for this Foreword to center on some of his earlier comments.

The electronics training program, developed as a solution to the crisis in naval radar mantenance at the beginning of World War II, was an important contribution, one that I completely missed in my book. It is an omission that I greatly regret, especially because it is the kind of history that I value, but I had not a hint of the Eddy Test or anything related. This well-researched book finally provides a thorough coverage of this very intense, long overlooked program.

The background material serves as a good, brief introduction to radar history, one that is free of the great radar myths that still fill many accounts: 'Before Rad Lab there was nothing.' 'We invented it in Britain and everyone copied it from us.' 'German radar was second rate and the Japanese did not have any.'

Particularly shown is the great, but generally ignored, importance of the Naval Research Laboratory and the Army Signal Corps in preparing the United States for that critical year, 1942.

The book fills a history gap between electronic foundations and hardware operations, the gap covering the selection and training of personnel for the highly critical maintenance function.

Louis Brown
Carnegie Institution of Washington

"Flying Bedspring" Antenna of the SK Radar –
The Navy's Standard Early-Warning System of WWII.

Prolog

In January 1939, the United States Navy was conducting maneuvers and battle practices in the Caribbean. Under darkness of night, several unlit destroyers were attempting a simulated torpedo run on the battleship USS *New York.*

A large bedspring-like antenna atop the battleship's conning tower moved slowly back and forth, scanning the dark sea like a lookout. A group of men in Air Plot peered intensely at a small fluorescent screen. Here was displayed a scratchy green line traced by a glowing point wavering along a scale showing distance.

A slight hump suddenly flickered on the line. The excited observers stopped the antenna and watched as the hump became taller and moved slowly down the scale. When the distance showed 5,000 yards, a searchlight was turned on in the direction faced by the antenna, illuminating the oncoming destroyers.

Operational radar had joined the U.S. Navy!

This occurred just 40 years after the first demonstration of radio in the Navy. In 1899, Guglielmo Marconi used his new wireless telegraphy to communicate between the cruiser USS *New York* and the battleship USS *Massachusetts,* some 35 miles apart. It was also only 24 years after the Navy opened the Radio Division (within the Bureau of Steam Engineering!) and started converting to electronic (vacuum tube) communication systems.

Radio had evolved amazingly during the four decades of its existence. While communications dominated its development within the Navy, electronics – the outgrowth of radio – also brought forth other important Navy equipment such as underwater sound detectors (sonar), navigational aids, and remotely controlled aircraft. Radio detection and ranging – radar – represented the climax of radio evolution, and by far the most technically complex.

In December 1934, Robert M. Page, A. Hoyt Taylor, and Leo C. Young at the Naval Research Laboratory tested the world's first "breadboard" radar, detecting a plane at a distance of one mile. And it was secret; the technology became tightly controlled in June 1937 and remained so throughout World War II.

Recognizing that the maintenance of electronic equipment required men with highly specialized training, in 1924 the Navy opened the Radio Materiel School (RMS) at the Naval Research Laboratory in Washington, D.C. There a few dozen carefully selected Electricians and Radiomen each year pursued an intense, six-month course in the theory and repair of this increasingly complicated electronic equipment. The training, however, was predominantly in communication systems – receivers and transmitters. When radar first entered the fleet, no instruction in this technology was yet given by the RMS.

As America prepared for war, enormous additions in hardware and manpower were planned. In 1939, the Navy had about 150,000 officers and enlisted men, and some 400 ships. At entry into war, this would need to quickly expand to at least 500,000 personnel and 2,000 ships. Industry would need to build aircraft at the rate of 50,000 per year. With the successful development of radar, it was ordered that all larger vessels be outfitted with this equipment, but when the war started, there were only 79 sets in the fleet.

How would the massive numbers of communication, sonar, radar, and other electronic equipment be maintained? Obviously, if the new marvels of electronics were inoperative, they had no value in warfare. Thousands, not dozens, of highly competent technicians must be trained, and on equipment that was mainly classified and still evolving. An additional complication was the introduction of microwave technologies – totally new to naval equipment, with the first microwave radars introduced in mid-1941. The existing RMS was totally inadequate for this effort, and rapid, large-scale replication of this school was not practical. The most urgent problem, however, concerned personnel, both instructors and students – how could the Navy obtain the large number of persons needed for this program?

There was a crisis in electronics maintenance, especially in radar.

The solution involved the development and operation of the most intense and intellectually difficult training program ever given by the military in America. Amazingly, there is almost no documentation concerning this activity. But this training, its creative originators, and the students involved deserve recognition in modern history. This book is an attempt to correct the deficiency.

The Naval Archives have essentially nothing on the overall training program. The program is also absent in periodicals and other media. The title of an article in the March 7, 2000, *Corpus Christi Caller-Times*, "Ward Island was a hush-hush radar school," indicates one reason for the lack of public documentation – secrecy. (Ward Island was a school in the program.) On the other hand, the development and use of radar itself was highly classified, but, nevertheless, there was much in the post-war literature on this accomplishment.

In the scant documentation, credit for this program is most often given to William C. Eddy, a medically retired (deafness) naval officer who came back to lead in its implementation. The examination (the Eddy Test) used for trainee selection carries his name. The program was conceived and implemented in just a few days by a small, *ad hoc* group while the Nation was reeling from the December 7th attack. In addition to Eddy, other key members of this founding group included Nelson M. Cooke, Wallace J. Miller, and Sidney R. Stock. Captain Louis E. Denfeld, just given the responsibility for all training in the Navy, approved the program initiation on January 7, 1942.

Radar *per se* in World War II has been extensively documented – often erroneously attributing its origin to Robert A. Watson Watt in Great Britain. While Watson Watt (along with Arnold F. Wilkins and Edward G. Bowen) did indeed independently develop radar and contributed greatly to this technology, the first demonstratio in Great Britain was several moths behind that at the Naval Research Laboratory.

Louis Brown's *A Radar History of World War II*, (Institute of Physics Publishing, 1999), is perhaps the most definitive book available on the evolution of radar, but it does not include even a mention of the maintenance and associated training. After reviewing an early draft of the present book, Dr. Brown included the following in his comments:

> I completely missed [the electronics training program] in my book. It is an omission that I greatly regret, especially because it is the kind of history that I value, but I had not a hint of the Eddy Test or anything related.

Was radar that important in the war effort? The atomic bomb is generally the U.S. development most identified with this conflict, with about two billion then-year dollars being applied. Radar development and implementation, however, was an even larger endeavor, with some *three* billion dollars expended. Almost a million radar sets of some 150 types were produced in America.

The Battle of Midway on June 6, 1942, first showed the value of radar. A superior Japanese naval force was devastated by the United States, with the enemy losing four carriers and 300 planes to a single American carrier and 100 planes. The Navy's radar contributed greatly to the success of this battle, and its continued use was a major factor in leading to victory. It has been said that radar won the war, and the atomic bomb brought the peace.

There is an interesting parallel between the evolution of radar and the atomic bomb. Einstein's $E = mc^2$, likely the best-known science equation in the world, was first published in September 1905. The atomic bomb, the most notable application of this equation, was first detonated 40 years later. As previously noted, it was also 40 years between the first radio demonstration aboard a U.S. warship and the first sea-based operation of American radar, certainly one of the most significant applications of radio.

The evolution of these highly diverse technologies followed similar timelines, with modest, open progress for over three decades – mainly with no understanding of what the ultimate result would be – followed by highly secret, intensive efforts that led to the two most significant wartime products. These technologies remained apart for most of the war, but finally came together on August 6, 1945, when a strapped-on radar provided the altitude trigger for the atomic bomb dropped on Hiroshima. It is interesting to note that *Time* magazine planned a cover story on radar – particularly the developments at MIT's Radiation Laboratory – for their first weekly issue in August, but this was upstaged by news of the atomic bomb and Hiroshima.

Like the atomic bomb, very few people knew about the development of radar. Work on both programs was carried out by close-knit teams whose sole objective was to make their product successful. The fruition of the atomic-bomb effort was event-oriented – success was the detonation of a few devices. Conversely, success in radar was measured by an operational continuum, and this required ongoing maintenance of this complex apparatus. The engineers and scientists who brought radar into being certainly knew how it should be maintained, but almost no one else had this knowledge.

To appreciate the nature of the electronic maintenance crisis requires going back and examining the four decades of radio and electronics evolution, particularly in the U.S. Navy. Chapter 1 describes activities during the first two decades (through World War I). In this period, the developments and applications of radio were essentially all in the communications field, starting with spark-gap transmitters and electro-mechanical receivers exchanging messages in Morse code at the turn of the century, and progressing into vacuum-tube, voice-modulated systems – including superheterodyne receivers – as the second decade closed. For America's fledgling radio industry, the Navy was the primary customer.

The period between the World Wars is covered in Chapter 2. The 1920s are sometimes called "The Golden Age of Radio," reflecting the emergence of broadcasting. The electronics industry flourished, fed by the public's fascination with this new entertainment medium. America's military benefited from the technical developments, but, no longer the prime market for industry, they opened their own research organizations – the Naval Research Laboratory and the Army's Signal Corps Radio Laboratory. This surge in commercial and military radio development essentially halted when the Great Depression began in late 1929.

Fortunately, the Navy had initiated a major upgrading of communication equipment in the late 1920s. In spite of the financial difficulties of the Great Depression, deliveries at the start of the 1930s enabled the Navy to soon boast the most modern communication system in the world. To a large measure, the "winds of war" cleared the residual effects of the depression; there was a worldwide resurgence in electronic development. Amateur radio operators became a large market for new components, and mechanical television became a novelty. A major section of Chapter 2 is devoted to the status of electronics technology as preparations for World War II began.

By the early 1930s, researchers throughout the world were aware of the basic concepts of target detection by radio. As the potential military applications of this technology were recognized, the development left the open scientific realm and went "undercover" in the government laboratories. Chapter 3 is devoted to the origin of radar in the United States and Great Britain. This goes back to the precursors of this technology, and then examines the early, very modest developments in the Navy and Army laboratories. Here the same scientists, engineers, and technicians who developed the equipment would also provide the maintenance. When serious preparations for war began, financial barriers were lifted and the development of radar accelerated. Recognition also began that specialty-trained maintenance personnel would be needed.

It is noted that the name "radar" only came into being in 1940, stemming from the acronym RADAR (Radio Detection and Ranging). Earlier, the technology was called radio detection, and in Great Britain it remained being called RDF (Radio Direction Finding) for some time. Although the word radar was never classified, the basic technology was secret, as was that of the RDF in Great Britain

In the pre-war years, there were no exchanges between the United States and Great Britain concerning classified technologies. After Great Britain went to war, it was vital for that nation to have access to the extensive natural resources and industrial capacity of America. President Franklin Roosevelt and Prime Minister Winston Churchill agreed that the time had come for full mutual disclosures of their national secrets.

The Tizard Mission came to the United States in September 1940, and, during exchanges of technical information, the two nations were amazed at the relative developments taking place in the other. It was in these exchanges that the resonant-cavity magnetron was revealed by Great Britain, providing the long-sought solution to microwave generation. The Radiation Laboratory was established at MIT for developing microwave radar.

While radar was the most recognized electronic system of World War II, it was just one of a number of very important electronic weapons contributing to the Allied success; Appendix I provides a summary of some of these. Modern radio communications emerged in the Navy during and just after World War I, but it was only in the 1930s that electronics technology really matured. Underwater sound detection (later called sonar) also started during World War I, and by the 1930s had evolved to a good level of application. Other wartime electronic systems included electronic countermeasures, recognition systems (IFF), direction-finders, radio-beacons, magnetic anomaly detectors, remotely controlled aircraft, radio-navigation aids, and proximity fuses.

For completeness, radar development throughout the world is summarized in Appendix II. The reader may be surprised to learn that in addition to the United States and Great Britain, there were significant radar programs in at least nine other countries: Germany, Japan, The Soviet Union, The Netherlands, Italy, France, Australia, Canada, and South Africa. Once the Germans had a demonstrated system – as in Great Britain, it followed that of the United States by a short period – they made rapid progress and were ahead of the Allied nations at the start of the war. Their lead, however, was not maintained; work in this technology had a low priority, primarily because radar was considered a defensive weapon and Hitler thought only in terms of aggression.

Very little of the material covered in Chapters 1, 2, and 3, and in Appendices I and II is original; it may be found in a huge number of sources. The contribution of the author to this information is to relate the individual developments to the whole and show how they led to the electronic maintenance crisis in the Navy at the start of the war, as well as the complexity of maintenance during the ensuing conflict.

Chapters 4, 5, and 6 cover the details of the maintenance crisis and its solution with the Electronics Training Program (ETP) – a named coined by the author in the absence of any official name for the program given by the Navy. Insofar as is known by the author, these three chapters present for the first time a concise and complete treatment of this very important activity of World War II. Unlike the previous described chapters and appendices, much information in chapters 4, 5, and 6 cannot be found in other, readily available sources.

A detailed description of the crisis and the approach to its solution is given in Chapter 4. Information on the founding team – William C. Eddy *et al* – is included. The ETP is fully described in Chapter 5, including the organization and changes over the brief time of its existence. The Eddy Test – the passport to this excellent training – is covered in this chapter, as are the operational details of Pre-Radio, Primary, and Secondary

Schools. For readers only interested in the training program *per se*, this is the chapter.

The schools, facilities, and leaders of the ETP are detailed in Chapter 6. Six engineering colleges distributed across the United States played an extremely important role in the Program. These were Bliss Electrical School (Maryland), Grove City College (Pennsylvania), Oklahoma A&M College, Texas A&M College, The University of Houston, and Utah State College of Agriculture. It was only through the outstanding cooperation of these schools in providing the instructional staff and facilities that the program was so successful. Radio Chicago and other Navy-operated schools are also covered in this chapter.

A brief Epilog covering some post-war events has been included. Also, to show the great diversity of backgrounds of people who completed the program, their Naval activities, and their following careers, brief biographical sketches on representative students are given in Appendix III. The reader might better appreciate the other sections by first examining this Appendix.

I was both a student and later an instructor in this Naval training activity, and have included much from my personal knowledge. I also found historical information hidden away in library newspaper and magazine files, college and university archives, and some from the Naval Historical Center. A major amount, however, has come from a multitude of former Electronic Technicians and Aviation Electronic Technicians, as well as a few other instructors, who have searched their memories and treasured collections of notes, letters, and other memorabilia, to provide me with invaluable information. To all of these persons, I am eternally grateful.

Thank God for the Internet, the World Wide Web, and e-mail, and for His allowing me to complete this endeavor.

Raymond C. Watson, Jr.
Huntsville, Alabama

THE NAVY NEEDS YOU IN
RADAR!
The war's newest and most
dramatic technical development

Chapter 1

RADIO AND RELATED TECHNOLOGIES

Following the discovery of electromagnetic induction by Michael Faraday in the 1830s, electromagnetic waves were mathematically modeled by James Clerk Maxwell, experimentally demonstrated by Heinrich Rudolf Hertz, and, in 1895, shown to be a practical means of communications by Guglielmo Marconi. This was truly an international accomplishment – Faraday was in England, Maxwell in Scotland, Hertz in Germany, and Marconi in Italy – but it was in the technical and business environment of the United States that radio truly evolved. The following sections cover this development in America, with a particular emphasis on the U.S. Navy.

RADIO BEGINNINGS

Guglielmo Marconi (1874-1937), whose mother was Irish and father was an Italian country gentleman, was privately educated and began electromagnetic radiation experiments at the family's estate near Bologna. In 1895, he built an apparatus that sent a wireless telegraphy signal a little over a mile. The next year, 22-year-old Marconi took his wireless system to England, where he met with the chief engineer of the General Post Office and successfully repeated his previous transmissions.

Guglielmo Marconi

The apparatus included a spark-gap transmitter and a coherer receiver, neither of which Marconi personally invented, but the combined system was granted a British patent for "improvements in transmitting electrical impulses and signals and an apparatus therefor." A short notice, "Telegraphy Without Wires," in the January 23, 1897 *Scientific American*, stated:

> A young Italian, a Mr. Marconi, has recently demonstrated to the London Post Office the ability to transmit radio signals across three-quarters of a mile, and if the invention was what he believed it to be, our mariners would have been given a new sense and a new friend which would make navigation infinitely easier and safer than it now was.

In July 1897, Marconi formed the Wireless Telegraph & Signal Company, soon re-named Marconi's Wireless Telegraph Company and having the stated purpose of attaining a worldwide monopoly in that field. In 1899, he placed his equipment aboard three ships of the British Royal Navy, and, in tests at sea, exchanged telegraph messages at a distance of 74 nautical miles.

Damped Wave Radio

Marconi's success came as no surprise to many people throughout the world; inventors at that time were searching for means of improving on the Morse telegraph by developing a wireless version. As the name implies, this would involve transmission of telegraph signals without the use of wires. Several means could be used to "send" information, including magnetic induction, conduction, electrostatic coupling, as well as "Hertzian" waves.

In the U.S., patents for elementary types of wireless communication apparatuses had been granted to a number of researchers, including Mahlon Loomis (1872), Amos E. Dolbear (1882 and 1886), Thomas A. Edison (1891), and Isidor Kitsee (1895). All of these inventions, however, involved means other than the generation and reception of electromagnetic waves.

Loomis, a dentist and amateur inventor, in 1866 claimed the transmission of signals 14 miles, using kites to hold copper wires aloft. His apparatus consisted of a telegraph key that connected the "transmitting" wire to ground, while the "receiving" wire was connected to ground through a sensitive galvanometer. Loomis reported that with the key open, the galvanometer gave a reading indicating the flow of atmospheric electrical charges into the ground, and when the key was closed, the galvanometer gave a downward swing. According to Loomis, this was because the atmospheric electricity was decreased, implying that this was a conductive device. Although he received a patent on the wireless telegraphy device in 1872, its operation was never independently verified.

Dolbear, a physics professor at Tufts College, performed experiments using an inductive system, transmitting a little over 1,000 feet. His 1885 patent, although not exploited into practical applications, kept Marconi from operating in the U.S. until after he bought Dolbear's patent. The patents of Edison and Kitsee were also for "induction" devices.

In keeping with its traditional scientific leadership, the U.S. Navy had early foreseen the potential impact of wireless communications on naval operations; only by such means could far-flung forces be effectively directed. Toward this realization, Lieutenant (later Rear Admiral) Bradley A. Fiske had experimented with induction wireless devices aboard American naval vessels in 1888.

Following Marconi's naval demonstrations in Great Britain, arrangements were quickly made for similar tests aboard U.S. Navy ships. In late 1899, Marconi demonstrated wireless telegraphy between the cruiser USS *New York* and the battleship USS *Massachusetts*, some 35 miles apart. These tests were conducted during the time of the Spanish-American War. In this conflict, Admiral George Dewey almost met with disaster by depending on signal lights for communications while invading the Manila Bay. *The New York Herald*, having followed the U.S. wireless tests, made the following prediction in its January 21, 1900 issue:

> The day of the flag and lamp signaling system in the Navy is drawing to a close. The Dewey of the next war, instead of signaling the course to be pursued by means of lights . . . will send out electric waves.

After the successful demonstrations, the Navy requested Marconi to quote for supplying 20 wireless sets. He responded by establishing the Marconi Wireless Telegraph Company of America (hereafter called Marconi America) and offering to lease the equipment to the Navy for an unreasonably high price. After rejecting Marconi's monopolistic offers, the Navy decided to carefully study the situation, seeing what new equipment might be available from European and American sources.

During this period, the French and German navies installed wireless equipment manufactured in their own countries, while the Italian and British navies used Marconi equipment. The British Army installed a Marconi system in South Africa, and the Russian Navy began using equipment obtained from France.

In America, a number of entrepreneurial inventors set up shop, including Harry E. Shoemaker, who in September 1899 established the American Wireless Telephone & Telegraph Company, the first radio firm in the United States. Other notable pioneers included Dr. John S. Stone, William J. Clarke, Dr. Lee de Forest, Reginald A. Fessenden, and Walter W. Massie. The well-known and controversial Nikola Tesla was also involved, but he primarily envisioned his system as a means for transmitting electrical power, not for communications.

The initial wireless transmitters used a high-voltage transformer to generate a spark, interrupted by a "chopper" a few hundred times per second. The spark current passed through a coil, the ends of which were coupled to the antenna and ground. The antenna and ground served as a capacitor (then called a "condenser"), providing a resonant combination with the coil. This resulted in a natural resonant frequency, but the inherent resistance of the circuit prevented this from being narrow (there was a low "Q"). Since the input was pulsed, the radiated wave was "damped" – dying out between pulses. The result was an electromagnetic signal centered at some frequency, but broadly spread over the spectrum. It is noted that the transmitters did not initially use internal capacitors for resonating, but they were soon added in the form of Layden jars – glass covered inside and out with foil.

To receive the signals at a distance, an electro-mechanical device, called a coherer detector, was connected between an antenna and the ground. This rapidly interrupted the arriving radiation, producing a "buzzing" audio signal in headphones (then called "telephone receiver") or closing a secondary electric-buzzer circuit. The coherer detector was invented by Frenchman Edouard Branly in 1890. Such a spark-transmitter and coherer-detector system was suitable only for Morse code telegraphy.

Perhaps the best-known early demonstration of long-distance wireless occurred on December 12, 1901. At Poldhu, Cornwall, Marconi used a transmitter 100 times more powerful than his earlier unit. He set up his receiver at the trans-Atlantic cable facility on St. John's, Newfoundland. Here Marconi repeatedly received the letter "S" (three dots in Morse code) sent across the Atlantic, a distance of 2,100 miles, using what he believed to be a 850-kHz transmission. In doing this, Marconi showed that electromagnetic waves could "bend" around the Earth, thus making this technology useful for long-distance communications. Prior to this, it was believed that these waves, like light rays, always traveled in straight lines and could not be detected beyond the horizon.

In 1902, Arthur E. Kennelly and Oliver W. Heaviside independently proposed that a layer of ions (charged atoms and molecules) was above the atmosphere and reflected certain radio waves back to earth. Successive reflections between this layer and the Earth's surface could bounce a signal across the Atlantic. Initially called the Kennelly-Heaviside Layer, this is now known as the ionosphere. Depending upon a variety of factors, there is a critical frequency above which the waves are not reflected. Under the time and conditions described for Marconi's first trans-Atlantic transmission, it is likely that an 850-kHz signal would have been above this critical number, indicating that the actual frequency was lower.

With Marconi's first transmitters, little was known about the relationships between the antenna and transmitter coil, which together set the transmitted central frequency. Receivers were untuned, responding to signals at any frequency; the best that could be done was to use an antenna and coil similar to those of the transmitter. There were no means of directly measuring the transmitted frequency. Consequently, the antennas – usually a parallel array of long wires – were likely very mismatched to the transmitter coil, and therefore markedly affected the subsequent frequency. Also, as noted earlier, the radiation was broadly spread over the spectrum.

In fact, the term "frequency" was not used at that time – the radiated signal was designated by "wavelength." Wavelength (λ) and frequency (f) are related in that their product equals the velocity of wave propagation – the speed of light. For electromagnetic waves in the atmosphere, the useful equation is then:

$$\lambda \text{ (meters)} = 300{,}000 \;/\; f \text{ (kHz) or } 300 \;/\; f \text{ (MHz)}$$

It should be noted that the earlier unit of measure for frequency was cycles per second (cps), leading to the terms kilocycles and megacycles, commonly abbreviated as kc and Mc, respectively. Hereafter, for consistency, frequency units in hertz (Hz) will be used (1 Hz ≡1 cps).

Fearing competition from wireless telegraphy, the Anglo American Cable Company, owners of the existing trans-Atlantic undersea cable, ordered the Marconi Company to shut down operations in Newfoundland. The Canadian government then offered the Marconi Company a location for a station at Glace Bay, Nova Scotia, and by the end of 1902, this station was ready to send and receive trans-Atlantic messages. Marconi convinced President Theodore Roosevelt to take part in a long-distance wireless demonstration, and the first two-way transatlantic wireless communication took place on January 19, 1903, between President Roosevelt and the King of England, Edward II.

As plans progressed to introduce wireless telegraphy systems into the Navy, consideration was given to the needs for operators and maintenance personnel for this new type of equipment. Seamen rated as Signalmen had traditionally operated the visual communication systems (flags and signal lights), and Electrician's Mates maintained the electrical equipment, but there were no personnel – officers or enlisted men – experienced or trained in wireless technology.

In October 1901, Commander F. M. Barber, USN (retired), then residing in Paris, was called to active duty for the purpose of studying and making a report on European radiotelegraph apparatuses. Two highly qualified enlisted men, Chief Electrician's Mates James H. Bell and William C. Bean, were sent abroad to study the European equipment and prepare instructions on its operation and maintenance. The knowledge and experience that they gained was to later prove of great value in installing the equipment and in instructing Navy personnel in its use, care, and maintenance. Thus, these two men have the distinction of being the first radio technicians in the U.S. Navy

Barber's initial report stated that all the continental navies maintained a high degree of secrecy and recommended that Navy should not waste more time in determining which apparatus to purchase. Based on this report, the Navy ordered for comparison purposes three sets each from Slaby-Arco, Braun-Siemens-Halske,

Ducretet, and Rochefort; the first two being German companies and the last two French. Marconi had again refused to sell equipment to the Navy.

As the equipment was being purchased, the Navy's Bureau of Equipment planned for test stations to be established at the Washington Navy Yard and at the U.S. Naval Academy, Annapolis, Maryland. In February 1902, the Bureau Chief informed the Secretary of the Navy that,

> While the fleet was adequately supplied with signalmen well versed in wigwag and in the operation of Ardois night-lights, and with able electricians, the problems involved in the care and operation of radio equipment were more complicated. . . . that it would be necessary to employ at each station a competent person to act as operator and instructor, who should be an educated electrician, skilled in the care and adjustment of delicate electrical apparatus.

It was also suggested that a special rating be created for radio operators, and that this rating be granted only after careful training and demonstration of competence. These recommendations, however, were not followed for many years; it was not until 1922 that the rating of Radioman finally came into being, and the first full school for training Electricians in radio was not opened until 1924.

By the spring of 1903, tests indicated that the German-built Slaby-Arco equipment gave superior performance, and the Bureau ordered 45 additional sets. This gave the Navy a total of 57 shipboard and shore-based wireless systems. Requirements were issued stating that vessels being constructed must accommodate wireless antennas. Over the next several years, sets were also purchased from Telefunken in Germany, and several U.S. firms. Pioneers de Forest, Fessenden, Massie, Shoemaker, and Stone eventually had their spark-transmitter designs being used by the Navy. Clarke built the Navy's first portable wireless set for the newly established base in Guantanamo, Cuba.

The early 1900s saw the commercial use of wireless telegraphy increasing rapidly throughout the world. The Marconi Company interests were fast establishing their monopoly by constructing shore stations in all the principal maritime countries. These stations were prohibited from handling messages from ships that did not lease Marconi Company equipment, thus making it undesirable for ship owners to use sets from other manufacturers.

To give credibility to his technology, Marconi obtained the consultancy of Dr. John A. Fleming, who in 1885 had established the electrical engineering department at University College London, the first such program in England. Through lectures and publications, Fleming declared the superiority of damped waves from the spark-gap units. This was particularly emphasized in his book, *The Principles of Electric Wave Telegraphy* (1906), the first full treatment of this subject.

By the end of 1903, in the U.S. there were 75 commercial stations constructed or in planning by Marconi Wireless Telegraph Company of America (Marconi America), De Forest Wireless Telegraph, International Telegraph, National Electric Signaling, Stone Telephone & Telegraph, and Massie Wireless Telegraph. A number of shipping companies had installed equipment for communicating with their vessels. *The New York Herald* had even placed a wireless system on a lightship to signal the arrival of steamers. Following the 1900 hurricane in Galveston – the greatest natural disaster in American history with over 8,000 persons killed – the U.S. Weather Bureau (then a unit of the Department of Agriculture) started an extensive wireless network for meteorological reporting.

Early Wireless Station

The Navy Department had 20 shore stations in operation and planned more than 5 times this number for the near future. The Army Signal Corps was establishing a network connecting the major posts throughout the country. Many of these commercial and government systems were concentrated along the east coast and had sufficient power as to present severe interference problems. Also, amateurs with their home-constructed equipment began increasing by the scores, and the interferences created by them in metropolitan areas posed additional problems.

From the start, operation of wireless stations in America was completely free and unrestricted. There was no regulation and no coordination, not even within the Government units using this equipment. Although some form of regulation had been advocated since the turn of the century, in this Nation, dedicated to the philosophy of free enterprise, it was slow in materializing.

In 1903, the problem of interference and need for control was expressed by the Chief of the Bureau of Equipment to the Secretary of the Navy, and from there it went to President Theodore Roosevelt, pointing out that foreign governments were already exercising such controls. An Interdepartmental Board of Wireless Telegraphy was appointed by the President to investigate the situation and make recommendations. Based on the Board's findings and report, on July 29, 1904, President Roosevelt issued an executive order, placing into effect the first well-defined radio policy of the U.S. Government.

The principal elements of the policy were (1) all coastal wireless stations operated by the Government were placed under Navy Department authority; (2) the Signal Corps of the Army was authorized to establish wireless stations as necessary, provided they did not interfere with the coastal wireless-telegraph system under control of the Navy Department; (3) Navy stations must handle, free of charge, wireless messages to and from ships at sea, but not in competition with commercial stations; (4) all private wireless stations, both commercial and amateur, in the continental United States would be regulated and licensed by the Department of Commerce and Labor; and (5) to prevent the control of wireless telegraphy by monopolies or trusts, any legislation on this subject would place the supervision of it in the Department of Commerce and Labor. This formed the basis of Government policy for the use of radio for almost two decades.

By the beginning of 1904, all the foreign-built wireless equipment purchased by the Navy had been installed. The Naval Radio Service (NRS), under a Superintendent and with headquarters in Arlington, Virginia, was formed to oversee all the related matters. The Navy's responsibility to provide wireless communication services to other Government departments led to considerable expansion and the development of new long-distance circuits. The NRS began broadcasting a noontime signal – originating at the Naval Observatory – in 1905, allowing ships with wireless receivers to more accurately determine their longitude.

Politically, the international situation had deteriorated and America's relationship with Germany was questionable. From a manufacturing standpoint, the United States was becoming more and more self-sufficient, and, with this, a stronger feeling of nationalism emerged. Transmitters and receivers from American manufacturers were added to the Navy's vessels.

The 1906 "Wireless Stations of the World," issued by the Navy, listed the numbers of naval radio installations of that year as: Slaby-Arco and Telefunken, 51; Shoemaker, 21; Massie, 13; De Forest, 9; Stone, 8; and Fessenden, 3; for a total of 105. This shows that about half of the equipment in use by the Navy at that time was of German manufacture. The operators regarded the Consolidated Wireless Company equipment (designed by Shoemaker) to be the best, but Shoemaker joined Marconi America and his talents were thus lost to the Navy.

About this time, the Navy apparently made a decision to cease, where practical, the purchase of wireless equipment from foreign countries. Such a decision would have been logical since the Navy was then dependent upon Germany for the necessary spare parts for a large part of the wireless equipment. In the event of a war with Germany, or with a country she favored, that supply could be cut off. Before this, most American firms had primarily manufactured wireless equipment for their own use – they were also operating companies. With increased opportunities, particularly from the Navy, many businesses, such as General Electric and Westinghouse, turned to being suppliers, with research laboratories devoted to developing wireless products. In addition, a number of smaller wireless-hardware companies were established.

Continuous Wave Radio

From the initiation of the wireless, many engineers recognized that the damped radiation from spark-gap transmitters should be replaced by continuous waves. In 1902, two new technologies for generating such waves emerged. Valdemar Poulsen at the Copenhagen Telephone Company in Denmark developed the arc transmitter, the first high-frequency generator without moving parts.

The technological ancestor of the Poulsen arc was William Duddell's "musical arc," an electric oscillator made from an arc lamp. Shunted by a resonant circuit, this produced a distinctive musical tone. Duddell had proposed that it might also be used as a generator for continuous electromagnetic waves, but his experiments in this were unsuccessful. Poulsen, who was already well known for inventing the magnetic-wire audio recorder (the Telegraphone) in 1898, used the "negative resistance" characteristic of the arc to keep a resonant circuit in constant oscillation. By 1904, the Poulsen arc was patented in Denmark and 14 other countries, and had been used for telephonic messages over considerable distances.

Also in 1902, Ernst F. W. Alexanderson at General Electric, seeking to improve the alternator that had been invented earlier by the renowned Dr. Charles P. Steinmetz, developed a high-speed, self-exciting alternator that was projected to be able to generate continuous waves at frequencies up to 100 kHz. Initially, the frequency was only about 10 kHz, too low for wireless waves, but within a few years this had been extended to generate much higher frequencies. While both Poulsen's arc and Alexanderson's alternator produced continuous waves and afforded significant improvements over the "dirty" spark-gap wireless telegraph, their major importance was that they allowed the potential of wireless telephony.

E. F. W. Alexandersen

Ernst Frederik Werner Alexanderson (1878-1975) was a native of Sweden and received his electrical engineering education there and in Germany before emigrating to the United States. In 1902, he joined General Electric in

Schenectady, New York, and was assigned to improve the basic alternator to meet requirements from Reginald Fessenden for generating electrical radio frequencies at kilowatt power.

An alternator running at a high rotational speed and with hundreds of poles was necessary, and Alexanderson eventually accomplished this with efforts involving major mechanical and electrical advancements. To control the output in Morse code or voice modulation, Alexanderson invented the magnetic amplifier, a transformer-like device with no moving parts that makes use of the saturation of magnetic materials to allow a small signal to control large amounts of output power. Over his long career with General Electric, Alexanderson contributed to all aspects of radio and was awarded 344 patents.

Earlier, the basic feasibility of wireless telephony had been shown by Reginald Fessenden. Then with the U.S. Weather Bureau in Washington, D.C., Fessenden was testing a spark-gap transmitter at a facility on Cobb Island in the Potomac River. To produce a more continuous signal, he built a high-speed interrupter – a rotating contact that increased the spark frequency to 10,000 breaks per second.

On December 23, 1900, using a carbon microphone inserted between the transmitter and antenna, Fessenden spoke loudly, "One, two, three, four. Is it snowing where you are, Mr. Thiessen? If so, telegraph back and let me know." Alfred Thiessen, his assistant with a receiver a mile away, replied by telegraph that it was indeed snowing. In great excitement Fessenden entered into his notebook, "This afternoon here at Cobb Island, intelligible speech by electromagnetic waves has for the first time in World's History been transmitted." Although of very poor quality, this message instigated wireless telephony. It would be several years, however, before this technology would be developed into a practical system.

Reginald A. Fessenden

Reginald Aubrey Fessenden (1866-1932), a native of Canada, was taken to see Alexander Graham Bell's first Canadian demonstration of the telephone. The 10-year-old asked, "Why do they need wires?" From then on, Fessenden was obsessed with developing the means for wireless transmission of the human voice. Fessenden had only the equivalent of about two years of college – mainly in language study – when he came to the United States, but after many inventions and publications while at the Edison Laboratory, at Westinghouse, and in several other employments, he was appointed to a professorship at Purdue University and followed this as the initial chairman of electrical engineering at Western University of Pennsylvania (later Pittsburgh University). His crude demonstration in 1900 at the Weather Bureau, followed by great successes in the next decade, certainly earns Fessenden credit for bringing into being amplitude-modulated (AM) radio.

Much work was also being done on improving receivers, particularly to find a replacement for the unreliable electro-mechanical coherer detector. Marconi invented the magnetic detector in 1902, thereafter using it in all of his receivers. Seeking an improvement in wireless receivers for handling continuous waves, in 1904 Fessenden patented an electrolytic detector, called a "liquid barrater," French for "exchanger." In 1904, Poulsen was one of the inventors of the "tikker," an electro-mechanical device that could detect the continuous-waves from his arc transmitter. The first "solid-state" detectors – galena (lead sulfide), carborundum (silicon carbide), and silicon crystals – were introduced around 1905, but their electrical contacts were difficult to adjust. Later improvements, however, turned them into the most used low-cost detectors.

Another Fessenden invention was the heterodyne principle. In this, an incoming wave is combined with a locally generated signal of a slightly different frequency to produce a lower-frequency "beat" signal that might be easier to handle. Fessenden used a small arc to generate the local signal, and detection of the beat depended upon the operator's ear. Although patented in 1902, the heterodyne circuit did not come into common use as a receiver until after the triode vacuum tube was perfected and used for the local oscillator.

In 1904, Dr. John Fleming in Great Britain invented the two-element vacuum tube (called the Fleming valve or diode). Two years later, Dr. Lee de Forest filed a patent application for a wireless detector called an Audion, also a two-electrode vacuum tube. This led to a controversy concerning the invention of the diode. Both devices were natural extensions of the 1883 discovery by Thomas Edison when he observed that an electrical charge would collect on a metal plate adjacent to a heated filament in a vacuum (the "Edison effect"). Fleming and de Forest added a positive potential to the plate and completed the circuit back to the filament. In January 1907, de Forest, then with the Radio Telephone Company, followed with for a patent application on a three-electrode Audion, containing a control grid inside the tube. This device, that could potentially both detect and amplify, ultimately revolutionized the wireless receiver, but found limited application when initially developed, apparently because de Forest did not fully understand its principle.

Lee de Forest

Lee de Forest (1873-1961) was raised in Alabama and earned his doctorate from Yale in 1999, writing a dissertation concerning Hertzian waves on an open-ended transmission line. He first worked briefly at Western Electric, where he invented the "responder" – another potential replacement for the coherer – then formed the Wireless Telegraph Company of America. In 1902, he joined the De Forest Wireless Telegraphy Company, a firm established by investors to exploit de Forest's inventions. This company gained great publicity by winning a gold medal for the best wireless system at the 1904 St. Louis World's Fair. During his career, de Forest worked at many locations, received over 300 patents, was almost continuously involved in lawsuits, and was defrauded by a number of business partners. His invention of the triode certainly justifies designating him as the "father of the electronic age."

Reginald Fessenden believed that the high-speed alternator, then being developed for him at General Electric, would provide the solution to continuous-wave wireless operation. While awaiting the alternator, however, he worked on a synchronous rotary-spark wireless transmitter that gave a more continuous output than the traditional spark-gap apparatus. In 1902, Fessenden left the Weather Bureau and joined the National Electric Signaling Company (NESCO). After the rotary-spark transmitter was patented in 1903, NESCO set up stations using this along the east coast. Operating at frequencies in the 50- to 200-kHz band, these were the first installations providing commercial wireless telegraphy services in America.

Fessenden also supplied the United Fruit Company with wireless telegraphy stations in New Orleans, as well as on their freight ships and plantations in Central America. In 1906, NESCO started two-way transatlantic wireless telegraphy between stations at Brant Rock, Massachusetts, and Machrihanish, Scotland. Shortly thereafter, Fessenden demonstrated transmissions as far as Cairo, Egypt, one-third of the way around the world.

By mid-1906, Alexanderson's alternators had improved to the extent that Fessenden could use one to generate 50 watts of continuous-wave power at 76 kHz. A smaller station was built by NESCO at Plymouth, eleven miles from Brant Rock, with voice communications regularly exchanged between the two stations. For his wireless telephone system, Fessenden developed two types of microphones: a condenser type and a water-cooled carbon-granule type, called a "trough

Brent Rock Station

transmitter," that could pass high currents. In November 1906, the operator at Machrihanish clearly heard a voice transmission from Brant Rock telling the operator at Plymouth "how to run the dynamo;" this was the first wireless telephonic transmission across the Atlantic.

On Christmas Eve, 1906, Fessenden used the alternator transmitter at Brent Rock for the first radio broadcast in American history, beaming a Christmas concert to the astonished crews operating receivers on ships of the U.S. Navy in the Atlantic Ocean and the United Fruit Company in the Caribbean Sea. It began with "CQ CQ CQ" sent in Morse code – a call to all receiving stations to expect an important message. Fessenden then switched to a microphone and gave a brief speech, followed by playing Handel's "Largo" on an Edison wax-cylinder phonograph. Fessenden himself also performed, playing his violin and singing "O Holy Night." Fessenden's wife and his secretary had intended to read seasonal passages from *The Holy Bible*, including, "Glory to God in the highest – and on earth peace to men of good will," but when their time came they stood speechless, paralyzed with mike fright. The broadcast concluded with Fessenden wishing his listeners, "A Merry Christmas."

The first time that the Poulsen arc transmitter was used in America was in 1906. Searching for wireless telephone systems for tactical use, the Navy purchased 26 sets from de Forest's newly formed Radio Telephone Company, using an arc transmitter and Audion (diode) detector. (It is noted that de Forest had not secured any rights from the Poulsen patent holder for offering this transmitter.) Unfortunately, the sets were improperly installed and inappropriately used in the fleet trials. Declared then to be a failure, this set back by many years the introduction of wireless telephony in the Navy.

The equipment installations for these trials had little internal communication with the bridge, where tactical signaling was needed. Experienced ship commanders were accustomed to maneuvering their ships in close formation by flags, and could readily see, in conditions of good visibility, the responses. Thus, they were reluctant to use a signaling method that they could not see, or even understand. Chief Electrician's Mate William C. Bean, who had been on the 1901 visit to Europe to examine wireless equipment, was basically responsible for the equipment in these tests, and he later described the final action:

> Admiral Evans [commanding the fleet] ordered the sets dismantled and the antennas taken down. This was because there was too much playing with the new toy; also, too much interference with the regular spark operation.

From that time until 1917, the Navy was without shipboard radiotelephone capability.

International Radio Regulation

Internationally, the need for regulation of wireless communication had escalated from the beginning of these services. The First International Radio Telegraphic Conference assembled in Berlin on August 4, 1903, with the governments of Germany, Great Britain, France, Russia, Austria-Hungary, Italy, Spain, and the United States participating. At the Conference, a protocol was proposed for intercommunication between all systems of radio telegraphy, including the caveat that any radio station should be compelled to accept messages from any ship, regardless of the equipment employed. Delegates of Great Britain and Italy – the Marconi Company strongholds – would not concur with this, and the protocol was not adopted.

The Second International Radio Telegraphic Conference was held in October 1906, again in Berlin. Where the 1903 Conference had representatives of 8 powers, this one was attended by delegations from 27 nations. The protocol from the First Conference was again introduced and this time adopted, insuring compulsory radio communication between ships and shore stations.

Since Great Britain could not accept this without violating agreements with the Marconi Company, an exception was made for them. Concerning ship-to-ship communication, it was adopted by a majority vote that every ship station "would be bound to intercommunicate with every other shipboard station, without distinction as to the radiotelegraphic system adopted respectively by these stations." Great Britain reserved the right to organize a separate system of shore stations in fulfilling this requirement. It was unanimously agreed that "all radio stations must accept, with priority handling, calls of distress from ships, and must answer these calls with priority dispatch."

Other important outcomes from the 1906 Convention were the adoption of an international code of signals and of call letters for shore stations and ships of various nations. It was also decided that the term "radio" would better describe "wireless." Transmitters for coastal and ship stations were to have call letters "distinguishable from one another and each must be formed of a group of three letters." Most stations were slow to make the switch, but Marconi-operated stations were some of the first, using three-letter calls starting with M in 1908. The U.S. Navy switched in late 1910 from a variety of two-letter calls to three-letters starting with N. As shown in the January 1912 edition of *Wireless Telegraph Stations of the World,* there were still duplicates and many stations continuing to use just one or two letters.

RADIO COMES OF AGE

The adjective "radio," apparently originating in America before the turn of the century, was sometimes used interchangeably with "wireless." Following the 1906 International Wireless Telegraph Convention, radio was intended to be used, but many people didn't change. In 1911, it was used in naming the Institute of Radio Engineers, and shortly thereafter adopted by the U.S. Government. At about this same time, it began to also be used as a noun.

Naval Radio Improvement

In the United States, by 1908 radio equipment had been installed in all naval surface vessels and most of the low-powered shore radio stations had been in operation for four of more years. In addition, there were a number of commercial and amateur radio stations. Most of the transmitters operated close to 750 kHz; consequently, there was considerable interference between signals. Since spark-gap transmitters dominated, there would have been little benefit from changing frequencies to get away from interference – the primary and secondary signals from old spark sets covered a wide portion of the spectrum. Thus, the station with the highest power and/or the best antenna dominated; others had to wait.

At that time, the Bureau of Equipment realized that the Navy should have its own radio scientists and engineers to improve the communications capability and prevent blunders such as in the earlier radiotelephone fiasco. The National Bureau of Standards (NBS) agreed to assist in this, providing space and certain equipment as well as the general use of the other laboratories. Dr. Louis W. Austin, a noted physicist who was already conducting research on radio for the NBS, was transferred to the Navy Department to head the newly established U.S. Navy Radio Laboratory.

Internal activities of the Laboratory would include establishing standards for high-frequency components, investigating dielectrics and other radio materials, developing receiving circuits, analyzing the propagation of radio waves from various types of transmitters and antennas, and testing the sensitivity of different types of receivers and associated headsets. The Laboratory would also conduct external tasks at radio shore stations, including studying the comparative efficiencies of various types of spark, arc, and alternator transmitters; analyzing atmospheric-electromagnetic wave interactions; and determining methods of reducing interference and counteracting natural disturbances.

A small group of civilian radio experts was established, primarily to supplement the efforts of Navy technical officers and, since the naval personnel were subject to billet rotation, to provide continuity to activities. The first civilian employee was George C. Clark, an engineer previously with Stone Telephone and Telegraph. Over the next decade, Clark contributed greatly to the improvement of equipment and gave valuable assistance in research studies such as on wave propagation. A number of Chief Electrician's Mates were assigned to the Laboratory and, although they lacked formal higher education, they made many valuable contributions.

The first project of the Laboratory was to reexamine Poulsen arc transmitters for shipboard use. Satisfactory results were obtained over distances up to 40 miles, but it was concluded that the equipment required more skillful attention than would be available at that time from Electrician's Mates, and that it was too bulky for shipboard installation. These conclusions were very unfortunate, particularly in light of long-distance tests conducted later. As a result, arc transmitters remained unused in shipboard applications for several years.

An early project of the Laboratory that had very positive results was in examining receivers developed by Dr. Greenleaf W. Pickard of Wireless Specialty Apparatus Company using silicon-crystal detectors. In the process, Chief Electrician's Mate B. F. Miessner greatly improved the detector through a wire contact known as a "cat's whisker." With Miessner's improvement, the resulting IP76 receiver performed better than other existing receivers. On recommendations of the Laboratory, many IP76 sets were purchased for replacements throughout the fleet. Miessner patented the cat's whisker in 1910 and sold it to Wireless Specialty for $200. It is noted that the cat's whisker and an inexpensive galena crystal formed the heart of first receivers built by thousands of young radio enthusiasts in the 1920s and 30s.

The Navy purchased its first airplanes in 1911. Their initial use was for spotting the fall of shot and for increasing the scouting ranges of ships, both applications requiring the use of radio for maximum results. Consequently, simultaneous with their delivery, efforts started to adapt radio sets for them. Ensign Charles H. Maddox was assigned this project. He prepared by briefly working at the Navy Radio Laboratory, then attended a post-graduate course in radio at Harvard University, the first naval officer to have such training. The equipment to be tested consisted of a repackaged quenched-spark transmitter and a 250-watt wind-driven electrical generator, limited to 40-pounds combined weight, and an IP76 crystal receiver from Wireless Specialty strung about the operator's neck and coupled to his headphones. Testing was conducted from aboard a Wright B-1 seaplane flying out of the Naval Experiment Station at

Annapolis. Not being a pilot, Maddox sat on a board along the lower wing with a telegraph key strapped to his leg.

On 26 July 1912, Maddox transmitted, "We are off the water, going ahead full speed on a course for the Naval Academy," which was received by the torpedo boat USS *Stringham* three miles away. Maddox wrote in his report, "These were the first radio messages ever received from an airplane radio transmitting set in the United States and probably in the world." Following this, however, little was done to improve aircraft radio until 1915 and then only under the pressure of necessity. Radio was not then popular with pilots who generally considered the additional weight a handicap to safety and its operation an undesirable personal burden.

Maddox on B-1 Seaplane

Commercial Radio Improvement

The arc transmitter, developed by Valdemar Poulsen, was first used in Great Britain by the Amalgamated Radio Telegraph Company who acquired rights in that country in 1906. They constructed stations near Newbury, England, and Lyngby, Denmark, about 500 miles apart. A 30-kW arc transmitter was built with the intent of starting a trans-Atlantic service. Before these were operational, however, Amalgamated's interests were bought by the Lorenz Company of Germany, and the Newbury equipment was used to build a station near Berlin. By 1908, the Berlin-Lyngby wireless link was being tested by both telegraphy and telephony. The system, however, had difficulties in reaching commercial readiness.

Disappointed with the ventures in Great Britain and Germany, in early 1909 Poulsen welcomed a visit from Cyril F. Elwell, an entrepreneurial engineer from California who wanted rights for using the Poulsen arc in the United States. Elwell returned to America convinced that the Poulsen system, with its continuous-wave output, was the solution to problems in wireless telegraphy, as well as being the system that would bring in wireless telephony. Elwell formed a small company and within a short time made a deal for manufacturing rights to the Poulsen system in America.

Australian-born, Elwell was only two years out of college when he founded his company and bought equipment from Poulsen to open several small wireless stations along the west coast. This was an immediate success, but to obtain large funding, a new firm, Federal Telegraph Company, was formed with Elwell as the chief engineer. His intent was to set up a network of wireless telegraph stations connecting the major cities on the Pacific coast and eventually extending eastward into the rest of the Nation. Located in the Palo Alto area, Elwell had ready access to the faculties of his alma mater, Stanford University. Within a short time, the original Poulsen arc transmitter was greatly refined by Elwell, with the output increased to 30 kW. The receiver, using Poulsen's tikker, was a weakness, and other possible detectors were examined.

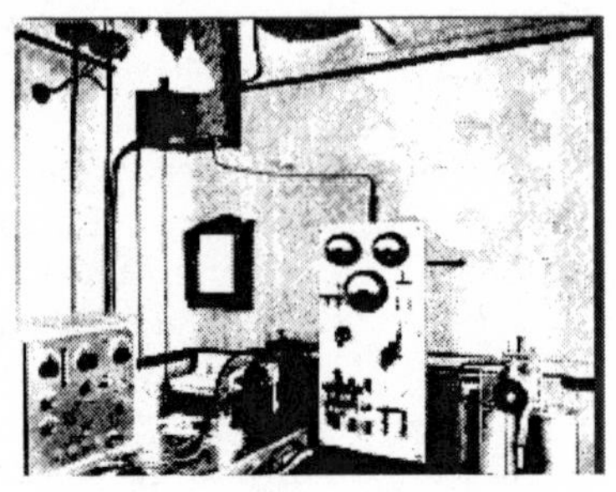

Poulsen-Ewell 30-kW Arc Transmitter

As this was going on, Lee de Forest was struggling with financial problems and approached Federal Telegraph about employment. He joined the company in 1911, and for the next two years conducted research in their Palo Alto laboratory. Building on his three-element Audion (triode), patented in 1907 but not put into use, de Forest set about developing a practical vacuum-tube amplifier. Success was achieved in August 1912, mainly by placing two or three triode stages in series using telephone transformers for coupling. Since this was limited to telephone (audio) frequencies, it was of immediate interest to AT&T but not of importance to Elwell's wireless telegraph.

While experimenting with the amplifier, de Forest accidentally coupled the output back to the input. This generated a "squeal," and this became an oscillator. Interestingly, two years would pass before de Forest applied for a patent on his regenerative oscillator, but Edwin H. Armstrong at Columbia University and Dr. Irving Langmuir at General Electric had also made similar discoveries at about the same time. (This led to a long-running patent conflict, with the U.S. Supreme Court eventually finding in favor of de Forest in 1934.) By October 1912, de Forest had increased the feedback circuit to much higher frequencies, and the first radio-frequency (RF) oscillator (the "ultra-Audion") came into being. This RF oscillator had an immediate application by replacing the low-power arc used in Fessenden's heterodyne receiver.

Technical and business development progressed rapidly for Federal Telegraph. By the end of 1912, the company operated 14 commercial stations using second-generation Poulsen arc transmitters developed by Elwell. The home station, at Point Bruno near San Francisco, soon had the largest antennas in the world – two towers 440 feet tall with 35,000 feet of strung wires. Regular communications were then demonstrated with Honolulu using a 30-kW transmitter. However, their venture into establishing an east-coast network was unsuccessful.

Regulation of U.S. Radio

On January 24, 1909, the R.M.S. *Republic* was rammed by the SS *Florida*, a ship transporting earthquake refugees from Italy to the United States. Six persons were killed in the collision, but subsequent rescue operations saved all of the others. Given credit for much of this success was the Marconi Company wireless system on the Republic and its 26-year-old operator, John R. "Jack" Binns, who stayed at his post throughout the slow sinking of the ship, using his wireless to direct rescue vessels. This incident, and the highly beneficial use of the wireless, led the U.S. Congress to pass its first laws concerning radio.

The Radio Act of 1911 required ocean-going vessels carrying 50 or more persons, including passengers and crew, for a distance of 200 miles or more from United States ports to have a radio operator and apparatus capable of working 100 miles. Following the sinking of the RMS *Titanic* on April 14-15, 1912, with the loss of over 1,500 lives, the law was revised in October. The Radio Act of 1912 required a second operator and an emergency auxiliary power source for the radio transmitter.

The original 1904 Radio Policy issued by President Theodore Roosevelt placed the responsibility of non-coastal and non-government wireless stations under the Department of Commerce, and the Bureau of Navigation within this Department handled these matters. The Radio Act of 1912 extended this responsibility to all non-government stations. This included radio amateurs, who had previously operated without restriction. Amateurs (later called "Hams") had to be individually licensed, and their transmitters were restricted to l-kW input power in the 200-meter band. The known usable spectrum at that time ran from about 300 to 3,000 meters (1 MHz to 100 kHz), and by restricting amateurs to the 200-meter (1.5-MHz) band, it was widely expected that most would cease operating.

The Bureau also mandated a uniform system of call letters. In accord with an International Agreement, regular stations in the United States would use K and W, the former in the west and the latter in the east. Amateur and special experimental stations did not come under this Agreement, and were initially assigned calls beginning with the numbers 1 through 9, corresponding to their location in nine radio-inspection districts. The first amateur Skill Certificate (license) was issued in 1912 to Irving Vermilya of Mount Vernon, New York, who operated as 1ZE (later W1ZE).

The role of radio amateurs in these early years should be noted. With the emergence of wireless telegraphy, many individuals immediately became enthusiasts of this technology, building their own receivers and transmitters. Articles before the turn of the century in the periodicals such *The Model Engineer and Amateur Electrician* and *American Electrician* gave descriptions of "simple-to-build" wireless equipment for an amateur audience. *Modern Electrics,* first published in 1908 by Hugo Gernsback, was the first magazine fully dedicated to wireless communication, and by 1911, its circulation passed 50,000 copies.

While amateurs often communicated with each other using Morse code, most of them concentrated on technical development, and it was in this activity that they made major contributions to the emerging field of radio. This is particularly true in increasing the upper frequency bounds, especially after being pushed beyond 1.5 MHz by the 1912 Radio Act. The new law, however, had a negative effect on the number of "legal" participants. In 1911, it was estimated that there were 10,000 amateur radio operators in the United States, but by the end of 1912 only 1,200 had obtained licenses.

Communication System Improvements

Despite the emergence of better technologies, widespread use of spark-gap transmitters continued. Wireless telegraphy, except in the U.S. Government, was dominated by Marconi who fought any divergence from that mode and introduced improvements such as the timed-spark transmitter. Federal Electric, however, had their share of the market with Fessenden's rotary-spark units. An article, "Around-the-World Wireless" in the September 1912 issue of *Popular Mechanics,* described plans for a British-American network of spark stations circling the globe. Congress appropriated funds to the Navy for a high-power chain extending southward to the Canal Zone and westward to the Philippines, expecting that spark transmitters would be used.

The U.S. Navy Department fully reorganized effective July 1910. All radio development activities were placed within the Radio Division of the new Bureau of Steam Engineering, and the existing Naval Radio Service was placed under the new Bureau of Navigation. Responsibility for all Government radio management and operational control had been placed with this Service in 1904. Added to this was an increasing requirement for intra-fleet and ship-to-shore communications. Then, the 1912 Radio Act required that certain naval shore radio stations be opened to commercial business, with charges being made for handling this traffic. The Service also maintained continuous watches at all naval shore radio stations listening for distress signal transmissions on the international calling frequencies of 500 and 1,000 kHz.

In 1909, the Naval Radio Service had requested bids for a new shore radio station at its headquarters in Arlington, Virginia. To generate innovation in the bids, the requirements far exceeded existing capabilities, including reaching ships at a distance of 3,000 miles at all times of the day and night, in any weather conditions, and during any season of the year. The system was also required to have wireless telephone capability within a range of 100 miles. NESCO, with the lowest bid, had won the contract over Radio Telephone, Marconi America, and Telefunken.

NESCO had proposed a 100-kW synchronous rotary-arc transmitter designed by Fessenden. While the power was sufficient for the coverage, it could not – nor could any other arc transmitter of that day – meet the operations condition. In attempting to meet these requirements, Fessenden improved the receiver by returning to his heterodyne method, patented in 1902 but dormant since then. In 1910, construction of buildings and one 600- and two 450-foot towers, comprising the station then known as Radio Virginia, was started in Arlington. The natural period of the antenna system was about 137 kHz.

Radio Virginia Towers - Naval Radio Service

After completing the Radio Virginia transmitter and following a major financial dispute, Fessenden left NESCO in 1911 and never again made any significant contributions to radio. It is ironic that the largest transmitter that he ever built was not a continuous-wave machine to which he had committed his career. Fessenden became a consultant to Submarine Signal Company, where he was very successful in using his oscillator to generate and receive underwater sounds (forerunner of sonar). NESCO continued to operate, but without commercial sales. Later, after losing a major lawsuit to Fessenden, the firm went into bankruptcy and their assets were acquired by International Radio Telegraph.

Besides building a commercial service, Cyril Elwell at Federal Telegraph wanted to convince the Navy of the superiority of the arc transmitter. Having just installed the 100-kW rotary-spark transmitter from NESCO at Radio Virginia, the Navy was reluctant to make a change. In late 1912, however, Captain William H. G. Bullard, newly appointed Superintendent of the Naval Radio Service and very knowledgeable in radio theory, allowed a 30-kW arc transmitter to be taken on a trial basis. To the surprise of most of the Navy personnel, the initial demonstrations in early 1913 were outstanding – the continuous-wave arc transmitter out-performed the more powerful spark system in communications with Panama and other distant stations.

On February 13, 1913, Radio Virginia, the Navy's first high-power station, was placed in operation. It included both the 100-kW rotary-spark and 30-kW arc transmitters. From initial testing of the coverage from the 100-kW transmitter, Dr. Louis Austin, head of the Navy Radio Laboratory, and Dr. Louis Cohen of NESCO empirically determined the mathematical relationships governing the strengths of received signals at specific distances. Known as the Austin-Cohen formula, this was so thorough that, with minor additions, it was used as the standard for over two decades.

George Clark, the civilian expert from the Naval Radio Laboratory, in early 1913 led a team aboard the USS *Salem*, sailing from Philadelphia to Gibraltar and back. A primary purpose was to make extensive comparison tests of transmissions from the arc and rotary-spark systems at Radio Virginia and of receivers aboard the ship. The receivers were a Fessenden heterodyne from NESCO, an IP76 crystal detector from the Wireless Specialty, and a tikker from Federal Telegraph. Clark's report on the test was that signals from both transmitters were about equally received. However, when the heterodyne receiver was used for the arc transmissions, there was doubt as to the superior system. Clark's report included the following:

> The combination of the heterodyne receiver and the arc transmitter constitutes the most noteworthy advancement in the development of practical radio-communications that has been made in the history of the art.

He went on to say that a chain of such high-power stations would "afford the first reliable radio service dependable by night and day and every day." From this time on, there was little debate as to the desirability of continuous-wave transmitters for Navy communications systems.

With the decline of NESCO, Leonard F. Fuller, one of the company's most promising young engineers, was hired by Federal Telegraph. Having devoted much of his senior year at Cornell to analyzing Poulsen's original arc developments and after working at NESCO on large rotary-spark machines, Fuller joined Federal Telegraph as their most knowledgeable transmitter engineer. Elwell resigned in 1914, and Fuller was made chief engineer. He soon made significant advancements to Elwell's second-generation arc equipment, particularly in increasing the power.

Early in 1914, the transmitters in the San Francisco home station and in Honolulu were replaced with Fuller's third-generation 100-kW arc systems, allowing regular day and night services to be established. In 1915, adding to their overland networks, Federal Telegraph went into the marine business, placing Fuller-designed sets on a number of passenger steamers and oil tankers operating out of west-coast ports to destinations as far as Australia.

Following the success of the arc transmitter at Radio Virginia, in 1914 orders were placed with Federal Telegraph for ten 30-kW arc transmitters for shipboard use and a 100-kW Fuller-designed arc transmitter for the Canal Zone. The Panama station would serve as a relay for far-flung naval stations in the Atlantic and Pacific, as well as for controlling the fleet in waters adjacent to the canal, which had just opened. This station was authorized to also handle commercial radio-telegraphy traffic. In 1916, orders were placed with Federal Electric for a 200-kW arc transmitter for San Diego and 350-kW units for Pearl Harbor and in the Philippines, capable of direct communications between these stations. These were completed in the 1915-16 time frame. The transmission distances were unprecedented: 5,300 miles from the Philippines to Pearl Harbor and 7,800 miles to San Diego.

At this time, commercial radio communications in the United States, primarily ship-shore, continued to be dominated by Marconi America. It succeeded in purchasing the assets of most of its competitors and, for practical purposes, possessed a monopoly. Although not yet providing reliable commercial radio communications between the United States and Europe, Marconi America was purchasing equipment for this and in 1914 announced its intention of using a 350-kW synchronous-spark system at New Brunswick, New Jersey. The next year, however, Marconi personally visited General Electric and was immediately convinced of the advantages of CW transmission. An order was placed for a 50-kW alternator for New Brunswick and Marconi America also commissioned Alexanderson to develop a 200-kW unit. In a step toward this, *The Electrical Experimenter* for August 1916 announced a 100-kW radio-frequency alternator under test at General Electric.

Alexanderson's Alternator

This action by Marconi America was clearly to stay ahead of competition. The Atlantic Communication Company, a Telefunken subsidiary, was constructing a station at Sayville on Long Island, New York. Homad, a German firm, was building a station at Tuckerton, New Jersey, for a French communications concern. Both of these stations intended to use European-built alternators and to provide trans-Atlantic radio-telegraphic services.

Although the Poulsen arc, which was very popular in America, had first been introduced in Europe, it never gained extensive use there. Shortly after Alexanderson and General Electric announced the alternator, this technology was taken up in Europe and became the dominant continuous-wave transmitter. It should be noted, however, that Europe's first wireless technology – the spark with its damped wave – also underwent improvements and remained the primary type of transmitter throughout that region.

Several types of alternators eventually emerged in Europe, all sufficiently different from Alexanderson's. In Germany, Dr. Rudolph Goldschmidt used multiple windings tuned to successively higher frequencies, resulting in a generator turning at 15,000 rotations per minute providing an output at 60 kHz. The Bethenod-Latour alternators, designed by Maurice Latour in France, had two or three sections ganged in a row, allowing higher power levels with moving parts of practical size. Another approach was to pass an alternator's output through one or more static frequency converters, producing radio waves at double or quadruple the frequency from the alternator.

The Telefunken alternator for Sayville ran at 9.6 kHz. Two static frequency converters were used to double this to 19.2 and then to 38.4 kHz, the station's transmitting frequency. It had a power output of 100 kW. In 1915, Telefunken used a combination of concepts for an alternator at its station in Nauen, Germany. Called the Schmidt system, it had a 500-kW alternator operating at 6 kHz and a tuning network that directly resonated the power into the station's antenna at 24 kHz.

As a transmitter, the arc was qualitatively inferior to the alternator. The arc required two frequencies for normal operation, it emitted many harmonics, and had a slight frequency variation. In contract, the alternator had no harmonics and transmitted but one sharply defined, stable frequency. The arc, however, was much simpler in construction, could be readily enlarged to almost any power, and was lower in cost. These two types of transmitters, each used for the purposes best adapted, would have remained supreme for years except for the introduction of the three-element vacuum tube.

Radio Goes Electronic

De Forest had patented a three-element vacuum tube, the Audion, in 1912 when he was with the Radio Telephone Company. The first triodes produced by this company, however, had serious weaknesses: they were expensive, short-lived, and lacked uniformity. This greatly inhibited their introduction into radio equipment, especially that of the Navy where spares would be limited. The Navy asked other industries to improve these devices, and Dr. Harold E. Arnold, Western Electric's director of research, responded with a tube with a high vacuum and an oxide-coated cathode that could be uniformly produced.

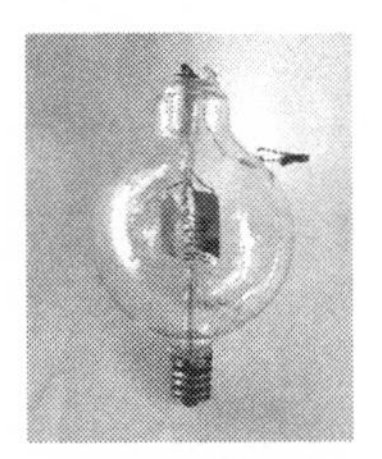

The Audion

In 1914, the Navy issued a procurement for tubes guaranteed to 2,000 hours and costing no more then $4.50 each. Western Electric, the manufacturing subsidiary of American Telephone and Telegraph

(AT&T), won the contract and a flow of tubes to the Navy began. This marked the beginning of a change to vacuum-tube receivers throughout the radio communications field. In addition, they were introduced by AT&T into their long-distance telephone equipment. Although the triode could also be used as the oscillator in a transmitter, this application would await several years until vacuum tubes for handling high power at radio frequencies could evolve.

From the beginning of radio, engineers had sought means of amplifying the received signal and improving the signal-to-noise ratio. Late in 1912, Federal Telegraph, where Lee de Forest was then working, had provided the Navy a crude "bread board" model of an amplifier using a triode vacuum tube. It performed well enough to allow the first cross-country transmission from the west coast to Arlington. De Forest left Federal Telegraph in 1913 to form his own firm, Radio Telephone and Telegraph, and submitted to the Navy the first model of a commercial amplifier. Ten units were purchased, beginning the use of an audio amplifier in a radio receiver. During 1914, amplifiers were installed on each shore station and major ship, but their expense only allowed one at each installation, with a switching arrangement for their use between receivers. These units, however, never operated satisfactorily and were later redesigned by the Washington Radio Test Shop.

Navy Shore Station

In 1914, both amateurs and commercial firms started to experiment with the new vacuum-tube transmitters for voice communications. De Forest, inventor of the triode oscillator (the ultra-Audion), took the lead, designing a vacuum-tube transmitter with sufficient power for short-distance applications. The July 18, 1914 *Electrical World* reviewed this, saying, "It is possible to [radio] telephone one to three miles, and the device is well adapted for use on small yachts, tugs, ferryboats, etc." De Forest later used an elementary power-amplifying stage and expanded into broadcasting activities, with nightly broadcasts from his experimental radio station, 2XG, located in New York City. The November 18, 1916 issue of *Electrical Review and Western Electrician,* announced this as "a remarkable step forward in the distribution of the world's news and music."

AT&T bought the U.S. commercial rights for de Forest's triode in 1914, primarily for amplifiers in long-distance telephone networks. It was also recognized as potentially useful for telephone radio links, but this would require much higher output power. Dr. Harold Arnold, research director at AT&T subsidiary Western Electric, immediately initiated development of such triodes, and within a year had devices suitable for transmitters.

In June 1915, AT&T and Western Electric installed an experimental vacuum-tube transmitter at Radio Virginia in Arlington, receiving permission from the Navy to use its extensive antenna systems. The successful tests were reported in "Wireless Telephony Now From Washington to Honolulu" in the November 1915 issue of *The Electrical Experimenter,* and "By Wireless 'Phone from Arlington to Paris" which appeared the following month. These tests marked the beginning of a ten-year period of increasing prominence for AT&T in the U.S. radio industry.

The commercial radio services were also interested in systems using vacuum tubes. The April 1915 issue of *The Wireless World* carried an article "Marconi's Wireless Telephone," describing a system under development with a "guaranteed working range of 50 kilometres between ships at sea." The transmitter used a valve (tube) designed by the Marconi Company handling 5 to 10 watts power.

New Ventures in Naval Radio

During the 1910s, the Navy had been responsible for most improvement in radio communications. By the middle of the decade, industrial research and development of radio equipment had almost ceased in this country. Commercial radio communications, primarily ship-shore, were dominated by Marconi America, but it had not succeeded in providing reliable radio communications between the United States and Europe and was dependent upon its British parent for technology improvements. NESCO was in poor financial condition and, having terminated Fessenden, had almost ceased research and development operations. None of the manufacturers held sufficient patents to meet Navy specifications and, moreover, were unwilling to provide equipment with the required ruggedness. This forced the Navy into designing and manufacturing its own equipment.

As the Radio Division was formed, the Bureau of Steam Engineering decided that this operation would, insofar as possible, provide the design and rigid specifications for future procurements of radio equipment and, if commercial manufacturers were unable or unwilling to meet the specifications, the Navy would manufacture its own equipment. By this time, essentially every radio circuit and system was patented, and no one company held all the patents for the most desirable combination of elements. In 1915, the Federal Court held that owners of patents infringed in the manufacture of equipment under Government contracts were limited to the recovery of damages from the United

States. Manufacturers, many of them just entering the field, believed themselves secure from litigation and thus sought Government contracts. Competition became intense; essentially all firms were willing to provide equipment based on Navy design and meeting Navy specifications.

Lieutenant Commander Stanford C. Hooper was named to lead the Radio Division of the Bureau of Steam Engineering in early 1915. Hooper had an excellent background in radio, having taught electrical engineering at the Naval Academy and served as the Chief Radio Officer on the staff of the Commander of the Atlantic Fleet. Hooper hired additional civilian engineers and in 1915 placed specialists in Radio Shops at Navy Yards in Boston, Brooklyn, Norfolk, Mare Island, Philadelphia, and Washington D.C. The first four Shops were devoted to specific components, and the Philadelphia Shop centered on transmitters. The operation at the Washington Navy Yard was designated the Radio Test Shop and concentrated on receivers; this was by far the largest and most productive of the Shops. A standard drawing number system, applicable to all Shops, was developed and put into use. Within a short time, 25 percent of the components of radio equipment that had type numbers assigned were of Navy design.

George Clark, who earlier had become the Navy's first civilian radio engineer, led the Washington Radio Test Shop. Dr. Louis Cohen was retained as a consultant, making a major contribution by introducing capacitor coupling between stages. Three standard receivers, incorporating the newly acquired triodes, were designed: types "A" (60-600 kHz), "B" (30-300 kHz), and "C" (1,200-3,000 kHz). To keep the cost low, the crystal detector was retained, and the heterodyne feature was used only in CW receivers. As rapidly as they could be produced, these receivers were placed in service at the shore stations and on the more important combatant ships.

Dr. Frederick A. Kolster, in 1915 while employed by the Stone Radio & Telegraph, had discovered that a loop of wire, connected to a receiver, could be rotated to determine the direction of a received signal. Kolster's patent for the radio direction-finder was obtained by the Navy in 1916 and, with Kolster as a consultant, the Naval Radio Laboratory (at the National Bureau of Standards) in association with the Pittsburgh Radio Shop developed this into the "radio compass," a system for shipboard use. Dr. Louis Austin, head of the Laboratory, designed an antenna to eliminate false bearings. These sets were immediately placed on most American warships.

An important component development was made at the Brooklyn Radio Shop. Large condensers (capacitors) were required for transmitters. They had evolved from the earlier Leyden jars to sheets of Bohemian glass plated with copper, but the war had interrupted the glass supply. In 1916, the firm Cornell Dubilier submitted a mica condenser for evaluation by the Navy, but it failed to meet requirements. Believing that mica might actually be superior to glass for condensers, Guy Hill at the Brooklyn Radio Shop drew up specifications for such a device in a metal container, the metal allowing generated heat to be better radiated. By December, with Hill's assistance, Dubilier had improved the product and the Navy immediately incorporated them in its new equipment.

MILITARY ERA OF RADIO

World War I (WWI) began in Europe in August 1914 and concluded with the signing of the Armistice on November 11, 1918. For most of this period, however, America had officially been neutral, only directly entering the conflict on April 6, 1917. The Military Era of Radio covers the period immediately preceding the war and extending to March 1920 when radio communications facilities that had been commandeered by the Government were released to their original owners.

Prelude to the First World War

As war clouds grew darker over Europe, the U.S. Government became increasingly interested in trans-Atlantic radio communications. Commercial companies, although they had pursued it for over a decade, had not succeeded in establishing reliable services between America and Europe. Most telegraph traffic, including essentially all Government messages, was through ocean-bottom cables, and there was fear that these cables might be severed during wartime. (That indeed did take place in June 1918 when the Germans cut two major cables 60 miles off the American shore.)

When the war began in Europe in August 1914, Great Britain, France, and Russia were allied against Germany and Austria-Hungary. America was officially neutral, and President Woodrow Wilson issued a proclamation of this neutrality as concerned radio services. The Executive Order, to be enforced by the U.S. Navy, prohibited radio stations within the jurisdiction of the United States from transmitting or receiving for delivery messages of an un-neutral nature, and from rendering to any one of the belligerents any un-neutral service. This enforcement was passed down to the Superintendent of the Navy Radio Service.

Instructions for enforcing the Executive order were issued to all Government and commercial radio stations, and to all radio operating companies. Navy censors were placed at major commercial stations, with instructions to prohibit the exchange of any coded messages, except enciphered messages between U.S. Government officials, unless the means of decoding was provided to the Navy. Marconi America refused to accept this prohibition and, after such a message was sent to their station at Siasconse, Massachusetts, the Navy closed it for several months.

The British severed all cables going to Germany, thus requiring communications from and through that country to be by radio. For services between the United States and European countries, the commercial station at Tuckerton was taken over by the Navy in January 1915. Although available for traffic with any European or United Kingdom station, the only one that continued exchanges was in Eilvese, Germany. It should be noted, however, that no radio circuits at that time provided continuous trans-Atlantic service; main reliance for communications between the United States and her European Allies continued to be placed upon the cables.

Great Britain first blockaded Germany, then Germany responded by declaring a war zone around the British Isles, giving notice that all ships of any nation would be sunk without warning and using their U-boats (submarines) for enforcement. On May 7, 1915, the British liner *Lusitania,* sailing from New York to Liverpool, was sunk off the Irish coast with a loss of 1,198 lives, including 128 Americans. The British deciphered a message from the Atlantic Communication Company station at Sayville, Long Island, and determined that it had been sent to a German U-boat, giving the location of the *Lusitania.* The Navy then took over the Sayville station and operated it under the same conditions as for Tuckerton.

To ensure that the Naval Radio Service could be quickly brought to readiness in the event of war, the Superintendent, Captain Bullard, ordered a mobilization of communication facilities for May 6, 1916. Telegraph and telephone connections were made between the Navy Department and all navy yards and naval radio stations in the United States. The radiotelephone transmitter, installed by AT&T the previous year at Arlington for long-distance communication tests, was still in place, and AT&T agreed that it could be used during the mobilization. Another smaller transmitter and one of the receivers used during the long-distance tests were installed in the USS *New Hampshire,* commanded by Captain Lloyd Chandler. The *New Hampshire* was to be in the vicinity of Hampton Roads during the mobilization tests, and another of the receivers was installed at the Norfolk Naval Radio Station. Upon commencement of the mobilization, two-way radiotelephone communications were quickly established, the first in the Navy – at least officially – since the failed trials in 1906. The station in Norfolk connected the conversations by telephone lines to the Navy Department and other stations.

All of the communication mobilization efforts were a full success. In reporting this, the AT&T, in its June 1916 issue of *The Telephone Review*, noted a conversation between Secretary of the Navy Josephus Daniels and Captain Chandler:

> I can hear you as well as if you were in Washington, Captain Chandler. It will not be long before the Secretary of the Navy will be able to sit in his office and communicate with vessels of the Navy all over the world by wireless telephone. That is something the captains may not like!

Radiotelephone Demonstration

After the 1910 reorganization, the Naval Radio Service in the Bureau of Navigation remained unchanged for several years. In July 1916, considering the growing use and importance of radio, Naval Communication Service under the Chief of Naval Operations was established, taking over all of the communication circuits. Captain W. H. G. Bullard, previously Superintendent of the NRS, was appointed the first Director, Communication Service.

William Hannum Grubb Bullard (1866-1927) graduated from the U.S. Naval Academy in 1886. His 20 years were spent in a wide variety of shore and sea tours, including an assignment in 1905 to evaluate a kite-lofted aerial photography system that had made huge panoramic pictures of San Francisco following the earthquake. In 1907, Bullard returned to the Academy to organize, and become the first Head of, the Department of Electrical Engineering. He served in this capacity for four years and prepared the textbooks on electrical engineering that were used by midshipmen for many years.

Having risen to the rank of Captain, Bullard was appointed Superintendent of the Naval Radio Service in December 1912. When the NRS was reorganized into the Naval Communications Services, with the responsibility for all radio operations in the Navy, Bullard became its Director, reporting directly to the Chief of Naval Operations. After completing his tour of shore duty, he was promoted to Rear Admiral and served at sea until early 1919 when he again became Director, Naval Communications. In 1921, Bullard had his last sea assignment, commanding destroyers and gunboats of the Yangtze Patrol in establishing radio communications along the Yangtze River and protecting American interests in that disturbed area of South China. He retired from the Navy in 1922. As will be noted later, Bullard played major roles in American post-war radio activities.

Communication Services

After the *Lusitania* incident, the American public was ready for war. President Wilson, however, tried to negotiate a peace. Although Wilson was reelected in 1916 with the slogan, "He kept us out of war," the Nation prepared for combat. The Navy Reserve was formed, and shortly thereafter many reserve officers and enlisted personnel volunteered for active duty. Major funding was released for military hardware, with the Navy receiving the largest share. Finally, after U-boats sank three American merchant vessels and Great Britain released an intercepted message from Germany asking that Mexico attack the United States to divert them from the European war, the President and Congress finally declared war against Germany.

On April 6, 1917 at 12:45 pm, Radio Virginia transmitted a signal directing all naval communications stations to "Cease all radio work and listen for rush signals." Every operator knew why and listened intensely for the historic message. At 1:00 pm, the President applied his signature. Radio Virginia broke its silence, and in seconds the fleet, all shore stations, and most of the entire world knew that the United States had entered the war on the side of the Allies.

The same day that war was declared, President Wilson issued an Executive Order for the Navy Department to take over all radio stations within the jurisdiction of the United States that were needed by the Naval Communication Service. All other stations were to be closed, including amateurs who were also directed to dismantle their transmitters. Fifty-five commercial stations, mainly owned by Marconi America, were affected, and 28 were closed. No commercial radio communications were to be through any channels other than Navy stations. Navy censors were stationed in all the open commercial stations, but most of the operators remained as government employees.

Trans-Pacific traffic was through the Navy-operated Marconi America circuits between California, Hawaii, and Japan, and the Federal Telegraph circuit between California and Hawaii. A major portion of the trans-Atlantic traffic was through the Sayville and Tuckerton stations, already operating under the Navy. The German-built alternators at these stations were replaced by arc transmitters. Since these stations were remote from Washington, traffic between them and the Navy Department was mainly relayed by telegraph over land lines.

The Government commandeered all large vessels (about 330) and placed them under Navy control. All radio equipment on these vessels was purchased and subsequently operated and maintained by the Navy. To block a potential acquisition of arc patents by British Marconi Company, the Government bought the patents and shore stations of

Federal Telegraph. Later, all of the coastal stations of Marconi America were also bought by the Government.

To improve the reliability of trans-Atlantic reception, special receiving facilities were set up at several coastal locations including the then-unused Marconi America station in Belmar, New Jersey. Led by Roy A. Weagant, chief engineer for Marconi America, the Belmar station had for several years conducted extensive static-elimination experiments to increase the hours during which trans-Atlantic traffic could be handled. When earlier taking over the Belmar station, the Navy allowed Weagant to continue, and with the conversion to a receiving facility, these very fruitful investigations continued. Weagant's premise was that much of the static came from outside the atmosphere (later shown to be true). This work was eventually reported in an article, "Weagant's Anti-Static Invention, Details of a Great Discovery Which Has Revolutionized Long Distance Wireless Communication," in the April 1919 issue of *The Wireless World*.

To further improve the trans-Atlantic services, a 500-kW arc transmitter was installed at a new station in Annapolis and a contract was signed with Federal Telegraph for a 1,000-kW arc transmitter for a station near Bordeaux, France; these stations were to serve the American Expeditionary Force and Allied Government affairs. When asked about the feasibility of the megawatt arc transmitter, Federal Telegraph's chief engineer, Leonard Fuller, said that if needed he could build one of five times that power.

A 50-kW alternator from General Electric was delivered to Marconi America's New Brunswick station in late 1917. In comparison tests with the already-installed 350-kW timed-spark transmitter, the alternator was clearly superior. The 200-kW alternator, commissioned earlier by Marconi America, was completed and installed at New Brunswick by mid-1918. For the first time, a reliable and continuous trans-Atlantic radio circuit was available. Throughout the war, this station handled the major portion of radio traffic between America and Europe, primarily for telegraphy but occasionally for telephonic transmissions. In November, this station would be used to transmit President Woodrow Wilson's ultimatum to Germany, bringing the First World War to a close. It was later used for radiotelephonic communication with the USS *George Washington* during President Wilson's trips to France for the Peace Conference.

200-kW RF Alternator

Personnel and Training

One of the major concerns of the Navy at that time was the availability of radio operators. The Bureau of Navigation, charged with recruiting and training personnel, provided a sufficient number to man the existing fleet and shore stations, but it was obvious that it would be extremely difficult to provide those with the necessary training in the event of war. Commercial companies were asked to organize their operators to agree to enroll in Government services should they be needed and the response was excellent. The National Amateur Wireless Association provided membership lists, and the various Naval Districts organized the members. In early 1917, the Bureau of Navigation created the United States Naval Reserve with the Naval Communication Reserve as an element.

The Naval Communications Reserve was called to active duty upon the declaration of war. They were immediately augmented by the enlistment of hundreds of commercial and amateur radio operators who had not previously joined but who now saw it as a patriotic duty. The closing of the commercial stations made additional hundreds of operators available for duty, and the immediate requirements for trained operators were well met by these people. As the war progressed, however, more and more ships were built and commissioned, causing a constantly increasing demand for qualified radio operators. The recruitment of amateur radio operators became critical. At the time stations were closed, there were about 6,000 licensed amateurs, and within a year some 4,000 were in uniform.

The primary skill brought by amateurs was in radio operation – the use of International Morse code, which still dominated radio communications. At that time there was not a Navy rating of Radioman (the Navy had only 13 different ratings during the war years). Petty officers serving as radio operators were usually rated Yeomen, while those performing equipment maintenance were rated Electrician's Mate or Gunner's Mate. Considering the nature of radio equipment at that time – particularly the spark transmitters – the "electrician" classification was appropriate. The "gunner" classification would appear to be inappropriate, but it likely resulted from the introduction of electrical gun controls.

To meet the need of thousands of new radio operators, the Bureau of Navigation established a four-month training program. Schools were set up in each naval district to give one-month preliminary training in radiotelegraphy and to eliminate those who lacked the necessary aptitude. For advanced training of three months duration, a school was established on each coast. In June 1917, Harvard University offered buildings for classrooms, laboratories, and dormitories. This offer was accepted by the Navy, and the school rapidly grew into a very large activity. A smaller advanced school was set up at the Naval Station at Mare Island, California. By the end of 1917, the two schools were graduating over 100 per week, and in early 1918 this increased to 400 per week. While amateurs, already possessing the basic skills, were the best students, many young men who had never before seen radio equipment were trained.

All commands were required to have a qualified communications officer. In addition, all Navy vessels of significant size were assigned a radio officer. The "U.S. Naval Communications Regulations, 1918," prescribed the specific duties of these officers. Radio officers on "first-rate" ships had this as a full-time duty. They were required to be knowledgeable of the equipment and stood a regular watch during which they personally did all of the sending. Each of these vessels had a portable radio set, and the radio officer was personally responsible for ensuring the operation and calibration of this important piece of equipment.

Several universities set up special engineering programs for training communication officers. Both the Army Signal Corps and the Navy Bureau of Navigation established limited training courses in radio maintenance for enlisted personnel, but information concerning these programs is essentially nonexistent. Some Electrician's Mates and Gunner's Mates attended special schools on radio and led the maintenance on capital vessels. Dr. John H. Dellinger and his colleagues at the National Bureau of Standards prepared two books on radio for Army and Navy personnel. One was *Radio Instruments and Measurements*, published as Bulletin 74 in 1918 and used as a basic reference for many years. The second, *The Principles Underlying Radio Communications*, was published as Pamphlet 40 in 1918 and gave an excellent treatment of the technology at that time.

The Secretary of the Navy took advantage of the fact that the 1916 law creating the Naval Reserve used the word "personnel" rather than "male" when referring to Navy Yeoman. The enlistment of women as Yeoman (F) was authorized in early 1917. Within a short time, over 10,000 "Yeomanettes" were recruited to "Free a Man to Fight." They were sworn into the Navy as the first officially recognized enlisted women in U.S. history. Most performed clerical duties, but some became radio operators and telegraphers. Some were used as instructors in radio operator schools. At the end of the war, all the Yeoman (F) personnel received honorable discharges and qualified for veterans benefits.

Upgraded Naval Radio

Immediately before and after entry into the war, the Radio Division under Commander Stanford C. Hooper in the Bureau of Steam Engineering greatly increased its work. The Radio Shops at various shipyards designed new equipment and evaluated systems commercially available, essentially all using continuous-wave technologies. Major advancements, particularly at the Washington Radio Test Shop, were made in incorporating vacuum tubes, turning "electrical" into "electronic" circuits. As an indication of the success of the implementation of vacuum tubes into military equipment, between 1914 and 1918 the Navy and Army Signal Corps purchased more than 500,000 type VT-1 (Navy) and VT-2 (Army) tubes from Western Electric and some 200,000 VT-11, -14, and -16 tubes from General Electric, copied from the design.

At the Washington Radio Test Shop, Lieutenant William A. Eaton led in making significant advancements in receivers. In cooperation with NESCO, they redesigned the Type A and B receivers, the first becoming SE-95 covering 30-300 kHz, and the latter SE 143 covering the most-used range of 100-1,200 kHz. Under designations CN-239 and CN-208, respectively, these were manufactured by several firms.

The loop antenna, developed by Dr. Frederick Kolster and improved by Dr. Louis Austin at the Naval Radio Laboratory, was used by the Radio Test Shop with a receiver as the SE-995 low- and medium-frequency direction-finding system. This was deployed on destroyers as well as around the harbor of Brest, France. After the war, ground stations were installed around approaches to the most important ports and used cooperatively with shipboard units for position determination.

In late 1918, the Radio Test Shop, with Dr. Louis A. Hazeltine, professor at Stephens Institute of Technology as a consultant, developed the SE-1420 receiver. This centered on Hazeltine's neutrodyne circuit that, for the first time, incorporated the detector as an integral part of the receiver. The neutrodyne, a tuned-radio-frequency circuit, effectively neutralized the high-pitched squeals that plagued early radio sets. The original audio amplifier from de Forest was modified into a two-tube unit by the Test Shop and put into service as the SE-1000. When an audio amplifier was incorporated with the neutrodyne circuit, the integrated receiver was designated SE-1440.

The urgent need of receivers for operating in the noisy environment of aircraft led the Test Shop to develop a six-tube receiver. This used two stages of RF amplification, followed by a detector and a two-stage audio amplifier. In the order that they were developed, these full receivers were designated SE-1613 (100-300 kHz), SE 1615 (30-100 kHz), SE-1617 (18-43 kHz), and SE-1405 (45-150 kHz), the last designed for use with aircraft radio direction finders.

The Navy made few improvements in damped-wave radio transmitters during the war years. Many spark transmitters of various types and sizes were purchased for ships, aircraft, and some shore installations, but this was primarily because they were readily available from existing manufacturing sources. There were, however, Navy-led developments in arc and alternator units, and in low-power vacuum tube transmitters. The arc, simpler in construction than the alternator, could readily be scaled to essentially any power level and was thus improved for shipboard applications. The alternator, with a superior signal, continued to be improved and became in great demand for shore stations. Frequency changers, designed by the Navy, allowed transmitters to be quickly shifted between several fixed frequencies.

It was in vacuum-tube transmitters that the greatest advancements were made. One of the first major applications was in radio telephone transceivers for submarine chasers – the "mosquito fleet." Western Electric developed this unit, the CW-936, having a five-watt transmitter. Ultimately, over 2,000 of these sets were built for the U.S. and British navies. Western Electric made the first 50-watt power tube, the VT-18, in 1918. This was quickly followed by a 100-watt tube, the "Oscillion," from Radio Telephone and Telegraph, then a 250-watt "Pilotron" from General Electric. The all-electronic transmitters used a regenerative circuit to generate the signal, followed by one or more RF amplifier circuits. The higher-power output tubes needed considerable input power; for example a typical Pilotron required 6,000 volts and drew 600 watts. A few transmitters using higher-power tubes were first used by the Navy in late 1918.

CW-936 Radio Transceiver

At the Western Electric research laboratory, Dr. Alexander M. Nicolson was developing high-frequency sound generators based on the piezoelectric effect. He found that quartz and Rochelle salt (potassium sodium tartrate) crystals made excellent sound sources, vibrating under electrical stimulation. He then reversed the process, applying mechanical stress on the crystal and found a highly stable, narrow-frequency electrical output. In 1918, a patent was filed for a crystal-controlled RF oscillator with a highly stable output frequency. This was one of the most important developments for transmitters in this period.

Dr. A. Hoyt Taylor, one of the Naval Radio Reservists called to active duty, was assigned as the 9th District Communication Officer with headquarters at Great Lakes, Illinois. Early in 1917, he established a temporary laboratory on the shores of Lake Michigan to determine directional and static-reducing properties of buried and submerged antennas. This research provided some interesting and unexpected results, not applicable to improving radio systems at that time but useful in the future for very-low-frequency communications. Following this, Taylor was assigned as Communications Officer responsible for the trans-Atlantic radio station along the northeast coast, and actively participated in the antenna research being done by Weagant at the Marconi America station in Belmar.

The Aircraft Radio Laboratory (ARL) had been established in 1916 at the Pensacola Naval Air Station and had a major role in evaluating new aircraft radio systems. Two types of aircraft were then used by the Navy: short-range crafts for spotting shot-falls and other observations, and longer-range "flying boats" for antisubmarine and convoy duty. Several types of wind-turbine-driven, low-power spark transmitters were developed in cooperation with a number of companies. These required the pilots and airmen to be proficient in Morse code; an ability to achieve 10 words per minute was included for pilot graduation. In association with Western Electric, a low-power radiotelephone transceiver, based on the CW-936, was developed and eventually put into production. At the beginning of 1918, the ARL moved to the Naval Air Station, Hampton Roads, Virginia, but a few months later it moved again to the Anacostia Naval Air Station in the District of Columbia. Dr. Hoyt Taylor was made director of the ARL when it transferred to Hampton Roads, and continued in this position at Anacostia.

Dr. Louis Austin at the Naval Research Laboratory analyzed the theoretical performance for various types of transmitters and antennas for aircraft. For the then-used transmitters, primarily spark, he recommended-trailing wire antennas, allowing the maximum possible communication distances. This being an obviously undesirable configuration, the search for alternatives was given priority. Frequency is inversely proportional to wavelength, and antenna size must be some appreciable fraction of the wavelength being received. Thus, if frequency is increased, the antenna length can be decreased. After joining the ARL, Taylor started an activity in high-frequency research, not only to reduce the antenna size but also to examine the general characteristics and merits of this relatively unexplored radiation band.

Army Signal Corps

Communications by electrical means in the U.S. Army dates back to the Civil War, during which the Army Signal Corps provided extensive telegraphy services, both in the field and over long distances. This included the first airborne electrical communications; Union observers were sent aloft in balloons and sent their messages by telegraph using wires strung to the ground.

When Marconi brought wireless to America at the turn of the century, the Signal Corps immediately recognized the benefit of this technology and initiated its own research in this field. In 1904, the Signal Corps opened the Washington-Alaska Military Cable and Telegraph System, portions of which included wireless links. At that time, the Signal Corps also had wireless links between all of the major forts. The Executive Order issued by President Theodore Roosevelt in July 1904 placed the responsibility for extended-range wireless communications in the United States under the Navy. The policy did authorize the Army Signal Corps to establish "such stations as necessary," provided they did not interfere with the wireless systems under Navy control. This led to a long-running rivalry between these services concerning radio.

The infantry, cavalry, and armor demanded better, lighter, more compact receivers and transmitters for tactical communications, and this is where the Army Signal Corps centered its developments. Most of the same commercial firms supplying to the Navy were also sources of Army radio equipment. Needed research was mainly performed in a special Army Research Laboratory at the National Bureau of Standards (NBS), with the equipment specified and procured by the Signal Corps' Electrical Equipment Division in Washington, D.C. The laboratory was under Major George O. Squier. Having earned a Ph.D. in physics from Johns Hopkins University, Squier made many personal contributions to electronics, including the technology of signal multiplexing, for which he was elected to the National Academy of Science in 1919.

The equipment carried the designation SCR, which initially meant "Set, Complete Radio," but later was an abbreviation for "Signal Corps Radio." This designation was used on essentially all types of electrical and electronic equipment – not just radios. Some representative equipment included the SCR-40, Fixed Station (3 –kW spark transmitter); SCR-41, Field Wireless Wagon Set (1-kW spark transmitter and crystal receiver); SCR-47, Field Radio in Artillery Caisson (range 75 miles); SCR-49, Pack Radio (quenched-gap transmitter and crystal receiver); SCR-52, Aircraft Receiver (with audio amplifier); SCR-57 Aircraft Intercom (the first); SCR-73, Aircraft Transmitter (200 watts CW); and SCR-78, Tank Radiotelephone Set. There was even a transmitter and receiver set for Balloon Observers – the SCR-62.

In October 1916, the Signal Enlisted Reserve was formed, with thousands of civilian telegraphers joining the ranks. As the war began and the subsequent draft started, in 1917 the Army established four training camps for new signal troops: Little Silver, New Jersey; Leon Springs, Texas; Fort Leavenworth, Kansas and The Presidio of Monterey, California.

Shortly after opening, the New Jersey facility became Camp Alfred Vail, named for a man sometimes given credit for developing the Morse code. Over the next two years, thousands of Signal Corps personnel were trained there in standard telegraph and telephone operations and in wireless telegraphy.

Signal Class at Camp Vail

As in the Navy, women also served in the Army during the war, particularly in the Signal Corps where they were used as telephone and radio operators. Among these, 223 women trained at Camp Vail as bilingual (French-English) operators and were sent to France. Unlike those in the Navy, however, none of the women were officially inducted into the Army but were treated as "contract employees." Because of this, after the war they were not eligible for honorable discharges or veteran benefits, nor were those who had served in France entitled to wear a campaign medal. In 1978, Congress finally passed a bill to recognize the 223 women with foreign service as veterans, although not retroactively.

"Hello Girls" in France

The existing Signal Corps capabilities at the Electrical Equipment Division and the NBS Army Research Laboratory were not sufficient for development of the new generation of electronic equipment. In late 1917, it was decided to establish the Signal Corps Radio Laboratories at Camp Vail. Quickly constructed

Col. Squier at Radio Laboratories

and placed into operations, activities in this new facility centered on the standardization of vacuum tubes and the testing of equipment manufactured for the Army by commercial firms. Under the direction of now-Colonel Squier, experimentation was also done on radio communication with aircraft and aircraft detection using sound. Two airfields were built, and up to 25 aircraft were involved in testing radio equipment. (It is noted that at that time the Army's air operations were a part of the Signal Corps.) Both ground and aircraft radio-telephones developed under the direction of the Signal Corps Radio Laboratories were introduced into the European theater in 1918.

The Army Signal Corps also had a development facility near Paris, concentrating on communications equipment used on the French battlefields. Signal Corps Reservist Captain Edwin H. Armstrong, who had earlier revolutionized wireless communications by inventing the regenerative receiver, was assigned there to develop a means of detecting enemy shortwave communications. In 1918, he designed a complex eight-tube receiver that greatly modified the existing heterodyne circuit. In tests from the Eiffel Tower, this superheterodyne receiver amplified weak signals to a degree previously unknown. The superheterodyne circuit was, unquestionably, the greatest advancement in radio technology made during World War I. Today, it forms the basis of receivers in essentially all applications.

Signal Corp's Paris Laboratory

UNDERSEAS ACOUSTICS

Parallel with the evolution of electromagnetic communications, the water around the globe was being researched as an acoustics medium. A brief history of these developments will be given, culminating in the developments during World War I in undersea acoustics technology.

Discovery of the excellent capability of water in conveying sound waves was well known in antiquity. Ancient boatmen signaled by striking the bottom of a pot held upside down in the water, and this could be heard miles away by persons with their ear against a hull. In the 15th century, Leonardo da Vinci described an instrument for hearing movements of vessels in the water. In 1826, J. Daniel Colladon and Charles-Francois Sturm, measured the velocity of sound in water (about 4,174 feet per second) by striking a submerged bell with a hammer and timing the receipt of signals 10 miles away using an ear trumpet held in the water.

The measurement of water depth was always important to boatmen, with playing out of a knotted cord the standard method through the centuries. (A "fathom" is six feet, the length of a man's outstretched arms in gathering up the measuring cord.) When transatlantic cables were considered, the need for measurements at considerable depths was needed. For installing the first such cable in 1857, the fathom cord was replaced by piano wire (which had great strength) and was hauled up by a power winch. This method was used until acoustic devices were introduced.

In the late 1890s, several researchers experimented with underwater bells and submerged carbon microphones to measure water depths. The Submarine Signal Company was formed by Arthur J. Mundy in 1901 to conduct such acoustical research and to develop equipment for assisting in navigational safety. For the latter, bells tuned to a ringing frequency of about 1.2 kHz were installed at a number of lighthouses, and carbon microphones driving headphones were placed on opposite sides of a ship's hull. The direction to the lighthouse was determined by steering the ship until signals of equal intensity came from the microphones. This system was successfully demonstrated to navigational authorities, and by 1912 bells were placed at dangerous coastal points around the world and ship owners quickly equipped their vessels with receiving equipment. The system functioned well to a range of about 10 miles.

Listening Device for Navigation Safety

Early Detection Techniques

Reginald A. Fessenden, the pioneer of continuous-wave radio, left NESCO after a financial dispute, and became a consultant to Submarine Signal. Starting in 1912, he applied his heterodyne oscillator in sound generating and receiving. Within a short time, Fessenden developed an acoustical device that revolutionized the use of underwater sound. It could potentially be used for warning of icebergs and taking depth soundings. Following the 1912 RMS *Titanic* disaster, the Government encouraged the development of all ideas that might enhance safety of life at sea.

Fessenden's system used a large circular plate that emitted undamped 540-Hz waves when electromagnetically driven by an oscillator. The plate and oscillator could then be switched to become a

Fessenden's Apparatus

sensitive receiver for detecting the reflected signal. The receiver could also be used for hearing the movement of distant ships.

In early 1914, the apparatus was tested aboard the Coast Guard cutter USS *Miami* on the Grand Banks, off Newfoundland, where echo ranging from an iceberg at about two miles distance and depth sounding were demonstrated. All of this was described in a paper by R. F. Blake, "Submarine Signaling: The Protection of Shipping by a Wall of Sound and Other Uses of the Submarine Telegraph Oscillator," *Transactions of the AIEE, 33 (1914).*

Over the next several years, Fessenden was awarded six patents in sound generation and detection, four in submarine signaling and detection, and the basic patent on the Fathometer (depth measurement device). His Fathometer, however, weighed about 1,200 pounds and was difficult to use. The transmitter also produced almost intolerable noise within the ship; thus, it did not gain wide acceptance.

Submarine Detection

As the threat from German submarines increased, considerable research in underwater acoustical detectors was done in Great Britain and France. In 1915, Paul Langévin, a distinguished French physicist, and Constantin Chilowsky, who was originally from Russia, developed an efficient device that they called a "hydrophone." Their device used thin quartz crystals glued between two steel plates and mounted in a housing suitable for submersion. The composite had a resonant frequency of about 150 kHz, in the ultrasonic frequency spectrum.

In Great Britain, a top-secret Anti-Submarine Detection Investigation Committee was formed, seeking a system code-named ASDICS. In this effort, a team led by Canadian Dr. Robert W. Boyle, often called the "father of ultrasonic research," developed the Langévin-Chilowsky transducer into an ASDICS listening device. The first-known sinking of a submarine detected by a hydrophone was the German U-Boat *UC-3*, in the Atlantic on April 23, 1916.

In this period, the U.S. Army also conducted work in underwater sound. In support of its port defenses, the Coast Artillery Corps had a research facility near Fort Monroe, Virginia, devoted to submarine detection. For the Army, Western Electric performed a fundamental study of the disturbances given off by submarines, and an analysis of other water-based disturbances of a similar nature. Later, the Army also established a Subaqueous (Underwater) Sound Ranging Laboratory at Fort Wright, New York

When Germany commenced an unrestricted submarine campaign during World War I, it became obvious that the U.S. might be drawn into the conflict. To strengthen preparations, in 1915 the Secretary of the Navy established the Naval Consulting Board (NCB) headed by Thomas A. Edison and consisting of the foremost scientists in the Nation. Recognizing the urgency of protection against submarines, the NCB set up a Special Problems Subcommittee on Submarine Detection by Sound.

Edison (center) with NCB

In early 1917, upon the recommendation of the Subcommittee and with funding from the Navy, Submarine Signal, General Electric, and Western Electric jointly established an experimental laboratory on the peninsula of Nahant, Massachusetts, jutting into the ocean outside the entrance to Boston Harbor. The basic equipment included microphones from Submarine Signal and the oscillator that Reginald Fessenden had developed for sending and receiving sound signals. The oscillator was improved by adding a Pilotron vacuum tube, and the resulting apparatus could detect ships and shallow submarines at distances of many miles. The equipment, however, was not yet capable of operating at the speeds required for tracking and destroying submarines.

Other work at Nahant included the development of the "C" (later designated "CD") and "K" sets. The C-tube was an aural device made of two rubber spheres mounted on the ends of an inverted T-shaped hollow pipe that terminated in a stethoscope. Hung over the side of a vessel or protruded through the bottom, the assembly could be rotated until the sound seemed directly in front of the listener. The K-tube set had three microphones arranged in an equilateral triangle, with the electrical outputs of two at a time fed to telephone receivers that were, in turn, coupled to the operator's ears through flexible air lines. These lines could be varied in length and the amount of variation was transferred to a dial calibrated in degrees, indicating the bearing of the originating acoustical signal. The third microphone was used to determine the sense of the bearing. The K-tube development team was led by Dr. Irving Langmuir, a scientist from General Electric and future Nobel Prize winner.

The K-tube set was used extensively during WWI for initially locating submarines, but when the ship was in motion the noise overwhelmed the submarine's signal. To overcome this, three types of towed devices designated OS, OK, and OV were developed. These operated in the same way as the K-tube set, but could be towed at high speeds without introducing noises created by water motion.

As a second approach to the submarine detection problem, in the summer of 1917 the Navy established the Naval Experimental Laboratory at New London, Connecticut, and also obtained the services of many leading physicists and engineers. Led by Dr. Albert A. Michelson, the 1907 Nobel Prize winner in physics, these consultants included Dr. Vannevar Bush, who would later have a major role in World War II defense research, and Dr. Harvey C. Hayes of Swarthmore College, who would ultimately lead the development of sonar. By the time the armistice was signed, the Station had a complement of 700 persons.

Working at New London, Dr. Charles Max Mason, later President of the University of Chicago, provided the creative genius behind several generations of the "MV" passive submarine sensor. The MV was an electrical sonic system of carbon button-type microphones with their output signals fed to a compensator that gave the direction of the received sound within a few degrees. To determine this direction, the operator needed only maximize the output on his earphones by turning a dial. This was one of many acoustic devices developed at New London during the war.

WWI Subchaser

The Navy established a Hydrophone School at New London in 1917 to train operators for the submarine detection equipment on "subchaser" vessels. Over 1,500 operators and 150 officers received training there by the end of the war. The destruction of at least six German submarines was credited to the use of American antisubmarine equipment.

Chapter 2

ELECTRONICS BETWEEN THE WORLD WARS

During the period immediately following World War I, the Navy was the primary user of radio. They were essentially the sole supporter of developments in vacuum tubes and electronic equipment, and spent considerable sums of money on encouraging manufacturers to extend radio technology. In a few years, however, the broadcast era began and industry turned to the commercial market.

The Radio Corporation of America (RCA), established immediately following the war and holding many of the essential patents, came to dominate commercial radio. From 1922 until the end of the decade, industrial research related to naval radio languished, but the boom in commercial electronics more than made up for this deficiency. In addition, the Naval Research Laboratory came into being in 1923 and was highly successful in improving fleet, shore, and aircraft radio communications. By the end of this era, therefore, naval radio facilities were fitted with the best equipment in the world.

The entire situation drastically changed in late 1929 when an economic depression swept over the world. Large firms found it necessary to have major layoffs. Consumer equipment sales fell rapidly as unemployment rose, thus forcing many of the smaller manufacturers to close. For several years, research and development by both the military and commercial interests almost ceased.

By the mid-1930s, the economic climate was somewhat improved. Although unemployment was still high, the political situation in Germany, and Japan's actions in China forced America to give new attention to defenses and ended the financial limitation of research in weapons systems. In defiance of the Treaty of Versailles that had closed World War I, Adolph Hitler and the Nazi party rebuilt the German army and embarked on an expansion of territory. This signaled to the world that another major war was inevitable. On September 1, 1939, German forces invaded Poland, and two days later Great Britain and France declared war on Germany, marking the start of World War II.

Activities as related to electronics during the period between the two World Wars are summarized in the following sections. The timelines for most of these activities are parallel; thus, they must be examined independently.

THE POST-WAR SITUATION

World War I officially ended with the signing of the Armistice on November 11, 1918, but real demobilization did not begin until after the Paris Peace Conference in June 1919. Draftees and reservists were then released as quickly as possible, and contracts for building military equipment were terminated. Within the Navy, the title of the Bureau of Steam Engineering was changed to the Bureau of Engineering in June 1920, but this had no effect on the organization of the Bureau. The Radio Division was retained, with Commander S. C. Hooper responsible for the development of communication equipment. Captain W. H. G. Bullard remained as Director of Naval Communication Services, responsible for the operational use of radio systems.

Stanford C. Hooper

Stanford Caldwell Hooper (1884-1955) has been called the "father of naval radio." He first worked as a telegraph operator, then graduated from the U.S. Naval Academy in 1906 and began a career that eventually led to the rank of Rear Admiral. After commanding a destroyer during WWI and receiving the Navy Cross, he taught electrical engineering at the Naval Academy for two years. For 11 years between 1914 and 1928, he was in charge of the Radio Division of the Bureau of Steam Engineering (later Bureau of Engineering). He received the IRE Medal of Honor in 1934 for his contributions to radio communications.

The patent situation just after the war was so confused that no manufacturer could safely sell equipment for either commercial or amateur use. Large quantities of surplus equipment were on hand, so the military requirements were very small. Further, the government no longer indemnified manufacturers against infringement. In testimony to the Federal Trade Commission, the Navy representative stated that there was not a single company among those making radio sets for the Navy that possessed basic patents sufficient to enable them to supply a complete transmitter or receiver without infringement.

The government, having indemnified manufacturers against patent infringement during the war, set up the Munitions Patent Board to determine its liabilities. Fourteen hundred contracts were examined, including over $40 million for radio apparatus purchases. Two-thirds of the equipment under consideration had been purchased or manufactured by the Navy – 90 percent since 1916 – and much was still in the original crates. The Government eventually settled practically all claims from suppliers.

In 1916, Congress had considered a bill that would give the Navy permanent, total control over all transmitters in the United States. Immediately after the Armistice, Congress again considered a bill that would, in effect, give the Government a monopoly on radio, leaving commercial operators and amateurs out in the cold. Much of this resulted from the continued attempts of Marconi Company, a British firm, to reestablish its monopolistic position. After great protests, the bill was dropped, but with the understanding that an American-controlled firm would be established to take over the major commercial communications operations.

The President approved the return of the commandeered radio communication facilities to their owners effective March 1920. However, since the Navy then owned most of the coastal stations, legislation was required to permit the continued use of these facilities for commercial purposes. Although new commercial stations were allowed, the use of naval radio facilities for commercial purposes continued at some level for a number of years

EMERGENCE OF THE ELECTRONICS INDUSTRY

The Marconi Company, wanting to regain its domination of wireless communication, recognized that it must have continuous-wave transmitters. Its officials approached Federal Telegraph to buy arc transmitters, but then switched to General Electric with an offer to buy 24 Alexanderson alternators – 14 for American Marconi and 10 for Marconi Company. The offer was with the condition that Marconi Company would have exclusive rights to using this technology. The transaction was nearly completed when a report of the proposal reached the Navy Department, raising their concern over loss of control of wireless communications to the British. Another issue concerned the ownership of American Marconi. During the war, radio had been considered such a strategic asset that it was decreed that no foreign company would be allowed to hold more than a 20-percent interest in any radio station on United States soil, and American Marconi did not meet this ownership requirement.

With the urging of the Assistant Secretary of the Navy, Franklin D. Roosevelt, General Electric turned down Marconi's offer and agreed to take the lead in establishing a new corporation to, among other things, take over American Marconi. Rear Admiral W. H. G. Bullard and Commander S. C. Hooper played major roles in the negotiations.

Radio Corporation of America

The Radio Corporation of America (RCA) was officially formed in October 1919, not as a General Electric subsidiary but as an independent U.S. firm, representing a "marriage of convenience" between private corporations and the government for the development of radio. A central feature of this deal was an agreement in which General Electric and RCA cross-licensed each other to use all radio technology they then owned or would develop in the next 25 years. This agreement became the template for the cross-licensing agreements around which other patent alliances would coalesce.

RCA immediately acquired American Marconi, bringing in all of its operations and U.S. patents, as well as hiring many of the its officials. Edward J. Nally, who had been vice president and general manager of American Marconi, became president, and David Sarnoff, Nally's right-hand man and commercial manager, became managing director. Dr. Ernst Alexanderson was named RCA's initial chief engineer and divided his time between RCA and General Electric for several years. To represent the Navy's interest, Admiral Bullard was approved by President Wilson for a seat on the new company's board of directors.

David Sarnoff (1891-1971) came to America from Russia in 1900. At age 15, he started work as an office boy for American Marconi. After studying at night, he became a wireless operator and gained experience aboard several ships. By 1910, he was a station manager, and in 1912 gained recognition for receiving the wireless distress message from the RMS *Titanic* and staying at his post for 72 hours during the disaster.

David Sarnoff

In 1913, Sarnoff became Marconi America's chief radio inspector and assistant chief engineer, then four years later, he was appointed commercial manager and also served as the primary advisor to the company's president. Soon after joining RCA, he prepared a ground-breaking memo entitled "Music Box Radio," giving glowing projections of opportunities in the entertainment field. He was promoted to general manager in 1922. Two years later, he received an appointment as a Lieutenant Colonel in the U.S. Army Signal Corps Reserves. Sarnoff ultimately became president and then CEO of RCA and is given major credit for the success of this firm.

Since no single company possessed sufficient patents to provide a complete system, other companies soon became RCA shareholders, bringing extensive cross-licensing. A cross-license agreement was executed between Westinghouse and AT&T (including Western Electric) in mid-1921, thus bringing their patents into RCA. Westinghouse, having acquired International Radio Telegraph (successor to bankrupt NESCO), had the rights to Reginald Fessenden's rotary spark and heterodyne circuits. However, realizing that General Electric's alternator was superior to that of Fessenden's spark system and that vacuum-tube transmitters – for which AT&T held the key patents of Lee de Forest – would dominate in the future, Westinghouse also became a stakeholder in 1921. Westinghouse had recently acquired rights to Armstrong's superheterodyne receiver, and this then came into the new company.

RCA also acquired from the Navy many German radio patents issued in the U.S., these earlier having been seized by the Alien Enemy Property Custodian. By the end of 1921, RCA either owned or possessed license rights to more than 2,000 patents, including the most important radio patents of that time. At the end of 1921, the ownership of RCA was General Electric, 30.1%; Westinghouse, 20.6%; AT&T, 10.3%; United Fruit, 4.1%; and original Marconi America stockholders, 34.9%. United Fruit was a participant through owning patents for the loop antenna and crystal detectors.

Because former Marconi America managers ran RCA, the company inherited the monopolistic and predatory characteristics of Marconi corporate culture. Aggressive, litigious, and monopolizing, RCA emerged as the dominant force in the industry, ready to sue, buy out, or collect excessive license fees from any fledgling electronics company in its path. RCA used its patent position to control commercial radio communications, forcing ship owners to lease or purchase RCA equipment and turn to RCA for maintenance assistance. Companies not affiliated with RCA did not hold sufficient patents to legally manufacture much of the radio equipment needed by the Navy; thus, it was not possible for the Government to purchase this equipment using competitive bidding. The situation resulted in a growing distrust by Navy officials in the company it had originally fostered. It was not until 1922 that RCA first permitted the use of its vacuum tubes on competitors' ship stations.

The agreements with stakeholders General Electric and Westinghouse was that RCA would be an operating company – continuing with the former communication facilities of Marconi America – and a marketing outlet for radio equipment manufactured 60 percent by GE and 40 percent by Westinghouse. It was expected that the communications activities would be dominant. RCA, however, soon grew in a different direction. Music and other entertainment had been transmitted by all types of stations on a sporadic basis for several years, but on October 27, 1920, experimental station 8XK in Pittsburgh began broadcasting (then called "radiocasting") on a regular schedule as KDKA. Owned by Westinghouse, it was the first licensed broadcast station in the Nation. This was followed by an avalanche of broadcast stations and a subsequent public demand for receivers – estimated at near 2.5 million by the start of 1923. Just six years after its formation, RCA's revenues from communication operations and equipment totaled less than one-tenth the sales from consumer products, particularly radio receivers and station equipment for the emerging broadcasting field – and the gap was widening.

RCA had research activities from its beginning. Alexanderson had brought a cadre of researchers from General Electric, led by antenna expert Harold H. Beverage. In 1920, they set up an experimental facility at Riverhead, Long Island, for research on antennas for an RCA communications receiving station. Beverage found that a single-wire antenna of great length gave highly directional characteristics over a broad range of wavelengths and required no tuning adjustments. Called a "wave" or "Beverage" antenna, this led to the development of a system operating at 20 MHz that enabled RCA to lead the world by handling short-wave traffic in daylight for the first time.

In 1922, RCA hired Dr. Alfred M. Goldsmith as a consultant to establish research for consumer products. The laboratory was originally in facilities at the City College of New York, where Goldsmith was a professor. Two years later, a team of engineers, scientists, and technicians under Goldsmith set up RCA's new Technical and Test Laboratory in the Bronx. There they maintained quality control and standardization of the vacuum tubes and home radios manufactured by General Electric and Westinghouse and marketed by RCA. Goldsmith coined the name "Radiola" for receivers sold by RCA during the 1922-29 period.

1922 Radiola

With radio for broadcast reception becoming a household necessity and the rising demands of burgeoning ranks of amateurs (Hams) and enthusiasts, many new radio manufacturers came into being. RCA and its affiliates, however, had a virtual monopoly through their patents, especially for vacuum tubes. Edwin Armstrong had licensed a number of firms to use his feedback patent prior to its sale to Westinghouse. These and other firms sold equipment without the necessary vacuum tubes. RCA, however, considered this to be an infringement on its tube patents, and in 1923 brought a suit against one of the small manufacturers. Radio publications joined the hue and cry raised by the public against RCA, with Congress eventually directing the Federal Trade Commission to investigate. Although the findings ultimately resulted in a satisfactory licensing program, the Department of Justice in 1930 instituted anti-trust proceedings against RCA, its affiliated companies, and its corporate parents. This was finally settled by a consent decree in 1932, under which General Electric and Westinghouse agreed to divest their interest in RCA. Despite these adversities, RCA continued to dominate the electronics industry for the rest of the decade.

Bell Telephone Laboratories

Also formed in this period was another organization that would profoundly affect electronics and communications worldwide – the Bell Telephone Laboratories (BTL). In 1869, Western Electric Manufacturing Company (first called Gray and Barton) was formed and in a little over a decade became one of the largest electrical hardware producers in the world. American Telephone and Telegraph Corporation (AT&T) was established in 1885 to run the first long-distance telephone network in the United States. It purchased a controlling interest in Western Electric in 1891, and made it the exclusive developer and manufacturer of AT&T equipment. In 1925, AT&T established the Bell Telephone Laboratories, Inc., taking over the work previously performed by the research division of Western Electric and research parts of the AT&T engineering department. The BTL was equally owned by AT&T and Western Electric. Dr. Frank B. Jewett was named BTL's president and Dr. Harold E. Arnold the director of research.

An initial major mission of the BTL was development of technologies for long-distance cable and radio networks that would be built by Western Electric, but, to do this, it also conducted basic research on a par with the best university laboratories. The BTL soon became internationally recognized as the world's preeminent industrial research facility and as a center of scientific and engineering excellence, with seminal papers published in *Bell System Technical Journal*. Within a decade, Dr. Clinton J. Davisson became the first of many Nobel Prize winners from the BTL for his experimental confirmation of the wave nature of electrons.

Some Smaller Radio Firms

General Electric, Western Electric, Westinghouse, and RCA were the giant American firms in the radio business during this period, but there were many smaller companies also making significant contributions to the technology. Representative of these were Philco, Raytheon, General Radio, National Radio, Hammarlund, Collins, Hallicrafter, and Aircraft Radio.

A small firm that started in 1892 was later reorganized as the Philadelphia Storage Battery Company, then adopted the trademark Philco in 1919. They did well as the radio boom started, first adding vacuum tube to their products, then full radios. In 1930, they introduced the Model 20 Cathedral – the home receiver that made Philco the number one radio maker in the nation. It would remain the leader in home entertainment electronics until World War II. [Author's note: In 1934, my parents bought a Philco Model 20 for me to listen to Admiral Richard E. Byrd's broadcasts being relayed into America's living rooms from the Antarctic, thus introducing me to radio.]

American Appliance Company was formed in Cambridge, Massachusetts, in 1922, by Laurence K. Marshall, Vannevar Bush, and Charles G. Smith. Their first successful product was a gaseous rectifier tube that for the first time allowed radios to be powered directly from standard AC power sources. The tube, marketed under the name Raytheon, was highly successful, and in 1926 the firm changed its name to Raytheon Manufacturing Company. Similar products quickly followed, positioning the company as a major contributor to the radio market.

Founded in Cambridge, Massachusetts, in 1915 by Melville Eastham, the General Radio Company concentrated on electrical test equipment. In a short while, it came to be recognized as the premier supplier of precision electronic instruments. Its Type C21H Primary Frequency Standard, introduced in 1928, was the most accurate instrument commercially available for the measurement of frequency and became the reference standard in laboratories throughout the world. Its publication, the *GR Experimenter*, was widely followed in the radio development field.

The National Toy Company was formed in Boston in 1914 as a manufacturer of items ranging from toys to power-plant components. In the 1920s, the company first manufactured a TRF receiver for the home radio market, then in 1930 switched to a line of regenerative short-wave receivers and registered as the National Company, Inc. It established a reputation with the amateur radio community and also developed a high-performance receiver for the Civil Aeronautics Authority (forerunner of the FAA). In 1935, the company introduced their top-of-the-line HRO (standing for Helluva Rush Order!) receiver, containing two preselection stages and a crystal filter. With few changes, this basic design survived for 65 years.

Hammarlund Manufacturing, organized by Oscar Hammarlund in 1910 in New York City, is one of a small number of firms that continued in business in the radio field without interruption down through the years. So pre-eminent were pioneering Hammarlund products that they became a part of the language of radio; early amateur radio operators affectionately referred to these as "Ham" products, and called themselves "Ham" operators. An emphasis on top quality helped develop a reputation for reliability that through the years was the hallmark of the Hammarlund name.

Just 15 years old at the time, Ham (9CXX) Arthur A. Collins won national acclaim in 1925 by maintaining reliable communications with Donald MacMillan's Arctic expedition using a home-built transmitter. In 1933, he formed the Collins Radio Company in Cedar Rapids, Iowa, and again entered the national spotlight, providing voice communications transmitters for Admiral Richard Byrd's historic second expedition to Antarctica. The firm prospered through supplying transmitters for the airline and broadcasting industries. Collins personally led the development of Autotune, an electro-mechanical device that automatically tuned radio equipment to pre-determined frequencies.

William A. Halligan, a Ham (W9AC) since a teenager, served as a wireless radio operator on the USS *Illinois* during World War I. After periods attending Tufts College, working as a journalist, and selling radio parts, he formed Hallicrafter Radio in Chicago in 1928. The company built handcrafted receivers (hence, the name) with state-of-the-art features at an affordable price. By 1938, it was considered one of the "Big Three" (National, Hammarlund, and Hallicrafter) manufacturers of amateur receivers and was selling in 90 countries.

In 1922, the booming radio industry caught the attention of Richard W. Seabury in Boonton, New Jersey, and he established the Radio Frequency Laboratories to exploit this new market. Aircraft Radio Corporation (ARC) was formed in 1927 as a subsidiary to develop aircraft receivers. It gained recognition in 1929 for providing the radio gear used by then-Lieutenant James H. (Jimmy) Doolittle in his famous "under-the-hood" landing at Mitchel Field on Long Island, proving that "blind flying" was practical. While civil aviation was its primary customer, by 1933, receivers from ARC were being installed in the first fighter squadrons of the Navy and Army Air Force.

ENTERTAINMENT MEDIA

The evolution of radio up through World War I was primarily engendered by naval and commercial communications needs. In the next two decades, however, radio as an entertainment media was the primary force driving developments. While broadcasting *per se* was not along the main stem of radio evolution, it had two very important side influences: in the overall development of the electronics industry and in the emergence of a "sister" field, television.

Radio Broadcasting

The first known amplitude-modulation (AM) "broadcast" in America was in 1906 when Reginald Fessenden used an alternator at NESCO's Brant Rock, Massachusetts, facilities to send out a Christmas Eve message. From that time on, experimenters and amateurs occasionally transmitted music and news item on an irregular basis. Many technical notables were involved in this type of activity. For example, Lee de Forest did this with an arc transmitter in 1909, and in 1916, he entertained listeners around New York City through a vacuum-tube transmitter on experimental station 2XG.

Electrical engineering and physics departments at many schools started early experimental broadcasting stations; one in particular should be noted. At the Bliss Electrical School – a college that would play an important role in the future radar maintenance training program – students started experimental station 2XH in 1913. Located in Tacoma Park, Maryland, just north of the District of Columbia, 2XH claimed to be the first broadcasting station serving the Nation's capital. The station was later converted to WBES.

The National Bureau of Standards was assigned the call letters of WWV for testing of broadcasting equipment. In May 1920, WWV began broadcasts of Friday evening music concerts. The station's 50-W transmitter operated at 600 kHz and could be heard about 25 miles. A news release on May 28, 1920, noted the significance of these broadcasts:

> This means that music can be performed at any place, radiated into the air by means of an ordinary radio set, and received at any other place even though hundreds of miles away. . . . suggesting interesting possibilities of the future.

An article in the August 1920 *Radio News* reviewed the Bureau's "Portaphone," a portable radio receiver designed to allow people to "keep in touch with the news, weather reports, radiophone conversations, radiophone music, and any other information transmitted by radio." The broadcasts of WWV were initially intended to provide a reference signal at a precision frequency, being stated as having an accuracy of "better than three-tenths of one percent." (Today, the station's many frequencies are controlled to within 1 part in 10^{13}. The well-known time announcement in telegraphic code was added in October 1945, and voice announcements began on January 1, 1950.)

Dr. Hoyt Taylor, when head of the Physics Department at North Dakota State University, built an arc transmitter and broadcast music and announcements in the Grand Forks area. In 1920, during developments in audio modulation at the Naval Aircraft Radio Laboratory, Taylor and Leo Young started part-time broadcasting to the public using call letters NSF. The U.S. Public Health Service became interested in this medium and, with the approval of the Navy Department, began public health lectures over the station twice each week. During 1922, broadcasts came from the House of Representatives and Congress, the latter including an address from President Warren G. Harding. There were so many special requests, however, that broadcasting began to interfere with research. In early 1923, this activity was transferred to Arlington, where it was restricted to programs in the public's or the Navy's interests.

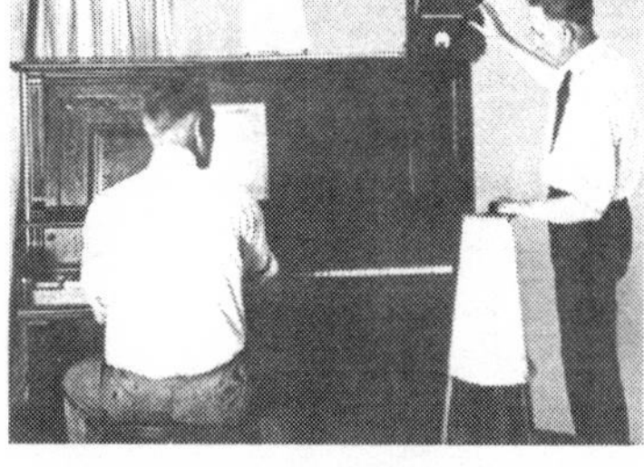

Navy Broadcasting – Young, right

Dr. Frank Conrad, assistant chief engineer of Westinghouse in Pittsburgh, personally received experimental license 8XK from the Department of Commerce. In 1916, operating from his home, he began "entertainment" broadcasts for persons with receivers in the area. Conrad eventually became deluged with mail from amazed listeners who asked for more broadcasts – more of the music and information he was sending over the airwaves. Recognizing the sales potential of receiving sets, Westinghouse applied for a license to provide a regular schedule of broadcasting. On October 27, 1920, station KDKA in Pittsburgh became the first "regular" broadcasting station in the United States, opening the door to what is often called the "Golden Age of Radio."

Other requests for licenses followed, and in early 1921, the Department of Commerce established a single common frequency of 833 kHz for this service. Later that year, 619 kHz was added for weather and market reports. In addition to broadcasters, communications stations and amateurs (Hams) were clamoring for new frequency assignments. A National Radio Conference, called by Secretary Herbert C. Hoover of the Department of Commerce, was held in early 1922, during which the radio spectrum between 50 kHz (6000 meters) and 3 MHz (100 meters) was partitioned and allocated for various types of users. A third frequency, 740 kHz, was allocated for "entertainment" broadcasting.

The three frequencies allocated for broadcasting were quickly swamped. This was followed by an avalanche of broadcast stations – 569 by the end of 1922 – and interference of signals became a major problem. To satisfy the demand for licenses, a second National Conference was held in 1923. In this, the Department of Commerce expanded the broadcast band to be from 550 to 1350 kHz, establishing channels between 550 to 1000 kHz in 10-kHz steps for "territorial" coverage. Forty-four frequencies were set aside for high-power, wide-area stations. Broadcasting interests, strongly supported by the public, demanded that the Navy give up the full 550-1500-kHz band. The Navy acquiesced and agreed to use this band only on a non-interference basis. In one way this was beneficial to the Navy – it provide an increased incentive to the commercial firms to develop improved vacuum tubes, transmitters, and receivers.

In 1924, a third National Conference extended the upper limit of the broadcast band to 1500 kHz, and recognized Canada's right to six of the high-power channels. There were concerns about the efficacy of expanding the band to 1500 kHz, "since few radios would tune that high."

During what became known as the "Chaos of 1926," the authority of the Commerce Department in controlling broadcasting was severely damaged by a Federal District Court's overriding a licensing decision. Secretary Hoover and the Department then gave up and approved essentially all applications with little regard to interference. About 200 new stations went on the air, and the resulting interference almost destroyed radio broadcasting. In addition, at that time there were no regulations on the content of broadcasts, and public complaints about obscene material arose.

Radio, originally developed as a means of communications, had become primarily a mode of entertainment, with its former use relegated to a secondary position, especially in the minds of the public. This had resulted in the necessity for tighter governmental control. Congress was stimulated to enact the Radio Act of 1927, superseding the Radio Act of 1912 and creating the Federal Radio Commission (FRC) to temporarily administer this new act. Rear Admiral (Retired) W. H. G. Bullard was nominated by President Calvin Coolidge to chair the commission.

FRC Members – Bullard, center

The FRC was given full authority to establish "avenues through the sky"– radio channels free of interference and providing reliable service over great distances – and to "preclude obscenities into the home" by enforcing rules of decorum on the licensees. "General Orders" issued by the FRC confirmed the standard broadcast band in the U.S. as 550-1500 kHz. This included the previously allocated frequencies for wide-area coverage, but limited their use to only one station each operating during nighttime hours (the "Clear Channel" stations) and a maximum power of 50 kW. By 1930, about 20 stations in the U.S. were operating at this level.

The Federal Communication Commission (FCC) was formed as a permanent regulatory agency in 1934. Its charter included regulating radio and television broadcasting, short-wave communications, and amateur radio operations. The FCC set standards for the equipment, issued licenses for stations, and examined persons who would operate and maintain the equipment. With the firm establishment of this regulatory board, the position of the Navy as Government spokesman for the industry drew to a close.

Television

The basic concept of transmitting moving pictures by radio waves came essentially with the introduction of the wireless. Paul G. Nipkow, a university student in Germany, received the first patent for an electrical imaging machine in 1884. His system used a rotating disk with a spiral of holes that scanned a viewed object onto a photocell, then sent the electrical signal to a receiver using a bulb behind a similar disc directly connected to the scan driver. The Nipkow disc was used by essentially all researchers in this field for three decades. Constantin D. Perskyi, a Russian scientist, introduced the name "television" in 1900 at a meeting in Paris to describe devices that make pictures using electricity.

Mechanical Television

Mechanical television systems were developed over the following years in many countries. Some of the best know foreign inventors were

Kenjiro Takayanagi in Japan, Boris Lvovich Rosing in Russia, Edouard Belin and Réné Barthelemy in France, Denes von Mihaly and August Karolu in Germany, and John Logie Baird in Great Britain. In the U.S., Charles F. Jenkins is perhaps the best known, but Ernst F. W. Alexanderson and Herbert E. Ives also deserve prominent places in mechanical television history.

The early television demonstration used the Nipkow disc for both image generation and reception, with the transmitter and receiver separated and connected by wires. The speed of disc rotation established the imaging rate. Since the human eye responds to individual images at about 18 per second, the rotation rate tended to be this value or above to make the images continuous. Synchronization of separately driven discs, however, was a major problem. The number of holes on the spiral determined the number of scanning lines (usually 20 or less). The image quality was limited by the number of lines and the responses of photocells and neon bulbs. Bandwidth of the connecting lines was no problem with short links, but longer telephone lines did increase the problem. When the wireless was used as the connecting channel, the bandwidth then became a significant limitation.

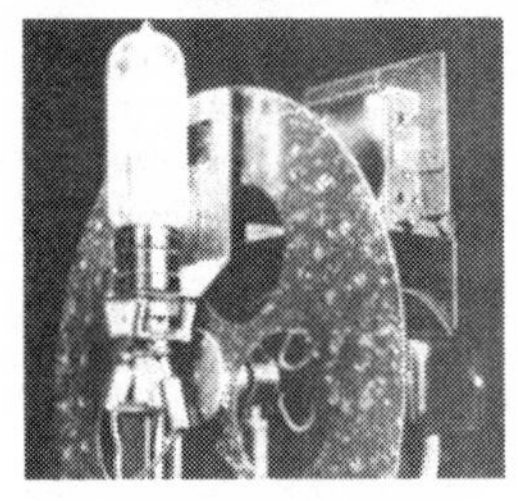

Nipkow Scanner

In Great Britain, Scotsman John Logie Baird developed a Nipkow-disc television system in the early 1920s. He first showed it to the public in early 1926, when it produced 30 lines of scan at 5 frames per second. The next year, he gave a demonstration over 438 miles of telephone lines, and followed this in 1928 with broadcasts over BBC, as well as a wireless channel from London to New York. From 1929 onward for several years, the BBC broadcast regularly scheduled television programs using the Baird camera, and receiving kits called Televisor were available in stores. Baird is generally credited in Great Britain with introducing television.

The most prominent system in America came from Charles Francis Jenkins, inventor of the modern motion-picture projector. Jenkins Laboratories was formed in 1921 for "developing radio movies to be broadcast for entertainment in the home." Jenkins developed a system using a Nipkow disc scanner and a rotating drum in the separate receiver; a patent application on this "electromechanical television system" was filed in March 1922. The system was demonstrated at the Naval Research Laboratory in June 1925, broadcasting over station NOF at Anacostia to the U.S. Post Office in Washington, D.C.

With further improvements in image quality, in 1928, Jenkins Laboratories built and sold receivers and formed experimental television station W3XK in Wheaton, Maryland. The short-wave station operated in the 26-meter (later 40-meter) band with regularly scheduled telecasts of "Radiomovies" produced by the Laboratories. These were at 48 lines and 15 frames per second. The Jenkins receiver, the Radiovision, was a standard multitube set with an attachment projecting a cloudy image onto a six-by-six-inch square mirror. It was estimated that an audience of 20,000 persons eventually bought Radiovision receivers or kits to build the imaging device for adding to a standard receiver. The U.S. Navy became a user of Jenkins's scanners, sending weather maps from Washington to ships.

Jenkins' Television

Realizing that television broadcasting was arriving in America, the Federal Radio Commission in February 1929 issued new "Rules and Regulations Governing Visual Broadcasting" that included the following:

> That visual broadcasting be designated to include both television broadcasting and picture broadcasting, or moving-picture broadcasting and still-picture broadcasting, and that all licenses issued to be of an experimental nature for a period of six months only.

The Commission authorized 2.00-2.20 MHz and 2.75-2.95 MHz for these experimental stations, giving a bandwidth of 200 kHz, which was much wider than the standard radio allocation of 10 kHz. In this it was recognized that television picture signals (later called video) were higher in frequency than audio.

Both General Electric and the Bell Telephone Laboratories also developed mechanical television systems similar to that of Jenkins. Ernst Alexanderson (of alternator fame) led the work at General Electric. By 1928, Alexanderson's system had improved to allow motion pictures, and General Electric began regular broadcasts on experimental station W2XB (later WGY) with a 48-line, 18-frame system, and sold home receivers with a 3-by-4-inch screen. Alexanderson also developed a projector system, with a screen 6-feet square. "Television for the Home," an article in the April 1928 issue of *Popular Mechanics,* gives details of Alexanderson's mechanical system.

Perhaps the most complex mechanical television system was developed at the BTL under Dr. Herbert E. Ives. The camera, designed by Dr. Frank Gray, used a Nipkow disc to sweep an intense beam of light over the object, with the reflected light collected by surrounding photocells. This produced 50 lines at a frame rate of 18. Two types of displays were used. One, a standard Nipkow disc display, produced a 2-inch by 3-inch picture. In the other, a Nipkow disc swept the output light across an array of 2,500 photocells, each connected to a neon lamp arranged behind a 1.5-by 3.0-foot glass screen. In April 1927, this system was demonstrated in sending by "picturephone" images and speech of Secretary of Commerce Herbert C. Hoover from Washington, D.C. to New York City. The press reported the image on the small screen as "exceeding clear," but on the large screen they were "not so good." There was also a transmission from BTL's experimental station W3XN in Whippany, New Jersey, to Washington, D.C. In 1929, Ives used this system in a demonstration of color television.

BTL Picturephone – Herbert Ives, left

A number of experimenters had also used a cathode-ray tube (CRT) in the receiver. For example, in 1922, Russians Dr. Boris L. Rosing and his student Vladimir K. Zworykin, devised a system that incorporated a cathode-ray tube as an imager. The basic CRT was invented in 1897 by the German scientist Dr. Karl Ferdinand Braun. In the CRT, electrons from a cathode are focused into a narrow beam that travels down the tube to strike a screen. The direction of the beam is determined by an electric or magnetic field along the tube. When used as a display, the CRT has a phosphorescent screen, and the intensity of the beam can be changed through a control grid.

Electronic Television

Analogous to spark transmitters in radio, mechanical television, even using the CRT for imaging, had severe limitations. The solution lay in an all-electronic system – from camera tube to the display in the receiver. The concept of television scanned, synchronized, and displayed by electronic means was originated in 1908 by Alan A. Campbell-Swinton, a radiologist in Scotland. In a letter published in the June 1908 issue of *Nature,* he described the "Distant Electric Vision." In 1911, Campbell-Swinton gave the presidential lecture to the Röntgen Society of London, amplifying his concept. This was eventually popularized as "Campbell-Swinton's Electronic Scanning System" in the August 1915 issue of Hugo Gernsback's popular magazine, *Electrical Experimenter.* (Here it should be noted that Campbell-Swinton's concept of television –

and it was no more than a concept – did not apply well to electronic television as it eventually evolved.)

Vladimar Zworykin

Vladimir K. Zworykin came to America in 1919 and joined Westinghouse a year later. Working in off-hours, he conceived the Iconscope – a CRT-based camera tube – and in 1923 submitted a patent application that was granted five years later. He proposed a formal television project at Westinghouse, but it was rejected by his supervisor, who suggested that he devote his time to more practical endeavors. Still working off-hours, Zworykin returned to experiments that he had done with Professor Rosing in Russia on a CRT-based display unit. At an IRE convention in 1929, Zworykin demonstrated a picture tube called the Kinescope. That same year, he completed his Ph.D. degree and shortly thereafter joined RCA to direct their new electronics research laboratory in Camden, New Jersey.

Philo Farnsworth

Philo T. Farnsworth, a highly intelligent farm boy living in Rigby, Idaho, became extremely interested in electronics, reading all that he could find on the subject. This very likely included the definitive article on television in Gernsback's *Electrical Experimenter.* In 1920, at age 14, Farnsworth conceived, and disclosed in writing to his high-school chemistry teacher, a television camera and display tube, both based on the CRT, forming a full electronic television system. It would be 1926, however, before he could obtain financial support for building his apparatus.

With funding arranged through the Crocker National Bank, a small laboratory was set up in San Francisco. At the heart of Farnsworth's system was his camera tube, called the Image Dissector. His receiver was called an Oscillite. In January 1927, patents were filed for a "Television Receiving System" and a "Television System." Farnsworth tested the full system on September 7, 1927. The image was a simple black line across a sheet, but as the sheet was rotated, the received image did likewise. A telegram was sent to the financial backers by one of the observers, "The damn thing works!" This event sounded the death knell of the mechanical rotating-disc scanner systems and initiated a rivalry between Farnsworth and Zworykin for the eventual title of "father of modern television."

By mid 1928, Farnsworth's system could generate full pictures at a rate of 10 frames per second and 50 lines per frame, comparable to images generated by mechanical systems. A front-page article of the *San Francisco Chronicle* on September 3, 1928, headlined: "SF Man's Invention to Revolutionize Television."

Television Laboratories, Inc., was established in San Francisco in 1929 to further develop the Farnsworth television system. Arch H. Brolly, a highly talented electronics engineer trained at the University of California, Harvard, and MIT, was hired as the Chief Engineer. Zworykin visited the laboratories the next year and was given full information on Farnsworth's system. When Zworykin reported back to Westinghouse, little interest was displayed and he was admonished to "work on something more useful." RCA, however, holding marketing rights to electronic products from Westinghouse, was interested and suggested that Zworykin continue with his research. Farnsworth's two patents were granted in 1930, and David Sarnoff, then President of RCA, personally came to San Francisco to see the Farnsworth system. On the spot, he offered $100,000 for the enterprise, provided that Farnsworth would then work for RCA. When turned down, Sarnoff dismissed the matter, saying, "Well then, there's nothing here we'll need."

In 1931, Philco agreed to a license arrangement with Farnsworth, and Television Laboratories was moved to Philadelphia, where the two companies shared research facilities. This arrangement, however, proved to be unsatisfactory, and Farnsworth Television, Inc., was formed in 1933, separate from Philco but still in Philadelphia. In this same period, Zworykin left Westinghouse and joined RCA to head their Camden Research Laboratory, devoted to developing all-electronic television. There an improved camera tube, the Iconoscope providing 120 lines of resolution, was publicly demonstrated in 1932. The stage was thus set for a long, intense battle between Farnsworth and RCA over television patent rights.

By 1934, British electronic firms EMI and Marconi had also created an all-electronic television system. It produced 405 scanning lines at 25 frames per second and used the Orthicon camera tube invented by RCA. This system was officially adopted by the BBC in 1936.

The U.S. Patent Office ruled in February 1935 in the case of *Farnsworth vs. Zworykin* that "Priority of invention is awarded to Philo T. Farnsworth." Naturally, RCA appealed, and with each decision for Farnsworth, appealed again. The legal fees consumed all resources of Farnsworth Television, essentially eliminating further development. On the other side, the giant RCA brought full legal force against Farnsworth while investing large sums in Zworykin's ongoing research. It was 11 years before the final settlement in full favor of Farnsworth took place. In this period, Farnsworth Television survived only through license agreements with British Gaumont, the firm that owned rights to Baird's mechanical television system. To further develop and promote the Farnsworth system, an experimental television station, W3XPF, was established in Philadelphia. This began operation in 1935 with improved Image Dissector cameras having 441 lines per frame.

The *Short Wave Listener Magazine* for April-May 1935 listed 27 experimental television stations scattered throughout the nation, some still with mechanical picture generation and the others all electronic. Zworykin continued improvements on his Iconoscope camera tube, increasing the lines to 343 and adopting a repetition rate of 30 frames per second. In 1936, RCA and its radio network, NBC, started experimental all-electronic television broadcasting from New York over W2XBS in the Empire State Building, the station used earlier by RCA for mechanically scanned television.

In 1937, the FCC made seven television channel allocations between 44 and 108 MHz. During this period, there was no national standard for television-picture generation, but most experimental broadcasts by RCA and others were then at 441 lines of resolution and 30 frames per second. Philco introduced a television receiver in 1936, followed by DuMont Laboratories in 1937, and General Electric in 1938. RCA began a regular schedule of experimental broadcasts from the Empire State Building in 1939, and also introduced the company's new line of television receivers. Although Farnsworth was eventually granted the basic patent for modern television, Zworkin and RCA must be given credit for bringing this technology to the practical stage.

In addition to entertainment, television also had military applications. In the mid-1930s, when America learned of the Japanese Kamikaze Corps being formed, Zworkin proposed that the same tactic could be accomplished by using television and remote control on an aircraft. The Navy, already developing radio-controlled planes, recognized the potential and contracted with RCA for a suitable television system. Called Block, this miniature (for its day) device combined an Iconoscope pickup tube and a transmitter. This technology, however, was far in advanced for the day and little was done with it until well into WWII, and then with only limited success.

Impact on Technology

Television required a wide bandwidth for picture and synchronization signals. The FCC frequency allocations allowed for a modulation bandwidth of 6 MHz. Designated as "video," circuits to handle wide-band signals primarily came into being as a part of television. Electronic scanning for the camera and picture tubes required circuits for pulse generation and accurate timing. Very-high-frequency transmitters with significant power were also required, as were antennas that radiated horizontal patterns. Some experimental television stations had mobile camera units with a relay transmitter for sending the picture and audio back to the main studio. These relays operated near 300 MHz at the upper end of the VHF band and used parabolic-reflector antennas.

Cathode-ray tubes (CRTs) were used for the picture display in receivers, as well as in monitors on the cameras and their controls. CRTs of the early-1930s were imported from Telefunken in Germany at high costs. They normally had screens only a few inches in size and burned out after 30 or so hours. Dr. Allen B. Du Mont deserves significant credit for developing low-cost, long-lasting cathode-ray tubes for television and other applications. In the middle of the decade, DuMont Laboratories developed CRTs into a relatively inexpensive product with a lifetime in thousands of hours, and by the end of the decade was producing a "standard" 12-inch tube.

Allen Du Mont

AMATEURS AND EXPERIMENTERS

The American Radio Relay League (ARRL), formed in 1916, was (and still is) the leading proponent of amateur radio, particularly through their magazine, *QST* (Ham code for "message to all amateurs"). Having discontinued publishing when amateur radio suffered a wartime shut-down, *QST* returned in November 1919, joyfully announcing that the President had reopened amateur radio. Many former service personnel had a first exposure to radio during the war and returned home to join the amateur ranks. In addition to licensed amateurs, there were many thousands of radio enthusiasts and general experimenters, not only from veterans but also from the public's discovery of this technology. Many were guided by magazines and books devoted to this new field; one 1920 book with world-wide popularity being *The Wireless Experimenters Manual* by Elmer Bugher.

Applicants for an amateur license, at that time issued by the Department of Commerce through a Radio Inspector in each of nine districts, were required to demonstrate technical expertise in adjusting and operating equipment and a knowledge of International Conventions and applicable U.S. laws. The code requirement was an ability to transmit and receive in the Continental Morse at least 10 words per minute and to recognize important distress and similar signals. Amateur stations were restricted to 200 meters and below (1.5 MHz and up), with input power not to exceed 1 kW. By 1923, Hams had increased to over 14,000 – up from about 6,000 at the start of the war. Their call letters continued to be started by the nine-digit identification of their radio district; it was not until 1928, however, that W and K prefixes were assigned to amateurs and experimental stations.

In July 1924, the Department of Commerce authorized new frequency bands for amateurs. They were 150 to 200 meters (1.5 to 2.0 MHz), 75 to 80 meters (3.5 to 4.0 MHz), 40 to 43 meters (7.0 to 7.5 MHz), 20 to 22 meters (13.6 to 15.0 MHz), and 4 to 5 meters (60.0 to 75.0 MHz). Spark transmission was prohibited except for operations in a portion of the lowest frequency band; then in 1927 it was totally prohibited. The Federal Communications Commission (FCC) was created in 1934, with amateur radio under its jurisdiction. Amateur licenses were reorganized into Classes A, B, and C. By 1936, there were about 46,000 licensed amateurs. Throughout the years, development of radio had been well served by pushing amateur operations to higher frequencies. In 1938, bands around 112 and 224 MHz were added. In late 1939, as America prepared for war, there were some 51,000 Hams

AVIATION RADIO NAVIGATION

With the invention of the loop antenna in 1915, this device was used on aircraft radios as the earliest form of radio navigation. Both military and civil aircraft first used fixed loops and a signal-strength meter on the pilot's control panel. Using commercial radio stations, and sometimes coastal marine stations, the aircraft was maneuvered to find the direction of maximum signal. This was soon followed by a rotatable loop with a mechanical indicator showing the direction. Known as a radio direction finder (RDF) or radio compass, this could also be used with two stations and triangulation to find the aircraft's specific position.

A further refinement introduced in the early 1920s was the automatic direction finder (ADF), a system to automatically rotate the loop antenna and continually show the azimuth to the radio station. Somewhat later, a more sophisticated approach to ADF was to use two loop antennas set at a right angle to each other, and measure the ratio of the signals from each antenna to obtain the angle to the beacon. The RDF and ADF systems were highly important to the Post Office in setting up a nationwide airmail service.

In 1907, Otto Scheller in Germany had designed a system of four antennas set at corners of a large square to generate an array of overlapping, very narrow radio beams. This antenna arrangement was used by Percival Davis Lowell in 1921 at the National Bureau of Standards for devising a system by which a pilot on an aircraft would hear either dots if on one side of a course, dashes if on the other side, and a continuous signal when on the correct course. This was demonstrated in 1924. In 1926, the Department of Commerce took over aviation safety and ordered a full development of Lowell's system.

Improvement led to the "four-course radio range" using "A" (dot-dash) on one side and "N" (dash-dot) on the other. Where the A or N signals meshed, the Morse code dashes and dots sounded a steady hum, painting an audio roadway for the pilot. The beacon transmitters operated in the 190- to 535-kHz band, with 1.5 kW power, and were spaced about 200 miles apart. Stations transmitted their own identification in Morse code every 30 seconds. Because of the antenna arrangement, each beacon provided four different pathways; the system was thus called a "four-course range." In 1928, this system was adopted as the navigational standard for air lanes in America. The system was later improved by marker-beacons installed every 20 to 30 miles along the airway to inform the pilot of his position. Each of these transmitted a distinct identification signal.

In the early 1930s, Sheller, then with the German firm Lorenz AG, used his antenna in developing a similar system called *Ultrakurzwellen-Landefunkfeuer* (LFF), or simply *Leitstrahl* (guiding beam). Intended as a night and bad-weather landing system, it was sold worldwide as the Lorenz guidance system. Commonly called the Lorenz System, it added a radio beacon near runways to assist in landing.

VACUUM TUBE EVOLUTION

General Electric began producing tubes for marketing by RCA in late 1920, and Westinghouse started supplying them a year later. Throughout the 1920s, these two firms made and distributed receiving and small transmitting tubes for the general consumer solely through RCA; however, both companies maintained a line of industrial types under their own brand. Western Electric had extensive tube-manufacturing facilities, dating back to 1914, but their products were primarily for AT&T's communications facilities. In 1930, RCA started manufacturing their own line of tubes, Radiotron; thereafter, General Electric and Westinghouse marketed their own products.

There were many other companies making vacuum tubes during this period, some under license and others operating illegally. Several of the better known manufacturers included Cunningham, Raytheon, Sylvania, Tung-Sol, Majestic, and Ken-Rad. Tubes were also available from European manufacturers, particularly Mullard in Great Britain, Phillips in The Netherlands, and Telefunken and Siemens in Germany.

While there had been major advancements during World War I in standardization, increased ruggedness, and longer life of tubes, little had been done in changing the basic design of de Forest's original triode. Up until the 1920s, vacuum tubes were used primarily for oscillators and amplifiers at low frequencies and low power. New designs were now pursued on vacuum tubes to improve the electronic performance of the triode, to increase their power-handling capacities, and to raise the operating frequency.

General Performance Improvements

In the early 1920s, two types of vacuum tubes dominated the market: the UV-200 "soft" (low-degree of vacuum) tube for detector service and a hard-vacuum UV-201 for amplifiers. The tungsten filament of these tubes required considerable power but produced a "hum" in the signal when operated from an a-c power source; thus, a husky "A" battery was required. As home radios swept into being, various tubes for these followed. In 1923, Westinghouse developed the WD-11, a tube with a lower filament power using an oxide-coated element. The UV-199 "peanut" tubes came next, offering an even more efficient filament and allowing a great saving of battery power. This same-style filament was installed in the 201 tube, which was renamed the 201-A and given a new lease on life as a result of the more efficient use of filament power.

Irving Langmuir

The "equi-potential" cathode was first invented in 1919 by Dr. Irving Langmuir at General Electric, but manufacturing problems delayed its introduction until 1927, when it was used in type 226 and 227 tubes. This type of cathode, immune to noise and hum, was heated directly from a step-down transformer operating from the a-c power line. Also introduced at this time was the type 280, a greatly-improved diode with two separate plates and a common cathode. The dual-diode was used in rectifying the a-c high-voltage from a line transformer, replacing the "B" battery. This combination of tubes resulted in the first hum-free a-c operation – batteries were no longer required.

The next major improvement came in 1929 with the type 224 tetrode – a four-element tube. The extra element, a screen grid, isolated the input grid circuit from the output plate, resulting in much more stable operation and delivering considerably greater amplification for each stage. Although invented by Dr. Walter Schottky at Siemens in Germany, the tetrode reached production status through development at General Electric. This was the first real advancement in vacuum tubes since the triode and quickly became the standard in both TRF and superheterodyne receivers.

During the 1930s, other types of tubes evolved, but, in general, these were simply improvements on the basic triode and tetrode, not new concepts. Included was the pentode, with a second grid used to suppress the reverse flow of electrons. There were tubes with 4, 5, or 6 grids, called hexodes, heptodes, and octodes, respectively; these were primarily used in frequency convertors with different signals applied to various grids. An important class was tubes having a variable amplification factor (commonly called "variable mu"), allowing cut-off to occur gradually; decreasing cross-talk and distortion. Still others were multple tube types – such as two triodes – in a single envelope.

One tube reversed the trend to multiple grids, returning to the diode without a grid. This was a gas-filled diode (often called a "glow tube") developed to serve as a voltage regulator. These contained inert gases that ionized at predictable cathode-to-anode voltages, thus holding a circuit to no more than that potential.

The desired power output from circuits, particularly from audio amplifiers and intermediate RF drivers for transmitters, brought gradual increases in the capabilities of ordinary tubes. This reached a high point in 1936, when RCA introduced the type 6L6 beam-power tetrode. This tube, capable of handling up to 30 watts, had angled plates to focus the electron stream onto a portion of the anode (plate) that could withstand the heat generated by the impact but also provided pentode behavior.

High-Power Tubes

In a transmitting system, the power radiated from the antenna is less than the input d-c power to the final tube, the difference being converted to heat within the tube. To increase the useful power – that which is radiated – tubes were needed with elements that could withstand the resulting high temperatures. Also, an external cooling means for the tubes was required.

Before General Electric had access to Lee de Forest's three-element Audion (triode) patents, they experimented on alternate methods of controlling the flow of electrons within a vacuum tube. Starting in 1916, Dr. Albert W. Hull researched the use of magnetic fields for this control, and by 1920, this resulted in what he called the "magnetron." This consisted of a cylindrical anode surrounding the cathode; around this was a coil that produced an axial magnetic field. The magnetron was tested as an amplifier (replacing the triode) and also as an oscillator, but the device did not show much promise in either application. Hull continued research on the magnetron, but for use in power conversion, not radio. By 1926, a magnetron was built that could generate 15 kW at a frequency of 20 kHz, but then little further work was done on this device at General Electric.

Albert W. Hull

With the advent of radio broadcasting in 1920 and subsequent increased commercial value in transmitters, both General Electric and Western Electric started intensive research in higher-power tubes. Rapid improvements quickly followed. By early 1921, General Electric had developed a 1-kW tube with a heat-conducting metal base, shortly followed by a 3.5-kW water-cooled tube from Western Electric. In 1922, General Electric released a 20-kW water-cooled tube, with its first use in a new communication station on Long Island. Over the next several years, the power-handling capacity of tubes steadily increased, so that by the end of the 1920s, over 20 broadcasting stations with 50-kW transmitters (using multiple tubes) were on the air in the U.S.

After RCA relaxed its patent positions in the early 1930s, a number of new companies entered the power-tube field. These included Eitel-McCullough, Heintz & Kaufman, Raytheon, Sylvania, Amperex, and Taylor. Eitel-McCullough changed its name to Eimac and soon came to be the leader in output tubes for Ham transmitters. As the amateurs bands moved higher, so did the Eimac tubes, and they also found extensive applications in the new high-frequency military equipment.

In addition to tubes for generating high RF power, vacuum tubes for controlling high d-c power were needed. In 1927, Albert Hull at General Electric developed the thyratron, a gas-filled triode-type tube used as a high-energy electrical switch. Various types of thyratrons were developed for specific applications. The most common was an argon-filled tube for switching relatively low levels of current. Larger mercury-vapor thyratrons were capable of handling several amperes. Hydrogen thyratrons could switch energy at kilovolts levels at very high speeds.

High-Frequency Tubes

Until the late 1930s, most radio systems, even those of an experimental nature, were limited to upper frequencies of about 500 MHz, primarily because of limitations in vacuum tubes. There was, however, a drive to higher frequencies for decreasing antenna sizes and in generating focused radiation beams.

Standard tubes operate with density modulation of the electrons, and the electron transit time between electrodes limits the frequency that they can handle. A few special tubes with very small inter-electrode spacing were developed to operate at higher frequencies. For receivers, RCA supplied "acorn" tubes starting in 1934. Dr. Arthur L. Samuel at the BTL developed similar receiver tubes, as well the "doorknob" tube for transmitters. General Electric developed the "lighthouse" tube, used in both receivers and transmitters. All of these tubes could function up to about 1000 MHz (1 GHz), but the output power of transmitter tubes decreased significantly beyond 500 MHz.

Acorn Tube

Lighthouse Tube

To overcome the inter-electrode limitations, special transmitter tubes were developed for velocity modulation of the electrons. These included the Barkhausen tube, the smooth-bore magnetron, and the klystron. None of these velocity-modulation devices, however, produced sufficient power for radio transmissions over significant distances. A triode can be made to function in the velocity-modulation mode by biasing the grid positive and the anode slightly negative. A resonant circuit, either external or internal, between these two electrodes completed what was commonly called a Barkhausen tube. Invented in 1920 by Drs. Heinrich G. Barkhausen and Karl Kurz in Germany, it operated at centimeter wavelengths and was used in experimental transmitters and receivers.

As earlier noted, Dr. Albert Hull developed the magnetron at General Electric in 1916. In theory, it could function at very high frequencies (determined by the size of its resonant cavity), but it produced insignificant power at these frequencies. The magnetron was independently investigated in 1921 by Dr. Erich Habann in Switzerland

and Dr. Napsal A. Zázek in Czechoslovakia. In a split-anode device, Habann was able to generate oscillations in the 100-MHz range. Zázek developed a magnetron with a solid cylindrical anode that generated frequencies up to 1 GHz. Papers on both of these devices were published in scientific journals. In 1927, Dr. Kinjiro Okabe in Japan developed a split-anode magnetron that operated at 5 GHz. Dr. Ernest C. Linder at the RCA Camden Laboratory built a split-anode device in 1932 that had an output of less than a watt at about 3 GHz. In The Netherlands during 1933, Klaas Posthumus at the Philips Physical Laboratory developed a magnetron with an anode split into four elements that generated 50 watts at 2 GHz (15 cm). This tube was publicly advertised and sold by Philips. Drs. Henri Gutton and M. R. Warneck in France improved the Philips tube, and by 1938 had developed a magnetron delivering 1 kW of pulse power at 2 GHz.

The first multi-cavity magnetrons appear to have been invented in 1936. That year at the BTL, Dr. Arthur L. Samuel patented a multi-cavity magnetron, and Dr. Hans E. Hollmann in Germany published the classic treatise, *Physics and Technique of Ultrashort Waves,* that included a description of his multi-cavity magnetron (he received a U.S. patent on this tube in 1938). Also in 1936, Russian engineers N. F. Alekseev and D. D. Malairov developed a 300-W, 3-GHz multiple-cavity magnetron, with a paper on this device published in 1940. The Japan Radio Company, cooperating with the Japanese Navy, in 1939 developed an 8-cavity magnetron with a continuous power of 500 W at 3 GHz. This used a water-cooled metal anode inside a glass envelope. Except for the Russian and later Japanese magnetrons, all of the early devices lacked sufficient power for practical applications. The Japanese tube was never put into production, and the Russians abandoned their microwave developments.

In 1935, Drs. Oskar E. and Agnesa A. Heil (husband and wife) published a paper in Germany on a linear-beam, high-power microwave tube. Likely without knowledge of the Heils' accomplishment, a similar tube was developed by brothers Russell H. and Sigurd F. Varian in 1937 at Stanford's Lawrence Livermore Laboratory. Called a klystron, this microwave device used a built-in resonant cavity that developed standing waves along the electron path. While the output power of the early klystrons was good, it was still insufficient for most transmitter requirements.

NAVY RADIO IMPROVEMENTS

When World War I ended, the Naval Communication System had the largest, most efficient, and best equipped radio communication capability in the world. Much of it was, however, rapidly becoming outdated, particularly with the rising application of vacuum tubes. A natural period of relaxation followed the war, and little was done to improve the equipment. Also, the economic and military results of a rapid demobilization were general disorganization and a loss of funding for needed improvements. The Secretary of the Navy warned, "The danger to every navy after a great war, indeed, the national tendency if history is any test, is for fleets to become stale."

The Radio Test Shop of the Washington Navy Yard was limited to assisting in the design of tube transmitters and the building of receiving systems. Research and development at other navy yards was drastically curtailed. The Naval Radio Research Laboratory at the National Bureau of Standards returned to its former work in fundamental research, particularly in studying the origin of the static caused by the earth's magnetic field. Under the direction of Dr. Hoyt Taylor, the work of improving aircraft radio communications was continued at the Aircraft Radio Laboratory. This Laboratory was also assigned the task of exploring higher frequencies for naval use.

In 1922, Taylor and his assistant, Leo C. Young, were making high-frequency radiation measurements using a transmitter on one side of the Potomac River and a receiver on the opposite side. When a ship on the river neared the transmission path, they observed that the signal faded in and out and concluded that this was due to interference between the radiation received directly and that reflected from the ship. Taylor reported this to the Bureau of Engineering with the suggestion that this phenomenon might be used for detecting passing vessels. Unfortunately, the Bureau took no action on this precursor to radar.

Naval Communications Versus Broadcasting

With budgets for military developments severely curtailed, a decision was made to encourage commercial production of standardized vacuum tubes and tube-type transmitters that could also be used by the Navy, and to limit internal developments to items and activities of special value to the Navy. Very shortly, however, industry became almost totally occupied in providing broadcasting equipment and home receivers for the rapidly evolving entertainment field – the so-called Golden Age of Radio. With the loss of industry support, essentially all research for new naval radio equipment was performed internally. This situation remained throughout the 1920s.

It should be noted that only by the Navy's earlier support of commercial manufacturers could the broadcast era have arrived in this country by 1921. In fact, most of the early household radios were essentially copies of receivers designed and developed by naval personnel. Early broadcast transmitters used the technologies developed during the war for or by the Navy.

An overall plan was developed in the Navy Department for best organizing their forces to meet the changing concepts resulting from the U.S. emergence from the war as a world power, as well as the development of aircraft as an integral arm of naval service. In this, the major fleets would be consolidated under a single command. Based on this, a fleet communication plan was jointly developed by the Bureau of Engineering and the Office of the Director of Naval Communications.

All ships would require additional receiving facilities to cover the Commander-in-Chief's circuit and the shore-stations broadcasting of time, weather, and hydrographic information. Ships would have to be fitted with narrow-band transmitters to reduce power to a necessary minimum, and with sharply tunable receivers needed to reject undesired signals. Installation would require the maximum possible separation of transmitters and receivers in each ship, as well as a means of multiple transmission and reception to reduce the number of antennas.

The Washington Radio Test Shop was assigned the task of standardizing shipboard receiving equipment with emphasis on the needed increased selectivity. On an interim basis, existing Navy-designed components were assembled and designated Models E, F, and R. Model E had two receivers to cover high-and low frequencies but was capable of simultaneous operation with a single antenna. Models F and R had a single receiver. All three models used a complicated and temperamental acceptor-rejecter circuit copied from the Royal Navy. Battleships were to have Models E and F; cruisers and some destroyers Model E, and other destroyers and smaller craft Model R.

None of these equipment assemblies were satisfactory for submarines. Earlier studies by Hoyt Taylor had shown that submerged antennas could collect low-frequency RF signals. This finding led to further experiments by the Naval Radio Laboratory at the National Bureau of Standards where a 24-kHz signal from Germany was received by a submerged submarine off the coast of Connecticut, a distance of over 3,000 miles. The Radio Shop then assembled the Model RA receiving equipment covering 16-1200 kHz for submarines, allowing both submerged and surface reception. Models RB and RC were developed for shore receiving stations.

In the new plans for communication facilities, receivers and their primary use would cover the following bands: 12-150 kHz, shore communications; 75-1000 kHz, long-range communications, including the Commander in Chief; and 1500-4000 kHz, tactical command and operations. Both arc and tube transmitters would cover 75-550 kHz, and tube systems would cover 1500-4000 kHz. The band between 550-1500 kHz had been given up by the Navy in 1923,

Vacuum-tube transmitters had been used to a limited extent during WWI, but it was recognized that ultimate replacements to the spark, arc, and alternator systems would depend on high-power vacuum tubes. While General Electric and Western Electric had made improvements to vacuum tubes during the war, little had been done to increase their output power. The British had developed a 2-kW vacuum-tube transmitter, but the most powerful one at that time in this country was 250 watts. At the end of the war, commercial companies were reluctant to continue the development of the tube transmitters, believing them to be of little commercial value. To gain interest in industry, the Bureau of Engineering released some of their limited funds to transmitter development. General Electric won a contract to develop two types of voice-modulated sets: the Model TC at 150 watts for installation on battleships and the 750-watt Model TD for use by air stations in communicating with aircraft. These were followed by the Model TF for submarines and newly introduced airplane tenders.

More powerful tubes soon followed, and in 1923, the Navy converted its 30-kW arc transmitter at Arlington to a vacuum-tube set using a 20-kW water-cooled tube from General Electric. With this, the eventual demise of Poulsen arcs, Fessenden rotating-sparks, and Alexanderson alternators in Navy communications was inevitable. It is noted, however, that the October 1920 issue of *The Electrical Experimenter* had information about a 1,200-kW alternator Alexanderson had built for transatlantic service, and several of these systems continued to be used in commercial facilities long after the vacuum-tube dominated.

Starting in 1923, Navy-designed vacuum-tube transmitters replaced the older spark units. These included Models TL, a test unit; TM, a 500-watt unit for submarines; TN, a 6-kW unit for shore stations; and TO, a 100-watt, voice-modulated unit for intra-fleet communications. High-speed (100 words per minute) keying and recording mechanisms came into being in 1924. These were installed at the shore stations, where they greatly increased the traffic-handling capabilities of point-to-point circuits.

The Naval Research Laboratory (NRL) began operations in 1923, with its Radio Division taking over the research and development activities of the Washington Radio Test Shop, the Aircraft Radio Laboratory, and the Naval Radio Laboratory at the National Bureau of Standards. The NRL Radio Division was responsible for new receiver and transmitter development, as well as research in high-frequency communications, while the Radio Division of the Bureau of Engineering continued with component and circuit development.

Initial equipment developed in the Radio Division included the RE, RF, and RG receivers collectively covering 10 kHz to 20 MHz. Transmitters were the TU, TV, TW, TX, XA, and XC transmitters covering the spectrum from 195 kHz to 4.5 MHz, omitting the 550 – 1500 kHz broadcast band. The XA and XC transmitters could also operate on harmonic frequencies, for the first time extending Navy communications into the high-frequency spectrum. Most ships and shore stations received this new equipment before the middle of the decade.

Starting with the Aircraft Radio Laboratory and continuing at the NRL, significant improvements had been made in high-frequency radio communications. Based on these, the Bureau of Engineering in 1926 decided to modernize all major ship installations during their next overhauls. At that time, however, the Navy was almost entirely dependent on its own research facilities for the development of radio equipment. Between 1925 and 1929, with the broadcast radio boom in full swing, the radio industry was occupied in providing millions of receivers for American homes and in the research on improvements to increase sales in this highly competitive market. Therefore, equipment for the modernization was primarily designed at the newly formed NRL, then, manufactured under contracts with Westinghouse, Western Electric, RCA, and several smaller companies. In this period, essentially no research or development of Navy communications equipment was performed by the manufacturers.

The Bureau of Engineering and the Office of Naval Communications developed a Naval Communications Frequency Plan that formed the basis for designs of new equipments. Very-low frequencies would continue to be used for shore stations and the Fleet for long-distance communications, but for all other applications, bands of frequencies between 2.000 and 4.525 MHz would be used. The plan also became the basis for establishing the Third International Radio Conference that was held in Washington in 1927. The Convention stemming from this Conference was of far-reaching importance in stabilizing the international use of the radio frequency spectrum and made it possible for all navies and merchant-marine shipping to operate on any part of the high seas without undue interference.

Looking toward a potential need for radio operators during future wartime, Captain Ridley McLean, Director of Naval Communications, initiated the Radio Naval Reserve in 1924. He requested the assistance of the ARRL in establishing a reserve force of skilled operators "constantly in the pink of condition." This would be in Class 6 of the Naval Reserve Force. By 1930, this force included 4,300 amateur radio operators. (In 1902, then-Lieutenant McLean had written the *Bluejacket's Manual*, the "sailor's bible" through the century.)

Depression Years and Winds of War

By the time the depression hit the country, the first receivers and transmitters of the 1926 modernization plan had been installed, and orders were already in place for many others. These included RAA, RAB, and RAC receivers. Transmitters were TAK, TAQ, TAR, and TAT for shipboard, and TAB, TAD, TAF, TAJ, TAQ, TAS, and TAT for shore basis. Most of these were received and installed by 1932. Transmitters TAU, TAZ, TBA, and TBC followed, and the Navy again had the most modern and reliable communication system in the world.

All of the "new" equipment, however, was based on 1920s technologies. With the depression, research and development within the Bureau of Engineering was reduced to the most urgent requirements, and with industry also crippled by the depression, improvements for the communication equipment did not exist. At the NRL, all available funds were used for the radio-detection work, improvements to the underwater sound equipment, and radio-controlled aircraft.

As the prospects of war increased, there were essentially no equipment reserves and few new designs under development. The Navy's need for communication equipment on new ships became so critical that it was necessary to reclaim receivers used in the 1920s and to modify commercial transmitters for ship installations. These stopgap measures and lack of reserves eventually proved to be a blessing; once funding was available at the start of the next decade, new and vastly superior equipment was developed to meet the requirements of the rapidly growing fleet.

While the fleet in this period could continue to use the old and modified radio equipment, aircraft radios were another matter. As the 1930s progressed, great improvements were made in aircraft, and their addition to the military created an urgent need for new, light-weight, reliable radio sets, particularly in the high-frequency region. This need also applied to the civil sector, and industry responded. After the first years of the 1930s, and with economic conditions improving, a number of firms developed greatly improved receivers and transmitters for commercial radio communications; notable in these were Aircraft Radio Corporation with its Type K receivers and transmitters (13.5 to 27 MHz), RCA's ARV-30 receiver (10.5 kHz to 9.05 MHz), and the Autotune device from Collins Radio, allowing transmitters to have up to ten pre-tuned frequencies (1- 18 MHz). These and others provided a head start for Navy aviation when the funding became available.

The Fourth and Fifth International Radio Conferences were held during the 1930s – in Madrid in 1932 and in Cairo in 1938. The first involved little of importance; the world economic depression had greatly slowed technological expansions. It covered further interference reduction, communication facilities for the rapidly expanding use of aircraft, and increasing the regulated spectrum upward from 23 to 30 MHz. The Cairo Conference primarily concerned the rapid increase in hemispheric and transoceanic aviation, limiting the frequency tolerance and decreasing bandwidth tables, and service use of bands between 30 and 300 MHz.

RESEARCH LABORATORIES

In the period before World War I, research supporting military technology, particularly in radio communications, had been relatively unorganized. This situation started to change when America's foremost inventor, Thomas A. Edison, made a far-reaching proposal. In an interview published in the May 30, 1915, issue of the *New York Times Magazine*, Edison said that the Navy should have an "inventions factory," modeled on those then existing in progressive American industries. Secretary of the Navy Josephus Daniels was impressed with this suggestion and immediately contacted Edison, asking him to take the lead and recruit a technical advisory group to assist in the endeavor. With typical enthusiasm, Edison gathered 24 of the best luminaries in the scientific-engineering community and formed the Naval Consulting Board (NCB) of the United States, whose charge was to identify and shepherd state-of-the-art military invention.

Acting on the recommendation of the NCB, Congress in 1916 appropriated funds for establishing a central naval research laboratory as a unit directly under the Secretary of the Navy. The onset of World War I delayed further progress on this development, but the NCB had a very active role in a number of wartime programs, including the emergence of underwater sound techniques and hardware.

Naval Research Laboratory

It was 1920 before construction of the Naval Research Laboratory (NRL) started. After more delays, it was officially opened on July 2, 1923. Thomas Edison had been very vocal in insisting that the Laboratory should be run by civilians, not Navy officials. Although most of his recommendations concerning the NRL were accepted, this one was not. America's best-known inventor resigned from the NCB and did not attend the opening ceremony in protest of placing the NRL under the direction of a naval officer. A bust of Edison is at the entrance of the present NRL main building, in recognition of his founding of this prestigious research organization.

Located on a campus along the east bank of the Potomac River at Bellevue in the District of Columbia, the NRL began life as a small-scale invention factory much as Edison and the NCB had envisioned it should. It had a few doctoral-level scientists, but most of the technical staff were engineers whose work was aimed at inventing new devices, rather than adding to what Edison believed was "an already enormous heap of untapped basic scientific understanding."

Naval Research Laboratory - 1923

As a unit directly under the Secretary of the Navy, the NRL's top management was composed of naval officers. Captain Edwin L. Bennett was the first official director, but the NRL was only one of his many responsibilities and his office was at the Navy Department, not the Bellevue campus. It was Captain Edgar G. Oberlin, the assistant director, who was the hands-on manager, making most of the executive decisions during his tenure that lasted until 1931.

The NRL absorbed most of the Navy's existing research and development activities, mainly forming the Radio Division. These included the Naval Radio Laboratory at the National Bureau of Standards, the Washington Radio Test Shop, and the Aircraft Radio Laboratory. Most of the initial personnel in this Division came from these three operations, including the superintendent, Dr. A. Hoyt Taylor, who previously headed the Aircraft Radio Laboratory.

The Sound Division was another initial activity at the NRL. After the war, the Naval Experimental Laboratory at New London had been transferred to Annapolis, becoming a part of the Naval Engineering Experimental Station. Dr. Harvey C. Hayes, who had led the sonics activities at the Experimental Station, was named superintendent of the NRL Sound Division. The Radio and Sound Divisions shared a common interest in many areas of electronics.

A short time after the NRL was formed, it established the Heat and Light Division, with Dr. Edward O. Hulburt as the superintendent. Hulburt had worked for the Army Signal Corps in Paris during the war developing radio sets for troops on the front lines. To attract and retain such highly qualified personnel, the NRL had established the policy of giving industrial rights to the inventors for patents resulting from work at the Laboratory.

The new buildings at the NRL would have been essentially empty except for 34 train-car loads of war surplus machinery and equipment that was provided by the Army. Also, all of the initial personnel brought with them their own research equipment. Congress gave an initial operating allocation of only $100,000, a sum that did not even cover the staff salaries and overhead. Thus, any Navy bureau needing work done by the NRL had to pay from its own budgets. Initially, only the Bureau of Engineering, through which the Laboratory reported, opened its purse and made use of the NRL.

One of the initial five buildings on the NRL campus was designed to be a training facility. Activities here began in 1924 with the opening of the Radio Materiel School (RMS). Although not under the NRL organization but operated by the Personnel Division of the Bureau of Navigation (then responsible for all training in the Navy), the RMS was developed to provide high-level maintenance training for enlisted personnel on the fleet communications equipment. Within a short time, the activities at this facility expanded to include a Radio Engineering program for warrant officers, as well as special courses for regular officers. Much more on the RMS is covered in following chapters.

The NRL organizational chart would undergo frequent changes in its early years, reflecting the various internal and external forces governing the perception of the role of the Laboratory in the Navy. Some additions, like the Division of Physical Metallurgy and the Division of Chemistry, would eventually grow into scientific and technological leaders. Others, like the Ordnance Section, initially funded by the Navy's Bureau of Ordnance to examine highly specialized problems, would eventually be absorbed into other units or go out of business. During the first decade of existence, the Radio Division under Hoyt Taylor was certainly the dominant NRL operation.

NRL Radio Division

The initial 18 professional members of the Radio Division were all well known to Hoyt Taylor, the superintendent. Essentially all were transferred from the merged organizations – the Naval Radio Laboratory at the National Bureau of Standards, the Washington Radio Test Shop, and the Aircraft Radio Laboratory. Taylor himself was ideally suited to the new appointment.

Albert Hoyt Taylor

Albert Hoyt Taylor (1874 – 1961), a native of Chicago, graduated from Northeastern University and became an instructor at the University of Wisconsin. He later studied at University of Goettingen, Germany, where he obtained his doctorate in electromagnetics in 1908. Returning to America, he accepted the position as head of the Physics Department, University of North Dakota. As a Lieutenant in the Naval Reserve, Taylor was called to active duty in 1917 and assigned as District Communications Officer, 9th Naval District, Great Lakes, Illinois. On his own initiative, he developed a research project in underground antennas. Following this, he directed the Navy's Trans-Atlantic Communications Systems along the East coast, then was assigned as director of the Aircraft Radio Laboratory when it moved to Anacostia. Eventually promoted to Commander, Taylor in 1919 returned to civilian status but remained as the director at the Laboratory until joining the NRL.

Taylor carefully selected the initial key personnel for the Radio Division. Dr. J. M. Miller was named assistant superintendent and director of the Precision Measurements group. Miller, a recognized expert in vacuum tubes, had previously been an associate of Dr. Austin at the Naval Radio Laboratory. Austin retired as the organizational transfer was made. Other initial directors and their groups were W. B. Burgess, Direction Finders; T. M. Davis, Receivers; L. A. Gebhard, Transmitters; C. B. Mirick, Aircraft Radios; and L. C. Young, General Research.

Burgess had worked with Dr. Kolster in developing the direction finder at the Naval Radio Laboratory. Davis, formerly a Gunner's Mate, had assisted Lieutenant Eaton in making major contributions to receivers at the Radio Test Shop. Mirick came from the Aircraft Radio Laboratory. Both Gebhard and Young had been members of Taylor's team at Great Lakes, then followed him to Belmar and the Aircraft Radio Laboratory.

Gebhard would eventually succeed Taylor as superintendent of the Radio Division. To this day, Taylor and Gebhard remain connected through the intersection of NRL campus streets named for them.

Leo C. Young

Leo Clifford Young (1891 – 1981) was closely tied to Taylor in all of his work. These two men had a strong friendship through their love for amateur radio, Taylor with call letters 9YN and Young with W3WV. In contrast to Taylor, Young's formal education stopped with a high-school diploma. He was, however, highly self-educated in radio technology. He grew up on a farm near Van Wert, Ohio, and built his first receiver when he was 14 years old. After high school, he worked as a railroad telegrapher for a number of years. Young was an avid amateur radio operator, and in 1913 set up the central control station for the Navy-Amateur Network. A member of the Navy Reserve, he was called up at the start of the war and assigned to the Great Lakes Naval Radio Station, where Hoyt Taylor was the officer-in-charge. Young followed Taylor to the Aircraft Radio Laboratory at Anacostia, and then joined him as a primary assistant when the NRL was formed.

Science writer Ivan Amato, in his 1998 book *Pushing the Horizon* that heralds the first 75 years of the Naval Research Laboratory, described the technical environment:

> In these first decades of the 20th Century, the anatomy of technology was more self revealing. Knobs turned, metal touched metal, dials swung from number to number, parts moved. The ways things worked were more obvious than in present day technologies inside thumbnail-sized chips. It was with their tools in hand and radio parts all asunder that many in the Radio Division had developed an intuitive relationship with the unseen electromagnetic waves.

A major project in radio-controlled aircraft was carried over to the Radio Division from the Naval Radio Laboratory. For a number of years, the Navy had been interested in a "pilotless torpedoplane," but the safe launching of these had been a major problem. Carl L. Norden, then a consulting engineer for the launcher and later inventor of the famous bombsight, recommended that the effort switch to a radio-controlled anti-aircraft target plane. In 1922, the project was assigned to the Aircraft Radio Laboratory, with C. B. Mirick as the lead engineer assisting Taylor. Mirick had served with Taylor as an Ensign during the war and had considerable experience with aircraft radio problems. By mid-1923,

Mirick had completed the radio-control gear that coupled with a motor-driven Sperry gyroscope. A surplus N-9 seaplane was flown out of the Naval Proving Grounds at Dahlgren, Virginia, with a check pilot along in the event of a malfunction.

The project continued under the NRL Radio Division. In September 1924, using radio control and without a check pilot, the N-9 taxied across the water, took off, and flew for 40 minutes. Various commands were successfully sent to the plane, then it was brought in to land on the water. Although damage was sustained in landing, this was considered as the Navy's first successful, completely unmanned, maneuverable, aircraft flight – the first true Navy UAV. Despite its success, official interest waned, and a year later flight-testing ended. The project lay dormant for over a decade.

Louis A. Gebhard with HF Transmitter

Most of the initial tasks of the NRL Radio Division centered on new transmitters and receivers. The Navy's loss of the 550-1500 kHz band resulted in the need for transmitters operating at higher frequencies. L. A. Gebhard and the Transmitter group developed the TU, TV, and TX series covering 2-3 MHz. Another initial task of the Radio Division was the continuation of the Naval Aircraft Radio Laboratory investigation of the use of higher frequencies. In this, Taylor and Young developed a series of transmitters, leading to the XC operating between 4.0-4.6 MHz and 2nd harmonic, with piezoelectric-crystal frequency control and producing 10-kW output power. T. M. Davis led the Receiver group in the design of two new type sets: the RE, a 10-100 kHz (low-frequency) unit, and the RF, a 75-1000 kHz (intermediate-frequency) unit. These receivers were widely used in the Navy for many years. The Receiver group also designed the Navy's first high-frequency receiver, the RG covering 1-20 MHz that met fleet operational conditions. It was initially proved on the dirigible, the USS *Shenandoah* (ZR-1), on a cross-country flight in 1924.

The Fleet was preparing for a cruise to Australia and New Zealand in early 1925. To examine the merits of high-frequency communications, a global experiment was planned by the Radio Division. With cooperation of the ARRL, amateurs worldwide would attempt communicating with the Fleet flagship, the USS *Seattle* that was equipped with the NRL XC transmitter operating at 5.7 MHz and an RG receiver. The same type receivers were also at the NRL and at San Francisco, Honolulu, and Balboa. Commander S. C. Hooper, now on sea duty, was the Fleet Radio Officer and was responsible for the equipment operation aboard the *Seattle*. The tests were very successful, with good communications between Washington and the *Seattle* even when the ship was in Melbourne, almost 10,000 miles away. Following these tests, immediate plans were made to equip ships and shore stations with transmitters capable of up to 18 MHz. Within a few years, this led to the U.S. Fleet as having far more reliable communications capability than any other fleet in the world.

It had been observed that communications between 2 and 4 MHz were better during the night but those between 4 and 12 MHz were better in daylight hours. Taylor arranged with a number of geographically distributed amateurs to gain data on this phenomena. These experiments resulted in the discovery of a "skip distance" – a zone of silence that varied with frequency. The existing theory of reflections due to the ionosphere (then called the Kennelly-Heaviside Layer) did not explain this. In 1925, Taylor teamed with Hulburt of the Heat and Light Division to develop a more generalized mathematical account of propagation based on refraction in the ionosphere. The resulting paper, "The Propagation of Radio Waves Over The Earth," was published in *Physical Review* 27 (1926). Hulburt later credited this work with putting the NRL on the scientific map because its significance went beyond military communication and into the realm of basic science. Much later, a panel from the American Physical Society would rate this work as one of the 100 most important applied research papers of the century.

Besides helping the Navy guide its use of high-frequency radio communication, the findings of Taylor and Hulburt were used to assist in the 1929 flight by Admiral Richard E. Byrd over the South Pole. For several legs of his flight, Admiral Byrd used a high-frequency aircraft radio system built especially for the mission by the Radio Division. The operating frequencies for each leg were selected using the new knowledge about the ionosphere.

With the onset of the Great Depression, the Navy's appropriations were cut to the core. At one time, closing of the NRL was considered, but, overall, the operations were only slightly reduced and all of the key personnel were retained. Some work in the Radio Division during this period included development of the decade frequency synthesizer and incorporation of the Dow electron-coupled circuit, devised to eliminate crystal controls in high-frequency transmitters. Work was also begun on a 60-MHz tactical radio transceiver.

An event of long-term significance to the NRL occurred in 1930. Lawrence A. Hyland was determining the directional characteristics of an aircraft receiving antenna by using a signal from a nearby transmitter.

He observed that the signal strength delivered to the receiver fluctuated when an airplane flew overhead. Leo Young recognized that this was similar to the "interference" that he and Hoyt Taylor had observed in 1922 while at the Aircraft Radio Laboratory. A report was duly sent to the Bureau of Engineering suggesting that this might be the basis of a radio-based target detection method. Although funds for continuing the investigation were not gained at that time, Hyland's chance observation did represent the NRL's first activity in what would eventually be their most important development – radar.

During the next several years, unfunded experimentation on signal interference as a potential for detection continued at a low level. Recognizing the advantages of higher frequencies for this application led to investigations with microwaves, but the lack of transmitters at these wavelengths soon brought this to a halt. By 1934, the economic situation had improved, and the NRL received more funding for research. In a short while, work on radio-based detection, and later ranging, was central to the NRL Radio Division. Other activities did include revival of the development of radio-controlled aircraft for anti-aircraft targets and limited research in high-frequency communications.

The radio-controlled plane project, discontinued in 1924, was revived in 1935 with the initial development of an anti-aircraft target designated the NT. The aircraft would be developed at the Naval Aircraft Factory in Philadelphia, and the necessary radio equipment by the NRL Radio Division. The Navy assigned Lt. Commander Delmar S. Fahrney to lead the project. The British had already developed such an aircraft, called the "Queen Bee," and Fahrney followed this by calling his aircraft a "Drone." The drone, converted from the Curtis N2C-2, was first flown in March 1937, with Fahrney riding on board to perform takeoff and landing.

Curtis N2C-2 Drone

In August 1938, the Navy first used the drone as an aerial target for the aircraft carrier, USS *Ranger*. Although the personnel of the firing battery were well trained, they failed to score a single hit on either of two runs over the ship by the drone at 12,000- to 18,000-feet slant range. In tests somewhat later, the gunners brought down the drone simulating a dive-bombing attack. From that time on, the Navy had a number of types of remotely controlled target aircraft.

NRL Sound Division

Following the conclusion of World War I hostilities, most of the Navy's research in undersea acoustics was terminated, except for a small group headed by Dr. Harvey C. Hayes at the Naval Experimental Station in New London. In 1919, this group was moved to Annapolis, Maryland, and consolidated with the Naval Engineering Experimental Station. When the Naval Research Laboratory was opened in 1923, the underwater sound activity was moved from Annapolis to the Naval Research Laboratory, forming the Sound Division. At the start, this involved only Hayes as the superintendent and one other scientist, but soon reached a half-dozen researchers. Their primary objective was the development of an ultrasonic echo-ranging system using electronic amplification. The ultrasonic frequency range (15 kHz and higher) was emphasized because this is above much of the noise from the vessel and ocean and it can more easily be focused into narrow beams.

Harvey C. Hayes

The first experimental sets of this ultrasonic echo-ranging apparatus, designated XL, used a transducer made with a quartz slab sandwiched between steel discs 16-inches in diameter. This transducer operated both as a projector (transforming electric pulses into sound vibrations and projecting them into the water) and as a hydrophone (picking up sound vibrations and converting them into electrical pulses). The 4-inch assembly was attached outside the hull. Upon application of a voltage impulse, it produced a one-quarter second duration of supersonic pulses in a sharp, cone-shaped beam. Upon striking a target, a small amount of this energy was reflected back where the assembly acted as a receiver, converting the one-quarter second "ping" into an electrical output that was heard on headphones. The target was in a direction normal to the transducer. The maximum range was only about three-quarters of a mile, and that only if the operating vessel was slowed to a few knots. Systems of this type were installed and tested in several naval vessels during 1927.

In 1928, the transducer was rebuilt using Rochelle salt crystals, which were much more sensitive than those of quartz. Designated the JK, it was placed into operation in 1932 as a passive (receive-only) device and could detect propeller noise at up to three miles. Improvements on the JK extended the range up to six miles, even while the operating vessel was traveling at nine knots. The JK was used during search operations to discover targets and bring them into range for the XL to take over and direct the attack. The Washington Navy Yard manufactured the JK quartz-steel echo-ranging equipment. By 1933, the NRL Sound Division integrated the XL and JK into a single system known as the QB. To reduce the turbulence and subsequent noise, Goodrich Tire & Rubber developed a spherical cover of sound-transmitting rubber. This permitted speeds to about 10 knots before water noise became excessive. Since this was the maximum speed of submerged submarines at that time, a fully satisfactory underwater detection system came into being. Under the personal supervision of Dr. Hayes, performance tests of the QB were made on the recently commissioned U.S.S. *Cuttlefish* (SS-171). Submarine Signal was given a contract to build 30 of these systems. Improved versions of the QB would be vital in the forthcoming Battle of the Atlantic.

Early tests on echo-ranging equipment showed significant variations in performance that were attributed to temperature gradients in the water. The Woods Hole Oceanographic Institution conducted experiments in association with the Sound Division to understand these effects. The velocity of sound propagation was a function of water temperature, thus affecting the range measurements. Also, with gradients in temperature, the propagation line would be bent, affecting the accuracy of the target direction. It was thus necessary to take these conditions into account in calibrating and the using the echo-ranging equipment.

At Harvard University, Dr. George W. Pierce, who had earlier worked in the New London group, developed an ultrasonic generator using the magnetostrictive effect. In this effect, certain materials mechanically deform in the presence of a magnetic field. In 1933, at the Sound Division, Hayes used Pierce's device to develop magnetostrictive tubes, replacing quartz and Rochelle salt-crystal transducers for some applications. These were small electromagnets consisting of a coil of wire wound on a hollow nickel-alloy tube. A changing flux from the coil caused the tube to elongate and produce a vibration in an attached diaphragm. In the reverse, sounds striking the diaphragm caused a change in the magnetic flux and generated an electric signal from the coil. The magnetostrictive tube was quickly adopted by Submarine Signal in improved QB systems. It was also adopted for the Fathometer.

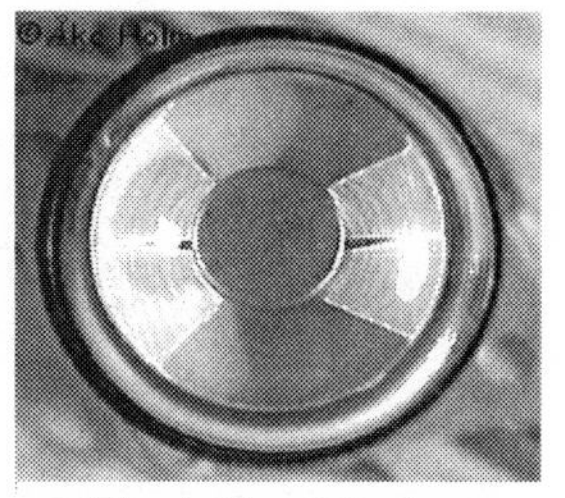
Magic-Eye Display

Because of the width of the radiated beam, it was difficult for the operator to determine the direction of the target using the intensity in the headphones. In 1935, the "Magic-Eye" or "Cat-Eye" indicator tube came into the market from RCA. Actually invented by Dr. Alan B. Du Mont, this was an ordinary-sized vacuum tube with a small fluorescent screen on the top. Electrons from the cathodes struck the screen, generating a glowing, green-ring display. The shape resembled a cat's eye, with the opening proportional to the voltage applied to a control grid. This was used by the Sound Division in a display for the operator, readily allowing a visual measure of the returned signal strength as a function of beam direction.

As the equipment evolved to provide 360-degree sweeps, the rotation of the beam pattern made it impossible for an operator to follow the changes with his unaided senses. In 1939, the Sound Division developed a radically improved display. A cathode-ray tube with a persistent screen had its spot driven in a spiral path, synchronized with the aim of the transducer. The returned signal was then displayed along this spiral, with the distance from the center proportional to the transmit-receive delay – the range to the target. The path of the spot on the screen was thus a map of the path of the active area. The spot brightness was controlled by the intensity of the received signal, rendering the bearing and range of the target visually perceptible.

This was called a plan-position indicator (PPI) and represented a major advance in display technology. If several targets were in the field, they would be displayed in their proper relative positions. Echoes from reefs or sand banks appeared on the screen as brightened areas. Thus, a scanning system with a PPI provided the operator with a complete map of the underwater situation. This technique was quickly adapted by the Radio Division for a similar display in its emerging target-detection systems.

The term SONAR was coined much later (in 1945) by the Navy. This was an acronym for Sound Navigation And Ranging, and soon came to be the name sonar. This name covers all types of underwater sound devices used for listening, depth indication, echo ranging, ship-to-ship underwater communication, and other uses.

Through the years, the activities of the NRL Sound Division were closely allied with those of the Radio Division, each centering on improvements in electronics. The respective superintendents, Dr. Hayes and Dr. Taylor, were the leading scientists in the Laboratory and worked together in promoting the NRL. In fact, Hayes was Taylor's main supporter in obtaining the first appropriated funding for radar research. Although the Sound Division had systems that used water for propagation and the Radio Division used electromagnetic waves, equipment for both involved transmitters, receivers, signal processors, and displays.

Signal Corps Laboratories

At the beginning of World War I in 1917, the Army Signal Corps had established a training facility for signal troops in east central New Jersey. Shortly after opening, it was named Camp Vail, after Alfred M. Vail, an inventor associated with Samuel F. B. Morse. Later that year, the Army determined that a need existed for a research laboratory devoted to radio and electronics, and established the Signal Corps Radio Laboratories at Camp Vail. Activities in the new laboratories centered on the standardization of vacuum tubes and the testing of equipment manufactured for the Army by commercial firms. Experimentation was also being done on radio communication with aircraft, and on aircraft detection using sound.

After the end of the war, aviation communication was transferred to the Signal Corps Aircraft Radio Laboratory at Wright Field in Dayton, Ohio. The Radio Laboratories continued at a low level, centering on design and testing of radio sets and field-wire equipment. The facility survived as an Army installation when all Signal Corps schools, both officer and enlisted, were moved to Camp Vail. The initial curriculum in the officers' division included courses in radio, telegraph, and telephone engineering, as well as signal organization and supply. For enlisted personnel, there were courses in radio, electricity, photography, meteorology, gas engines, and motor vehicle operation. In the 1920s, the facility was elevated to Fort Vail. The training operation there was named the Signal School and expanded to accommodate the ROTC and National Guard, as well as personnel from other services. In the two decades between the World Wars, over 4,600 enlisted men graduated from the school, about half being Signal Corps personnel and the remainder representing 16 other branches or services and foreign nations.

In 1925, Fort Vail was renamed Fort Monmouth. Although overshadowed by the Signal School and at a reduced scale due to budget restrictions, the Radio Laboratory remained an important activity at Fort Monmouth. Developments included a variety of radios for telephone and telegraph communications. Coupling capabilities in electronics and meteorology, in 1929 the laboratory developed and launched the first radio-equipped weather balloon.

Going into the 1930s, continuing decline in economic conditions led to the consolidation of the Signal Corps' widespread laboratories. The Electrical and Meteorological Laboratories and the Signal Corps Laboratory at the NBS were moved from Washington, D.C., to Fort Monmouth, as was the Subaqueous (Underwater) Sound Ranging Laboratory from Fort Wright, New York. The Aircraft Radio Laboratory, however, remained at Wright Field. On June 30, 1930, the consolidated operations became the Signal Corps Laboratories (SCL). The initial SCL had a personnel strength of 5 officers, 12 enlisted men, and 53 civilians. Dr. (Major) William A. Blair was named the director.

This Laboratory was responsible for the Army's ground radio and wire communication development and for improvement of the meteorological service. The next year, the SCL was also made responsible for the detection of enemy surface vessels and submarines from coastal stations by under-water sound, as well as the detection of aircraft by electromagnetic radiation. While the number of personnel was evidently inadequate for major work in these many and diverse areas, it must be noted that Blair, the director, was personally knowledgeable in all of them.

Blair's 1906 doctoral dissertation from the University of Chicago centered on experimental studies of microwave reflections. After graduation, he was an aerological specialist with the U.S. Weather Bureau. In 1917, he had prepared a report, "Meteorology and Aeronautics," for the NACA (National Advisory Committee for Aeronautics, predecessor of NASA) that was widely circulated as a basic handbook. When World War I started, Blair was commissioned as a Major in the Aviation Section of the Army Signal Corps Reserves. Following the war, he remained in the Army as a meteorologist and participated in planning the first round-the-world airplane flight in 1924. While attending the Command and General Staff College, he developed an interest in acoustical direction-finding for antiaircraft artillery. Later, he was involved with sounding-balloon observations at extreme altitudes.

During the 1920s, the Army Ordnance Corps at Frankfort Arsenal had made tests detecting infrared emitted from airplane engines or reflected by their surfaces. When the SCL was formed, this work was transferred to that Laboratory. Carrying this forward, in 1931 Blair initiated Project 88, "Position Finding by Means of Light." Here "light" was used in the general sense of electromagnetic radiation, including infrared and the very-short radio waves with line-of-sight transmission characteristics. Blair's primary assistant in these studies was Dr. (Major) Harold A. Zahl.

Initially, emphasis was placed on special detectors with tuned amplification for detecting reflected infrared from a pulsed searchlight. Civilian physicist S. Herbert Anderson led this effort. In 1932, a blimp was tracked using this equipment to a distance of over a mile, but then further pursuit of active detection techniques was abandoned because of the limit of infrared energy available from searchlight sources. Although Zahl and Anderson continued research in the passive detection of infrared emitted from aircraft motors, Blair became convinced that practical detection systems would involve reflected radio signals. No doubt he was influenced by his earlier doctoral research, and was also aware of work in this that had been done at the NRL.

In early 1932, the report from the NRL on airplane detection by radio was passed on to the SCL by the Secretary of War with the comment that "the subject is of extreme interest and warrants further thought." Blair's return comment was that he was already aware of the Navy experiments and "the possibilities of this method of finding airplanes are being considered." From that time on, there was a complete exchange of information on radio detection projects by both services.

University and Private Laboratories

Some universities in this era had organized research work that advanced the applications of engineering and physics to military technology, but few had major efforts in this area. One notable exception was MIT. As early as 1926, Dr. Edward L. Bowles of the Communications Division of the Electrical Engineering Department started experimental studies of radio antenna patterns and infrared radiation with specific application to the military. Dr. Vannevar Bush, who would later head the U.S. Office of Scientific Research and Development, participated in the research. Bowles later turned to microwaves for blind-landing techniques. In 1932, Dr. Wilmer L. Barrow initiated major research in millimeter and microwave radio transmission and hardware, providing the foundation of what would eventually become MIT's Radiation Laboratory. There was also outstanding work in MIT's Instrumentation Laboratory, started by Dr. Charles S. Draper in 1932. Dedicated to scientific instruments for accurately measuring and studying motion, in its early years the laboratory pioneered the development of gyroscopically based gun-pointing and firing systems for shipboard and fighter-aircraft use.

By the 1930s, very few private, non-commercial laboratories still existed. One that must be mentioned was financed and operated by Alfred Lee Loomis at Tuxedo Park, an exclusive residential area about 60 miles north of New York City. Educated as an attorney and very wealthy from years as a Wall Street tycoon, Loomis also had an insatiable appetite for scientific research. He formed the Loomis Laboratories at Tuxedo Park in the 1920s, and personally paid for a cadre of highly qualified researchers to pursue studies in a wide variety of areas, including ultrasonics, physical optics, and electromagnetics – foundation stones of sonar and microwave radar.

Alfred L. Loomis

Loomis gained international scientific attention in 1927 when he and Robert W. Wood, a renowned experimental physicist from Johns Hopkins University, published "The Physical and Biological Effects of High-Frequency Sound-Waves of Great Intensity" in the respected British *Philosophical Magazine and Journal of Science.* In the Wall Street crash at the beginning of the Great Depression, Loomis was one of a few shrewd investors who actually improved their wealth, and the Loomis Laboratories thrived. In the early 1930s, Loomis did personal research in the accuracy of time-keeping devices, coordinating with scientists at the Bell Telephone Laboratories and using a dedicated telephone line to compare time signals. He later turned to research on signal processing of brain waves, publishing a number of papers in association with Dr. Hallowell Davis of the Harvard Medical School that led to the development of electroencephalography.

Tuxedo Park

As the U.S. mobilized for war, Loomis was anxious for his Laboratories to make significant contributions to these efforts. In early 1939, Dr. Karl T. Compton, president of MIT, suggested that Loomis Laboratories join in a project on distance finding by radio. Loomis was immediately intrigued by this and discussed it with Dr. Vannevar Bush, then president of the Carnegie Institution of Washington but still a professor at MIT. Loomis suggested using microwaves in a plane-detector for anti-aircraft weapons that would "include a fairly simple computer which would control the gun directly."

A team at Loomis Laboratories was assembled and began research in microwave technology. Within a short time, a simple Doppler microwave object-detection system was demonstrated. Bush was named chairman of the National Defense Research Committee in June 1940, and appointed Loomis as the chairman of the Microwave Committee. In the remaining pre-war period and throughout the war, Loomis and his private laboratory made important unpaid contributions in advancing military science and technology.

Another private organization that should be mentioned is the Hammond Radio Research Laboratory, established by John Hays Hammond, Jr. The son of a wealthy family and a protégé of Thomas Edison and Alexander Graham Bell, Hammond graduated from Yale and started his privately funded research activities in 1911.

John H. Hammond, Jr.

In the years leading up to World War I, Hammond concentrated on developing an "automatic course stabilizer" for the remote guidance of naval vessels. Called "Iron Mike" and used in association with a Sperry gyroscope, the system was satisfactorily demonstrated to the Army Coast Artillery in 1918, piloting a small craft from an airplane at five-miles distance. That year, however, the Navy decided that a radio-controlled underwater torpedo would have better tactical value. Hammond then redirected his efforts toward controlling a submerged torpedo without the antenna projecting above the water. In 1925, a successful run at six-feet depth was made with the controlling station about three miles distant. Further work was discontinued for a decade.

Hammond Castle

During the 1930s, Hammond continued with his private research efforts at his seaside "castle" near Gloucester, Massachusetts. With many assistants, he concentrated on military projects including a multi-channel radio, altitude determining system, secrecy radio communications, a paravane device for mine protection, submarine sound transmission, and incendiary shells. His significant non-military developments included synchronization of motion pictures and sound, radio dynamic controls, television communications, the "dynamic amplifier" (today's stereo), gaseous detector of radiant energy, a cosmic ray detector, and the regenerative piano and harmonic organ. Producing 437 patents and ideas for over 800 inventions, he was one of America's premier inventors. His most important work remained the development of remote control by radio, earning him the title, "The Father of Remote Control."

The Carnegie Institution of Washington should also be noted. Beginning in 1895, Andrew Carnegie, then often called the richest man in the world, contributed his vast fortune toward establishing a number of organizations to carry on work in art, education, international affairs, peace, and scientific research. Carnegie founded the Carnegie Institution of Washington (CIW) in 1902 as an organization for scientific discovery. His intention was for the institution to be home to exceptional individuals – men and women with imagination and extraordinary dedication capable of working at the cutting edge of their fields. Some of the early researchers included Dr. Edwin P. Hubble, who revolutionized astronomy with his discoveries concerning the universe, and Dr. Charles F. Richter, who created the earthquake measurement scale. The Department of Terrestrial Magnetism (DTM) was founded in 1904 to map the geomagnetic field of the Earth. Over the years the research direction shifted, but the historic goal – to understand the Earth and its place in the universe – remained the same.

From its start, the DTM had a close relationship with Johns Hopkins University. Students from Johns Hopkins regularly used facilities at DTM for their graduate research. After the formation of the NRL, there was considerable interaction between these two organizations. In 1925, at the same time that Hoyt Taylor and Edward Hulburt were developing their theory of the ionosphere at the NRL, Dr. Gregory Breit and Merle A. Tuve, researchers at the DTM, were attempting to measure the height of this layer. Breit was already a well-known experimental physicist, and Tuve, an avid Ham, was a graduate student at Johns Hopkins. They borrowed from the NRL one of Louis Gebhard's crystal-controlled transmitters for conducting these experiments, but had problems in extracting height information from the continuous returned signal. Leo Young suggested pulsing the transmitter, and built a device using a square-wave generator for this purpose. This solved the problem, and it was determined that the ionospheric height varied both during the day and the time of year within a range of 55 to 130 miles.

Gregory Breit & Merle Tuve

Unfortunately, Breit and Tuve did not receive full credit for their discovery. In England starting in 1924, Edward V. Appleton had used the signal from a BBC station in London and a receiver in Cambridge to make ionospheric height measurements. His results were published before those of Breit and Tuve, and he ultimately received the Nobel Prize for this and other discoveries concerning the ionosphere.

Tuve used this ionospheric-probing work for his 1926 doctoral dissertation and continued with the CIW, making major contributions in electronic equipment for particle accelerators. One of his assistants in this effort was Dr. James A. Van Allen, discoverer of the radiation belt around the Earth. Tuve would later gain major recognition as the leader of the proximity fuse program and eventually become director of the DTM. Dr. Vannevar Bush, then dean of engineering at MIT, was appointed president of CIW in 1937, and strongly influenced the development of programs in advanced electronics.

THE STATUS OF ELECTRONICS TECHNOLOGY

Significant advancements had been made in radio technology during the 1920s, particularly driven by the broadcast industry, radio amateurs, and the Navy. While slowed by the economic depression, these advancements continued into the 1930s. Although coined earlier, *Electronics Magazine,* founded in April 1930, popularized the term "electronics." The magazine subheading read "Electron Tubes - Their Radio, Audio, Visio and Industrial Applications." A brief review will be provided of electronics technology during this era and its status just before World War II. The Navy's electronic systems at that time incorporated this technology, and their technical breadth and relatively sudden evolution greatly contributed to the crisis in maintenance.

Circuits

After the introduction of the triode vacuum tube as an amplifier in 1912, electronic circuits had been devised to perform many additional functions. Special circuits included feedback, both positive and negative, with the former leading to oscillators and improvements in receivers, and the negative bringing improvements in frequency response and in automatic gain control in amplifiers. The cathode-follower became popular in non-amplifying, isolating applications. The application of electronics in television and underwater acoustics, as well as industrial controls and physics research, gave rise to many new special circuits for timing and control functions.

Amplifiers

Early amplifiers used in audio devices, such as telephone repeaters and modulators in transmitters, were designed to be as linear as possible, with the output voltage following the wave shape of the input signal. Low- and high-frequency cutoff was mainly limited by the a-c characteristics of the coupling transformers. Introduction of capacitance coupling greatly extended the high-frequency characteristic, and direct coupling (confusingly called "d-c") made bandwidths from zero to tens of kHz possible.

Attention was also given to reducing noise and signal distortion, the former largely as "hum" coupled from the power source and "microphonics" from mechanical vibration of the vacuum-tube elements. Where the input was very-low-level signals, thermal and "shot" noise were natural limitations in the background. Distortion was caused by phase shifting, harmonic generation, and cross-modulation, all reducible by creative design. Invention of the negative-feedback circuit in 1927 by Harold S. Black at the BTL greatly improved the overall characteristics of amplifiers. By the mid-1930s, particularly with the advent of multi-grid vacuum tubes, audio-frequency amplifiers were readily available with low noise, low distortion, and essentially flat over the 20-20,000 hertz range.

Typical amplifiers had an efficiency of only about 30 percent. To increase the efficiency, beginning around 1930 tubes were sometimes biased to intentionally make them non-linear, actually giving an output signal only during a fraction of the input sinusoidal signal. Operation over the full cycle was designated Class "A," while Class "B" was cut off for up to half the cycle and Class "C" for half or more of the cycle. A "push-pull" amplifier used two tubes operating at Class "B," but with the input 180 degrees out of phase, the combined output being essentially linear and the efficiency greatly increased.

At radio-frequencies, a Class "C" amplifier could use a resonant circuit (then called a "tank circuit") in the output to reproduce the missing half cycle, giving a sinusoidal signal at an excellent efficiency. This intermittent driving, however, produced highly undesirable harmonics in the output. At the BTL in 1936, Dr. William H. Doherty designed a circuit using two tubes (one called the "carrier" and the other the "peak") that provided an efficiency of up to 60 percent but with greatly reduced harmonics. The Doherty amplifier was particularly beneficial in microwave transmitters where the available tubes had low power outputs.

The efficiency under any class of operation decreased in inverse proportion to the current through the tube. For a given power input, it was thus desirable to have the tube plate voltage as high as practical. The linearity of tubes was also better at higher voltages. Thus, even receivers operated with plate supplies (called "B" supplies) at more than 100 volts, and it was common for high-power transmitters to operate with 1,000 or more volts. This was not only dangerous (particularly during maintenance) but also required substantial power supplies having heavy transformers, capacitors, and chokes (inductances). By the end of the 1930s, improvements in these components substantially decreased the weight of both transmitter and receiver power supplies.

Televising brought the need for another type of amplifier – the video. Picture signals and synchronization pulses required bandwidth covering both traditional audio- and radio-frequency ranges, commonly 50 Hz to 6 MHz. These were basically audio amplifiers with compensation circuits – usually shunt or series inductances – to extend the upper frequency limits

Oscillators

Any amplifier circuit that can supply its own input signal with the proper phase and amplitude will generate oscillations. In general, this is done by having a portion of the voltage in the output load fed back approximately 180 degrees to the input. Many circuits can be used for this purpose, some using components that result in oscillations at audio frequencies (nominally 20 to 20,000 Hz), and others resulting in radio-frequency oscillations. Shortly after making the first vacuum-tube amplifier in 1912, Lee de Forest "accidentally" invented the oscillator when the output was coupled back to the input and a "squeal" resulted. Edwin Armstrong at Columbia University and Irving Langmuir at General Electric had also made similar discoveries at about the same time, and it was not until 1934 that the courts awarded de Forest patent rights on the regenerative oscillator.

Although the initial squeal was at an audio frequency, the first practical oscillator – de Forest's ultra-Audion – operated at radio frequencies by using a resonant "tank" circuit between the plate and grid for feedback. Several types of RF feedback oscillators evolved during the 1910s, including the Messiner, Hartley, Colpitts, and tuned-grid-tuned-plate. The quartz crystal with various cuts was introduced in 1921, substituting as the tuned circuit at the grid of a tuned-grid tuned-plate oscillator. The crystal provided precision frequency control, especially when the crystal was placed in a temperature-controlled "oven."

In 1931, Lt. Comdr. J. B. Dow at the Bureau of Engineering's Radio Division invented the electron-coupled oscillator (ECO), a circuit using the recently developed tetrode as both an oscillator and power amplifier. The ECO, which provided stability near to that of a crystal-controlled circuit while allowing the frequency to be varied, revolutionized transmitters.

Also in the early 1930s, the grounded-grid oscillator was developed, increasing the frequency up to about 200 MHz. This was a variation of the Colpitts oscillator with an inductor between the grid and the plate of a triode and the inter-electrode capacitance completing the resonant circuit. A circuit of this type was developed using a quarter-wave transmission line in the resonator; this particularly found applications as the local oscillator at up to about 300 MHz in VHF superheterodyne receivers.

As operating frequencies increased, vacuum tubes correspondingly decreased in size to reduce the inter-electrode capacitance. This also decreased their individual power-handling capacity. Two circuits were developed to provide multi-tube oscillators: the push-pull, a two-tube circuit giving twice the power of a single tube, and the ring oscillator, using any even number of tubes. The ring-oscillator circuit was invented in 1938 by Robert M. Page at the NRL. This circuit also reduced the effect of inter-electrode capacitance, allowing operation at up to 400 MHz.

At ultra-high frequencies, conventional oscillators failed because of the electron transit time within the vacuum tube, and the Barkhausen and split-anode magnetron tubes were adopted to provide low power at microwave frequencies. While these oscillators were used with some improvements during the 1930s, the only real breakthrough in this period was the microwave klystron tube invented in 1937 by the Varian brothers (Russell and Sigurd). Then in 1940, the resonant-cavity magnetron, developed by John Randall and Harry Boot, came into being, changing forever high-power oscillators in the microwave bands.

At the other end of the frequency scale, oscillators at audio frequencies were also developed. The first of these was the beat-frequency oscillator (BFO), wherein signals from two RF oscillators were mixed using Reginald Fessenden's heterodyne principle to produce a third frequency equal to their difference. Starting in the 1920s, the BFO was commonly used in superheterodyne receivers to produce an audio tone from the keyed CW signal, beating a second oscillator against the IF signal.

The BFO lacked stability because of temperature-induced changes in RF circuits, particularly capacitors. A concept originally developed near the beginning of the century by Max Wein at Telefunken was used in the early 1930s to devise a vacuum-tube audio oscillator without tuned circuits. Called the Wein-bridge oscillator, the frequency was solely determined by resistor and capacitor combinations; thus, it was also called the R-C oscillator. This circuit was much simpler than the BFO and, in addition, it was considerably more stable. The Wein-bridge circuit quickly became the standard for audio oscillators.

Modulation

The function of modulation is simply the superposition of a signal upon a carrier signal of higher frequency. In this era, it was usually amplitude modulation (AM) of an RF carrier by an audio signal, and was most often accomplished by adding a high-level audio signal to the plate input d-c power of the final Class C radio-frequency amplifier. The modulation could also be accomplished by injecting the audio signal onto the grid of the final RF stage; this required little audio power, but the efficiency of the Class C was greatly reduced. The original AM technique used plate modulation of the RF oscillator, but this resulted in an unwanted frequency variation and was soon discontinued.

The spark-gap made a comeback in 1940 as a means to pulse-modulate high-power RF output tubes, particularly the resonant-cavity magnetron. A spark-gap, placed across the high-voltage source, is discharged by a relatively low-voltage trigger pulse. As in the original spark-gap transmitters, this discharge produces an electrical impulse into a resonant circuit – in this case a series of such circuits – that forms the intensive pulse delivered to cathode-modulate the RF power tube. The spark gap has an advantage over the vacuum tube in that it does not require filament power and dissipates very little heat energy, resulting in an efficiency approaching 90 percent. It can be designed to handle currents of 100 amperes or more at very high voltages.

Under amplitude modulation, information is solely contained in carrier harmonics – the side bands. Under full (100 percent) modulation, 50 percent of the power is in the carrier and 25 percent in each side band. The carrier component, therefore, might be suppressed without affecting the information being transmitted. Further, both side bands contain the complete information, and one might be suppressed without degrading the information. Single-side-band (SSB) communication involves suppressing the carrier and one side band and results in a signal covering only half of the normal bandwidth and also saves two-thirds of the power. The most common technique was to use two tubes in a balanced arrangement (the van der Bijl amplifier).

In 1914, the carrier and side-band concept was mathematically expressed, and the concept of SSB transmission quickly evolved. The next year, John R. Carson at the Naval Radio Station in Arlington conducted experiments that substantiated the SSB concept, and applications of the technology followed. AT&T quickly realized the advantages of SSB for carrier-based telephone systems, and the first trans-Atlantic SSB telephone communication took place in 1923, the year Carson's patent was awarded. The technology, however, was very slow in penetrating the radio communications market, likely because of lack of interest in frequency-spectrum conservation. During the 1930s, there were few users of SSB, but later it came to dominate radio communications. It might be noted that SSB was an exception to technology improvements by Hams; although some amateurs used it starting in 1934, they did not really take it up until it was accepted on the commercial side.

Demodulation involves the extraction of the modulation information from the carrier. Often called detection, this occurs in the receiver through the rectification and filtering of the carrier, or, in the case of the superheterodyne receiver, the intermediate frequency. Through the years, rectification was most often accomplished by a diode, but the silicon-crystal detector – the first solid-state device dating from about 1910 – was and is still used, particularly at microwave frequencies. For SSB demodulation, it was first necessary to re-insert the suppressed carrier; this was usually done by using a beat-frequency oscillator.

In 1933, Edwin H. Armstrong announced frequency modulation (FM), a totally new technology. This was Armstrong's third major contribution to radio, the first being the regenerative circuit in 1912 while he was still a student, and the second being the superheterodyne receiver, developed in 1918 while he was serving with the Army Signal Corps in France. FM involved changing the carrier frequency in proportion to the information being transmitted while the carrier intensity remained constant. Such a transmitted signal contains all of the side bands of the AM transmission plus higher-order side bands that are multiples of the modulation frequency, resulting in a much wider bandwidth.

Edwin H. Armstrong

Armstrong started an FM broadcasting station in the middle of the Great Depression, and the technology was quickly used in transceivers for police and similar applications. However, FM was not adopted by the military for almost a decade, and it was still another decade before it impacted the entertainment media. Despite his inventive genius, Armstrong, holder of 42 basic patents and a respected professor at Columbia University, spent the latter part of his life in patent lawsuits and eventually committed suicide, believing that he was a failure.

Special Functions

Shaping of signals was important for many applications. Clipping circuits were used for changing sine waves to square waves, and involved clipping diodes, clipping amplifiers, and clipping with cathode followers. Simple resistor-capacitor (R-C) and triode circuits were developed for generating saw-tooth, trapezoid, stepped, and spiked signals. They were also used for clamping signals to offset values.

One of the most used circuits was the multivibrator; either free-running or driven by an external signal, these produced nearly rectangular signals of a specific, possibly adjustable, duration. This basic circuit, called the "flip-flop," was invented in 1919 by Dr. W. H. Echols and F. W. Jordan, then "reinvented" by Dr. C. E. Wynn-Williams at Cambridge University in 1931 for the first "scale-of-two" counting circuit. It soon found application as a timing device in many electronic systems and much later was at the heart of electronic computers.

Microwaves

Although microwaves (loosely frequencies at 1,000 MHz or 1 GHz and higher) had been researched since the beginning of radio, at the start of the 1930s they were still a laboratory curiosity. This was primarily due to the lack of a power source at these frequencies. The magnetron was invented by Dr. Albert W. Hull at General Electric in the early 1920s, but was not perfected for practical use at that time.

In 1937, Russell H. and Siguard F. Varian, brothers working at Stanford University, developed the klystron, an amplifier tube that could produce microwave energy at levels significantly higher than any other device at that time. It was quickly adopted at the Radiation Laboratory (later to become the Berkeley National Laboratory) of the University of California where there was extensive research in microwaves for driving its linear particle accelerator. It was also adopted in experimental microwave transmitters. Sperry Gyroscope Company underwrote the klystron development and in turn received the patent rights. A miniature version, the reflex klystron, was developed in 1940 by Robert W. Sutton in Great Britain, and became standard as the local oscillator in superheterodyne microwave receivers.

A number of special vacuum tubes capable of operating at low microwave frequencies came on the market in the mid-1930s. These included the "acorn" tube (type 955) developed at RCA by Browder J. Thompson and George M. Rose, Jr. and a similar tube from Dr. Arthur L. Samuel at the BTL. Samuel also developed the "doorknob" tube (type 316A) used in transmitter, as well as the first low-power, multi-cavity magnetron. General Electric released the "lighthouse" tube (type SC46), developed by Dr. C. Guy Suits for both receivers and transmitters.

In 1936, the waveguide was independently developed by Dr. Wilmer L. Barrow at MIT and Dr. George C. Southworth at the BTL. Barrow also invented the Magic T, a device for splitting microwave signals. Another significant contribution to transmission lines came from DuPont, where, in 1938, Dr. Roy J. Plunkett developed polytetrafluoroethene (later trade-named Teflon), making flexible coaxial cable practical. These cables are suitable at wavelengths greater than 20 cm; at shorter wavelengths, their loss becomes excessive and waveguides are required. Philip H. Smith of the BTL developed a circular presentation of the complex impedances of microwaves transmission lines. The Smith Chart, first published in the January 1939 issue of *Electronics* Magazine, is still in common use.

Since the beginnings of radio, attempts were made to concentrate the emitted electromagnetic radiation into narrow beams. This required antennas, or arrays of antenna elements, comparable in size to the radiation wavelength. As ultra-high frequency operations evolved, antenna elements centimeters in length were possible and the parabolic reflector became common for forming transmitting and receiving beams. With the advent of waveguides, many special devices were developed for coupling to the transmitter and receiver and to the antenna itself, including provisions for allowing the parabola to sweep and for rotating an off-center antenna element to produce conical scanning. Somewhat later, Dr. Arthur C. Crawford at the BTL invented the horn antenna that coupled directly to waveguides. Of particular importance in microwave systems was the transmit-receive (T-R) switch that protected the receiver from the transmitted signal. Developed by Dr. Arthur H. Cooke at Oxford University, the spark-gap T-R switch was used in several configurations for this function.

Displays

Although the cathode-ray tube (CRT) had been a laboratory curiosity for many years, in the early 1930s Dr. Allen B. Du Mont contributed greatly to their evolution by developing low-cost, long-lasting CRTs for experimental television and other applications. Television pictures were then in black and white and changed at the rate of 30 frames per second; thus, the phosphor used inside the screen glowed white and had a low persistency. As the CRT came to be used for other types of displays, various phosphors were used, especially one giving a green glow and with a long persistency.

CRTs were (and still are) identified by a tube number; common ones in the late 1930s included 2AP1, 3BP1, 3FP7, 5BP4, and 7BP7. The first number identified the diameter of the tube face and the first letter designated the order in which a tube of a given design was registered. The letter-digit combination indicated the type of phosphor: P1 producing a green light at medium persistence, P4 a white light with short persistence, and P7 a green light with long persistence. If a modification was made after the original design registration, a letter was placed at the end.

Display CRTs used either magnetic or electrostatic deflection of the electron beam, the former involving coils outside the neck of the tube and the latter using internal plates. In addition, either a coil or permanent magnets was needed for focusing. The electron beam was accelerated toward the screen by a high potential between the cathode and an anode located inside of the forward portion of the tube; this potential was typically several kilovolts. The brightness of the spot was dependent on the intensity of the electron beam, and this was controlled by a grid as in ordinary vacuum tubes. While the anode voltage was high, the supply could provide only very low current so it was not as dangerous as it sounds. In fact, the greatest danger associated with CRTs was from dropping or striking them, resulting in an implosion when the high vacuum was broken; the internal metal elements then fired out as from a gun.

CRTs used for radar and sonar displays required phosphorescent screens of long persistency, holding until new information arrived. Those displaying information on target range and direction had three general types of scanning. The A-scan was a horizontal line indicating range, with a vertical displacement showing target echo strength. Type B-scan used a vertical line for range and a horizontal displacement indicating the azimuth angle. The P-scan had a line rotating about the center of the screen, showing range as a radial displacement and the angular position indicating the azimuth angle; this was commonly called plan-position indication (PPI). Less used was C-scan with azimuth and elevation displayed horizontally and vertically, respectively, with no range indication, and J-scan, similar to type A but with range displayed circularly.

In addition to the high-voltage power supply, electronics for these displays included a timing circuit to determine range, a video amplifier for the echo signal, and special circuits for developing the sweep voltages going to the vertical and horizontal displacement coils or deflection plates. Much attention was given to designing highly linear sweep signals and their synchronization with the radiation emitter. For the PPI, mechanically rotating coils could be placed around the CRT neck, or the rotation could be produced electronically.

Just before entrance into the war, military CRTs and their associated electronics were primarily built by RCA, but other manufacturers included Dumont (note spelling simplification) and Sylvania. The state-of-the-art CRT for PPI-scanning was the 12DP7 with magnetic deflection and requiring 7,000 volts at the anode.

Controls

Various types of control systems had been used since machines were first invented. Isaac Watts' original steam engine had a spinning governor to prevent the machine from "running away," and James Clerk Maxwell developed differential equations to model this control. As electrical machines entered industry, they were accompanied by suitable electrical control units. Electronics entered the controls industry in the late 1920s with the development of the thyratron vacuum tube for safely switching relatively large currents.

During the 1930s, there were significant advancements in synchros and servomechanisms, electrical devices for remote positional control. A synchro is a small a-c machine used for electrical transmission of angular-position data; Selsyn and Autosyn were two of a number of trade names. Synchros operated in pairs, one the generator and the other the receiver. They do not involve torque amplification, so they are normally used to drive dials and move control valves. The servomechanism, commonly called servo, is a somewhat similar device. Two common trade types were the Ward-Leonard and the Amplidyne. Servos have amplifiers to handle large loads from small torque inputs, and include feedback to sense and automatically correct the positioning. The servo amplifier might be hydraulic or pneumatic, but electronic amplifiers became dominant in that era. Simple analog computers were introduced in these systems, automatically handling the associated differential equations for automatic control. Positional control systems found early military applications in ship steering and gun-aiming, and in the late 1930s became important for positioning antennas

.

Power Sources

Prime power for most electronic equipment in the United States had long been standardized at 60 Hz (cycles per second) and 120 volts single-phase or 208 volts three-phase. This usually came from the commercial power grid, but large ships had their own central power source. Where neither of these sources was available, prime power might be supplied by a generator driven by a combustion engine. These auxiliary generator sets often had higher output frequencies – some over 1,000 Hz – allowing much less iron in the generators and transformers, and thus lighter weight. Prime power on aircraft and small ships usually came from the main battery and engine-driven generator, normally producing 24 to 28 volts d-c.

By the 1930s, most electronic equipment had built-in transformers to change the voltage of the prime power to various a-c levels, mainly those for vacuum-tube filaments and inputs to the high-voltage supply. Power-rectifier vacuum tubes had long been available for converting the a-c to d-c, and these were used in the high-voltage supply. The pulsating d-c from the rectifier was smoothed by chokes (inductors) and condensers (capacitors). The advent of the electrolytic condenser in 1932 greatly reduced the physical size of these supplies. This component was invented at the Ergon Research Laboratories of Magnavox by Dr. Julius Edgar Lilienfeld, a prolific physicist who had immigrated to America due to the increasing persecution of Jews in Germany. The fundamental theories and practice laid out by Lilienfeld remain in use to the present day for aluminum capacitors (a $6B/year business worldwide).

For airborne electronic equipment, conversion from the d-c voltage of the prime source to the voltages needed by the equipment was more complicated. The simplest method was to use a vibrator to interrupt the circuit, then use a conventional high-voltage supply to convert this back to d-c at the desired level. Vibrator supplies, however, were only useful at power levels of some 100 watts and lower. For higher power, the dynamotor was an early device used for directly converting between levels of d-c voltages and was simply a d-c motor and a d-c generator built into one unit. This device had no means of regulating itself, however, and suffered from an inconsistent output when the load varied.

A rotary converter is a device to change a-c to d-c. A single winding produces a steady d-c voltage, but separate input and output windings are necessary for different d-c voltages. An inverter is a rotary converter operated in reverse, with a d-c input and an a-c output. As airborne electronics increased in usage, the inverter gained widespread acceptance, and it was in this device that most developments of power sources took place during the 1930s. Inverters used multiple windings to increase stability, and units of varying power and voltage levels were designed to be mounted together with the electronic units being served.

Personnel maintaining electronic systems that depended on auxiliary power, such as motor-generator sets and inverters, often found that more time was required in tending to the power sources than in repairing the electronics

Instrumentation

Precision electronic instruments were indispensable to the developments in the 1930s, but at the beginning of that era many such instruments had not evolved to be commercially available and were usually constrained to basic research laboratories. They were usually very expensive, physically difficult to transport, and functionally inadequate for use in developmental and test environments. Many electronic instruments were introduced or improved during the 1930s. Three of the most important were the oscilloscope, the vacuum-tube voltmeter, and the precision audio-frequency generator; these are briefly described below. Others included transconductance vacuum-tube testers, frequency standards and measurements equipment, audio distortion analyzers, and radio-frequency field-strength meters.

The cathode-ray tube dates from 1898, but it was not until 1931 that the full oscilloscope was developed by General Radio. Believing that the product was too expensive, difficult to manufacture, and with a limited market, GR sold the design to Dr. Allan B. Du Mont in 1934. At the DuMont Laboratories (slight difference in spelling), improvements were made by Dr. Thomas T. Goldsmith, including the use of Du Mont's recently developed low-cost, long-life CRT. The DuMont Released in 1936, the Type 208 oscilloscope was a landmark in the history of electronic instruments. It was the first truly commercially available oscilloscope and was purchased by laboratories throughout the world. It had a stable time base, a signal amplifier with a previously unheard of bandwidth of nearly 1 MHz, and was rugged and dependable. It was to prove to be the most significant product of the DuMont Laboratories.

DuMont 208

Accurate measurements of d-c and a-c voltages had always been difficult because of the loading effect of the meter. The triode vacuum tube, with a very large input resistance, was first used by Dr. Eric B. Moullin of Cambridge University to improve such measurements, but his circuit was very non-linear and lacked the stability needed for reliable measurements. The negative-feedback amplifier, designed by Harold Black at the BTL in 1927, gave the needed linearity. Alan D. Blumlein of EMI in England invented the differential circuit in 1936, giving the needed stability. All of these were incorporated to produce a practical Vacuum Tube Voltmeter (VTVM) that by the end of the 1930s was commercially available from RCA, Hewlett-Packard, General Radio, and Ballantine Instruments.

Various types of audio-frequency generators had been commercially available for many years, but the Model 200A, released in 1939 as the first product of the Hewlett-Packard Company, immediately set the industry standard. Designed by William R. Hewlett while in graduate school at Stanford, it had greater stability, but was about half the price, of previous instruments of this type. This was a resistance-capacitance (Wein-bridge) oscillator and used an incandescent bulb as a temperature-stabilized resistor in a critical portion of the circuit. The oscillator covered 20 to 20,000 Hz in three ranges. The Hewlett-Packard Company, founded jointly by Hewlett and fellow student David Packard at the suggestion of their mentor, Dr. Frederick W. Terman, was the first electronics firm in the Palo Alto Research Park, an area that eventually spread and became known as Silicon Valley.

By the end of the 1930s, catalogs of electrical and electronic instruments showed the availability of a wide range of types, quality, and price. To ensure reliability and precision, the government found it necessary to set standards and specifications for equipment to be used by the military.

ELECTRONICS EDUCATION

During the 1930s, there was not a large number of people in the United States knowledgeable in the detailed maintenance of electronic equipment. The Federal Communication Commission (FCC), formed in 1934, was responsible for regulating radio and television broadcasting, short-wave communications, and amateur radio operations. Among other things, the FCC examined persons who would operate and maintain the equipment. By the end of the 1930s, there were about 51,000 licensed amateurs (Hams) and an estimated 8,000 licensed commercial operators.

In addition to commercial and amateur operators, other people were involved in the ever-increasing consumer, industrial, and military radio products. Electronics maintenance was at the component level – chip, board, or module replacement did not exist. When equipment failed, technicians systematically traced signals, checked voltages, tested components, and repaired defects. While this usually did not require four-year college training, maintenance personnel did need a fundamental understanding of radio theory. The FCC's examinations for both amateur and commercial licenses strongly emphasized theory.

Formed in 1912, the Institute of Radio Engineers (IRE) was by the 1930s the dominant professional organization for persons working in radio communications. The *Proceedings of the IRE* and the *Bell System Technical Journal* were the two primary publications for advanced theory, while *Electronics Magazine* was the main source of up-to-date information for most practitioners.

On the Civilian Side

Bachelor's degree programs in electrical engineering were offered at many colleges and universities by the 1930s, but the typical curriculum was significantly different from that following World War II. Before the war, the emphasis was on the highly practical side of this field. Few programs included mathematics beyond introductory calculus. There were no digital computers, and all calculations were made using one of the many types of engineering slide rules. One of the more enduring textbooks was *A Course in Electrical Engineering* (McGraw-Hill 1922, 1928, & 1934) by Chester L. Dawes of Harvard. In two volumes (*Direct Currents* and *Alternating Currents*) and with no mathematics beyond elementary algebra and trigonometry, these books covered about everything needed up to specialized topics in the last two years of an undergraduate curriculum. Dawes also had a less-detailed two-volume set, *Industrial Electricity* (McGraw-Hill, 1924 & 1937), which was also very popular.

Most electrical engineering programs included a communications or radio option. Perhaps the most popular author of textbooks in this option was Frederick E. Terman of Stanford University. His *Radio Engineering* (McGraw-Hill 1932, 1937) and companion *Measurements in Radio Engineering* (McGraw-Hill, 1935) were broadly used. Terman covered the same material in *Fundamentals of Radio* (McGraw-Hill, 1938) but with less mathematics; this book was used at many colleges, including the U.S. Naval Academy.

On the upper side of formal education, only a few colleges and universities offered radio engineering graduate study. The more theoretical (read "mathematical") aspects of this field were mainly found in courses of physics departments; a good example of a textbook was *Theory of Thermionic Vacuum Tubes* (McGraw-Hill, 1933) by E. Leon Chaffee, a professor of applied physics at Harvard. Actually, more students earned higher degrees in physics than in electrical engineering.

The limited extent of advanced theory in undergraduate electrical engineering programs in this decade is shown by the continuing training offered by several of the larger industries. The BTL had extensive offerings in what they called "out-of-hours" courses. Academic credit could not be earned, although notes from many of these courses became university textbooks. An example was *Network Analysis and Feedback Amplifier Design* by Hendrik W. Bode, originally prepared in 1938, then later published as a graduate-level textbook (Van Nostrand, 1945). Perhaps the most extensive of the industrial programs was the Advanced Course in Engineering at General Electric; here a select number of recent graduates participated part-time in their first two years of employment. A well-known textbook to come from this program was *Mathematics of Modern Engineering*, by Robert E. Doherty and Ernest G. Keller (Wiley, 1936).

Operating between trade-school and senior-college levels was a number of excellent schools in the United States during that period. One of the best known was Capital Radio Engineering Institute (CREI), which offered very good home-study programs. RCA Institute had an outstanding two-year resident program in radio and television engineering. The Melville Aeronautical Radio School provided electronics maintenance training primarily for the airline industry. Bliss Electrical School, located just outside Washington, D.C., in Tokoma Park, Maryland, had intense, one-year programs in electrical engineering and radio communications.

For persons who wanted to study on their own, *Practical Radio Communications* (McGraw-Hill, 1935) by Arthur R. Nilson and J. Lawrence Hornung provided an excellent preparation for the very tough FCC commercial radio license examination. Another popular 950+ page tome was *Radio Physics Course* by Alfred A. Ghirardi (Radio & Technical Publishing, 1933). There were also the many editions of the *ARRL Handbook* (at $1 per copy) and the *Radio Handbook* (Editors & Engineers Publishers), primarily for persons studying for their FCC advanced amateur license.

As the national defense built, the need for personnel in engineering and similar fields far exceeded the normal output of schools. Further, the military draft was significantly affecting the college enrollment. Recognizing the urgency for training highly qualified personnel for defense industries, Congress in 1940 created the Engineering Defense Training (EDT) program within the Office of Education. EDT later expanded by adding Science and Management to become ESMDT. Overall, 214 institutions of higher education participated, providing college-level evening or extension instruction to help allay shortages of engineers, chemists, physicists, and production supervisors.

With all expenses borne by the Federal Government, many thousands of persons received advanced courses through the EDT / ESMDT program. In California alone, the program prepared about 100,000 individuals for "higher-echelon industrial positions." Records at several major schools, including Princeton and Drexel, indicate that they first admitted women to engineering studies through this program. (It might be mentioned that the author completed courses in Radio Electrical Engineering during 1941-42 through EDT from the Alabama Polytechnic Institute, predecessor of Auburn University.)

Navy Radio Materiel School

Advanced training in radio maintenance had a high priority in the U.S. Navy during this era. In 1924, shortly after the Naval Research Laboratory was opened, space in one of the initial buildings was allotted to the Radio Materiel School (RMS), providing maintenance instruction for enlisted personnel on the NRL's radio products.

There was also a Radio Engineering School for Warrant Officers in the NRL facilities. Established in 1927, each class had six Radio Electricians – about ten percent of the total Warrant Officers with this classification – devoted to a period of intense study and potentially leading to their appointment as Chief Radio Electricians. In addition, the NRL school provided special courses for Communication Officers, but for longer training these persons usually attended one of the military radio programs set up by Harvard and similar universities.

The RMS was six-months in duration. Admission required passing a rather stringent examination covering mathematics, electricity, and elementary radio theory. A manual, "Preparation Required for Candidates," provided example problems from these topics, and it was recommended that well-known textbooks also be studied. The third edition (October 1934) of the preparation manual warned that even after passing the entrance examination, approximately 30 percent of the men in classes to that date had failed to graduate.

Students were normally Seaman First Class or Third-Class Petty Officers with Radioman or Electrician ratings, all having a number of years duty in the fleet. Nelson Cooke, who later played a major role in the RMS, graduated in about 1928. One of the best-known students from the early days was future entertainer Arthur Godfrey, who finished in 1929. Top graduates who also had other qualifications were potentially in line for appointment as Warrant Officers, with subsequent promotion to Radio Electrician. Instructors at the RMS were all petty officers or warrant officers – commissioned officers were never used for direct instruction in this or other schools for enlisted men. A Radio Electrician was usually responsible for the various areas of instruction.

Although from its start the RMS was the most technically demanding school for enlisted men in the Navy, it was further upgraded in the mid-1930s. The entrance examination was made tougher; it was recommended that persons preparing for the exam devote two to three hours in private study each day for an extended period. An intensive "review period" was given at the beginning of the course, and more theory was added to the curriculum. The slide rule and calculations were given greater emphasis, and textbooks to supplement the school notes were adopted. These included the highly regarded publication *Principles of Radio* by Keith Henney (John Wiley, 2nd ed. 1934).

As the Navy's electronic equipment became even more complex, it was recognized that maintenance personnel needed to better understand the theory behind the hardware and, for this, they needed a curriculum somewhat like that in college electrical engineering programs. In 1938, Nelson Cooke returned to the RMS as a Chief Radio Electrician to develop more stringent mathematics instruction.

By 1940, the RMS had increased to eight months in length, with the curriculum divided into Primary and Secondary parts. The three-month Primary still included a review, but emphasized the theoretical portions of the program; this was followed by the five-month Secondary, centering on specific equipment and its maintenance. Cooke's notes became the "bible" for the Primary instruction, and he designed a special slide rule that was manufactured by K&E for the RMS. Cooke's notes were eventually published as *Mathematics for Electricians and Radiomen* (McGraw-Hill, 1942), possibly the most used textbook ever on applied mathematics through trigonometry and vect-or algebra.

Chapter 3

THE RADAR AVALANCHE

By the mid-1930s, the foundations of remote detection of objects by radio waves were possessed throughout the world. Researchers in the United States, Great Britain, Germany, Japan, Russia, Italy, France, The Netherlands, and several other nations had independently developed rudimentary detection systems. The potential value of these systems by the military was quickly recognized; therefore, essentially all further developments were closely held as national secrets.

The initial detection systems gave only indications of the presence and direction of a target, often using interference or Doppler effects. More practical detection systems, however, needed to also indicate range or distance to the target. There are many conflicting claims as to where and when radio-based detection with ranging was first developed. In preparing this book, care has been given to citing accomplishments and dates, and it is shown that this was first accomplished at the Naval Research Laboratory (NRL) in the United States.

The following sections give a brief summary of radar development in America and Great Britain prior to the U.S. entering World War II, while developments during the war are summarized in Appendix I. This Appendix also gives a summery of other wartime developments in electronics. Radar developments in other countries are briefly described in Appendix II.

PRECURSORS OF RADAR

The fundamental concept of reflecting electromagnetic waves predates the invention of radio. As far back as 1886, Heinrich Hertz gave laboratory demonstrations to show that electric inductors could reflect his waves. This led to many activities in which radar was "almost" invented. A few of these precursors that are frequently mentioned in the literature are briefly recounted – some have already been mentioned.

Twenty-two year old Christian Hülsmeyer formed a company in Dusseldorf, Germany, to apply Hertz's reflections to detecting the presence of ships. In 1904, Hülsmeyer registered German and foreign patents for an apparatus, the *Telemobilskop* (Telemobiloscope). An article, "The Telemobiloscope," was published in a 1904 issue of *The Electrical Magazine*. This was described as an anti-collision system using a 50-cm wavelength spark-gap transmitter and a coherer detector. The radiated signal was beamed by a funnel-shaped reflector and tube that could be aimed. The receiver used a separate vertical antenna with a semi-cylindrical movable screen.

Hülsmeyer

Hülsmeyer gave a public demonstration of his system, receiving reflections from a ship nearby on the Rhine River and causing an electric bell to ring as long as the ship was in the transmitted beam. Later improvements increased the detection distance to 3,000 meters. Although a number of patents were granted, Hülsmeyer was never able to gain financial backing or to sell his patents to industry, so the effort was dropped.

During the First World War, another German, Richard Scherl, apparently without knowledge of Hülsmeyer's previous work, designed the *Strahlenzieler* (Raypointer), a wireless device using echoes for detection. Assisted by a well-known science-fiction writer and engineer, Hans Dominik, they successfully produced an experimental set working at about 10-cm wavelength. Scherl sent details of his apparatus to the Imperial German Navy in February 1916, but his offer was rejected as "not being of importance to the war effort."

In 1922, Dr. A. Hoyt Taylor and Leo C. Young at the Naval Aircraft Radio Laboratory were testing a 60-MHz transmitter and receiver that they had built. In driving the receiver around in a car, they noticed that the buildings would cause the signal from the transmitter to fade in and out. To get away from this interference, they placed the transmitter and receiver at fixed sites on opposite sides of the nearby Potomac River. At first, the signal was strong and steady; then, as Young later stated,

> We began to get quite a characteristic fading in and out – a slow fading in and out of the signal. It didn't take long to determine that that was due to a ship [the wooden SS *Dorchester*] coming up and around Alexandria.

This was reported by Taylor to the Bureau of Engineering with the following suggestion:

> Destroyers located on a line a number of miles apart could be immediately aware of the passage of an enemy vessel between any two destroyers of the line, irrespective of fog, darkness, or smoke screen.

No action was taken on this at that time, but this would later lead to one of the most important developments in Navy history.

In Germany, Dr. Heinrich Löwy conducted extensive research on distance-measuring devices. In July 1923, he filed a U.S. patent application for connecting an antenna alternately between a transmitter and a receiver for "measuring of the distance of electric conductive masses and for ascertaining of the height of flying vehicles." His proposed apparatus was described as radiating very short trains of waves and using the transit-time between transmitted and returned (reflected) signals to measure the distance. Although there is no indication that such a device was ever built, the description is close to that for the first practical radar system.

Dr. Gregory Breit and Merle A. Tuve, researchers in the Department of Terrestrial Magnetism at the Carnegie Institution of Washington, were studying characteristics of the ionosphere (then called the Kennelly-Heaviside layer). Using a crystal-stabilized, high-frequency transmitter developed for them by Leo Young and Louis Gebhard at the NRL, in 1925 they devised a technique for measuring the varying height of the ionosphere by pulsing the transmitter. The experiments were successful, demonstrating distance measurement using pulsed transmission.

At General Electric in August 1928, Jetson O. Bentley filed a patent application for an airplane altitude-indicating system. This involved transmitting a signal with a linearly varying frequency (frequency modulation) from an aircraft toward the ground and a receiver in the aircraft to detect the signal reflected from the ground. A mixing of the transmitted and received signals would give a beat frequency proportional to the two-way radiation time, from which the altitude of the airplane could be determined. Although patents were readily granted in France, Belgium, and Germany, and eventually (1935) in the U.S., it does not appear that actual implementations were made of this precursor to FM radar.

While conducting outdoor tests on a receiving antenna at the NRL in June 1930, Lawrence A. Hyland observed interference between the direct signal from a high-frequency transmitter and the signal reflected from a passing airplane. Leo Young confirmed this phenomenon, and noted the similarity to the observations by himself and Hoyt Taylor in 1922. This was reported it to the Bureau of Engineering, and the Bureau responded by directing the NRL to continue with the investigations. No funding was provided for the continued effort, so it was given a low priority.

In Great Britain, William A. S. Butement, a scientific officer at the Woolwich research station of the Signals Experimental Establishment, considered the potential of using pulsed 50-cm signals for detection of ships. In January 1931, Butement and his assistant, P. E. Pollard, prepared a memorandum on this subject, and it was entered in the Inventions Book maintained by the Royal Engineers. Since there was no source of 50-cm radio power at that time, the War Office did not give it consideration. (Butement later applied their concept in developing the first radar projectile fuse.)

Dr. Carl L. Englund, Arthur C. Crawford, and William W. Mumford at the BTL conducted studies on propagation phenomena, including the effect of single trees, woods, wired houses, fluctuations from moving bodies, and aircraft-to-ground links. In 1933, they published a paper, "Some Results of a Study of Ultra-short-wave Transmission Phenomena," in the *Proceedings of the IRE*. In discussing the field fluctuations from moving bodies, re-radiation from aircraft was highlighted and the following was noted:

> For ordinary airplane heights a high-energy transformation loss in the re-radiation process can occur and still give marked indications in the receiver meter. The airplane re-radiation was noticed at various subsequent times, sometimes when the airplane itself was invisible.

In the summer of 1934, a small group of engineers from the U.S. Army Signal Corps gathered at a promontory on New York's lower bay to watch an experiment by Drs. Irving W. Wolff and Ernest C. Linder of the RCA Camden Laboratory research staff. The two scientists had brought with them a microwave transmitter using a low-power magnetron, a receiver, two four-foot parabolic antennas, and an audio amplifier with a loudspeaker. As the assemblage watched and listened, the antennas were aimed toward a small boat passing some 2,000 feet away. An audio tone emerged from the speaker as the boat went by. As the antennas were turned to follow the boat, the tone continued until the boat passed out of range.

Wolff had been investigating microwaves at RCA since 1930. Linder joined him two years later and developed a magnetron tube that could generate about a half-watt of power in the 9-cm (3-GHz) range. Wolff described the 1934 experiment in Kenyon Kilbon's book, *A Short History of the Origins and Growth of RCA Laboratories, 1919 to 1964* (RCA, 1964):

> They [Signal Corps engineers] asked if it would be possible to detect a boat? Would these microwaves reflect off a boat? And so we said we might just as well try. There was a boat coming in the harbor, and sure enough, we pointed it towards the boat and we could get a [Doppler shifted] signal off the boat, which combined with a little bit of the signal coming directly off the transmitter to give us a beep tone, and I guess that was our first radar experiment. It wasn't radar as we know it; it was just direction, whereas radar is distance and direction, but it was the first experiment.

Using low-power microwave equipment, researchers under Dr. William R. Blair at the Signal Corps Laboratories (SCL) at Fort Monmouth also observed interference phenomena, detecting a moving truck at a distance of 250 feet. In subsequent experiments using an SCL-built copy of a Hollmann tube from Germany with a 5-watt output at 50 cm (600 MHz) and a receiver placed some 12 miles away, reflected Doppler signals were generated from a large boat passing through the path. In his 1934 annual report, Blair commented on this method, which was only effective where there was relative motion between target and detector:

> It appears that a new approach to the problem is essential. Consideration is now being given to the scheme of projecting an interrupted sequence of trains of oscillations against the target and attempting to detect the echoes during the interstices between the projections.

Thus, the principle of pulse detection was stated. The Army researchers, however, did not reduce this to demonstrative equipment for two more years.

THE U.S. MILITARY AND RADAR

The early 1930s were a financially difficult period for research in America's armed services. Only a short time after the American financial crash and still several years before the rise of Hitler, the nation's domestic economic worries were pressing and any foreign menace seemed remote. The depression resulted in further reductions to military budgets, which had never been large following World War I. Nevertheless, a few dedicated researchers at the NRL and the SCL initiated some of the most important developments of the twentieth century – those related to radar.

In the period from 1933 to 1936, both the NRL and the SCL pursued research toward radio detection of targets. These laboratories communicated with RCA and BTL concerning industrial attempts in using microwaves for this application, and, in fact had microwave projects of their own. Hoyt Taylor and William Blair, leaders at the NRL and the SCL respectively, both had doctorates in electromagnetics and readily recognized the advantages of microwaves for this purpose. However, work at these frequencies was abandoned by the Government because of the lack of suitable transmitter tubes; both the NRL and the SCL turned to state-of-the-art components at lower frequencies.

It was not until 1936 that any significant interaction between the NRL and the SCL occurred. At that time, arrangements were made for mutual visits and exchanges of technical information, mainly to avoid duplication of effort. This also allowed them to better compare their work with that being done in industry.

In October 1936, Lt. Colonel Louis E. Bender of the Office of the Army Chief Signal Officer visited laboratories of RCA, BTL, and General Electric to assess the state of their research applicable to radio detection. He found that while they had made progress in some of the necessary components, none of these firms was prepared to take on the development of a complete aircraft detector with a reasonable assurance of meeting the requirements at an early date. His findings were summarized as follows:

> Comparison of the work done by commercial concerns applicable to this problem with that done by the Laboratories of the Army and Navy leads to the conclusion that the latter are further advanced, showing more practical results, and definitely more promising.

Navy Radar

When the Naval Research Laboratory opened in July 1923, Dr. A. Hoyt Taylor was named superintendent of the Radio Division. Under Taylor's outstanding leadership, and with strong assistance from and an outstanding staff that included Leo C. Young and Louis A. Gebhard, the science and technology of Naval radio blossomed.

In 1930, tests were being conducted by Lawrence A. Hyland at the NRL on directional characteristics of an aircraft receiving antenna. A 32.8-MHz transmitter was placed at a distance, and it was noted that aircraft passing nearby produced variations in the signal strength at the receiving antenna. Young verified these observations and realized that they were similar to those that he and Taylor had found in 1922 prior to the formation of the NRL. Concluding that this resulted from interference between the direct and reflected radio signals, in November 1930 Taylor and Young submitted to the Bureau of Engineering a detailed report entitled, "Radio-Echo Signals from Moving Objects." The Bureau responded by directed the NRL to "investigate the use of radio to detect the presence of enemy vessels and aircraft." Although the report was immediately given a security classification, no specific funding for this work was allocated and the project, by necessity, was given low priority at the NRL.

The report from Taylor and Young made its way slowly through the bureaucracy in Washington. In January 1932, the report was eventually forwarded by the Secretary of the Navy to the Secretary of War with the following suggestion:

> Certain phases of the problem appear to be of more concern to the Army than to the Navy. For example, a system of transmitters and associated receivers might be set up about a defense area to test its effectiveness in detecting the passage of hostile craft into the area. Such a development might be carried forward more appropriately by the Army than by the Navy.

This was, in turn, forwarded to the SCL with the comment that "the subject is of extreme interest and warrants further thought." The response of the SCL director, Dr. (Major) William A Blair, was that he was already aware of the Navy experiments and "the possibilities of this method of finding airplanes are being considered."

Research with "interference" of signals continued at the NRL on a part-time basis. The experimentation was expanded to cover frequencies up to about 100-MHz and used portable receivers and transmitters operated in various locations. By 1932, a complete system was devised for the protection of an area around Washington. Sufficient components of the system were built and installed to prove that the presence of aircraft could be detected and their approximate location given when they were within 50 miles of the center of the protected area.

This system involved separated transmitting and receiving stations, and was thus not adaptable to shipboard use. In late 1932, the Secretary of the Navy suggested that such a system might be beneficial to land-based military operations, and the NRL was directed to share its finding with the Army Signal Corps. Like other projects of this era, lack of funding soon brought this activity to a close at the NRL, and the SCL did not take it up.

Based on the microwave work at RCA and the BTL, in 1933 NRL researchers turned their attention to reflections of ultra-high-frequency radio waves. Some basic laboratory experiments were made, but it was soon realized that power from existing and potential transmitter tubes from American and foreign sources at these frequencies would be a major obstacle, and the microwave project was abandoned.

In 1934, Young returned to the pulsed-transmission technique that he had developed for Brett and Tuve at the Carnegie Institution for their ionospheric measurements. He had earlier discounted the possibility of this technique for target ranging because of the continuous nature of the ionosphere as contrasted with small reflecting areas of airplanes and ships. Now, however, he realized that pulsed techniques were the mainstay of work in the NRL's Sound Division. There, the director, Dr.

Harvey C. Hayes, and his associates had developed a technique in which an echo from an underwater sonic pulse could be detected and shown on a cathode-ray screen to measure the distance to a relatively small target.

Robert M. Page

With a new technique having a strong potential for solving the radio detecting and ranging problem, the activity was placed under Robert M. Page, charged with turning Young's concepts into reality. Page had joined the 126 members of the NRL in 1927, just after receiving an undergraduate degree in physics from Hamline University, a small Methodist-supported school in Saint Paul, Minnesota. Although he had no experience in radio engineering, he quickly gained the confidence of Taylor by providing very creative solutions to a wide variety of problems.

Page, assisted by LeVern R. Philpott and Robert C. Guthrie, first gave attention to the transmitter. In a short while, they had designed and built a transmitter that could emit 10-microsecond pulses with a wait-time between pulses of 90 microseconds. This allowed the initial pulse to radiate out, then be reflected and received during the dead period. With the velocity of radiation at 300,000 kilometers per second (186,000 miles per second), the range could be determined from the transmit-to-receive time interval.

A "breadboard" system tested in December 1934 detected an airplane at a distance of one mile. The transmitter operated at 60 MHz and used separate antennas for transmitting and receiving. Although the detection range was small, the experiment was termed a success. Based on this, Page, Taylor, and Young are generally credited with initiating radar development in the United States.

Leo Young with First Radar

The next year was spent on improving the receiver to solve the problems peculiar to reception of microsecond pulses with an extremely high-gain receiver adjacent to a transmitter radiating in pulses on the same frequency. The pulsing reflected signal required the receiver to have an output bandwidth far beyond that of conventional receivers, essentially the same as that of video amplifiers in the emerging television sets. Drawing on the sonic system, a cathode-ray tube was adopted to display the range to the reflecting target. New equipment was built to operate at 28.6 MHz; this allowed the use of an existing large antenna built for that frequency.

In considering the time required for the development, it is noted that Page, as well as others on the team, made this progress despite the lack of financial support earmarked for the project. They worked on the project "between" more conventional radio problems that had direct funding. Moreover, the Bureau of Engineering took no action toward securing needed funds. In early 1935, after receiving permission from the NRL Director, Taylor and Hayes (from the Sound Division) made a direct appeal to the Senate Subcommittee for Naval Appropriations. This resulted in a special $100,000 appropriation earmarked for the project.

Finally, with adequate funding, the improved system was completed in late 1935. Tests during the following months showed a detection range up to 16 miles. The following June, the system was demonstrated to officials from the Bureau of Engineering, and an NRL report was submitted detailing the accomplishments. Based on this report, Rear Admiral Harold G. Bowen, Chief of the Bureau, directed the NRL to "give the highest possible priority to the development of shipboard equipment," and also wrote:

> It is requested, upon receipt of this letter, that the subject problem be placed in a secret status, that all personnel now cognizant of the problem be cautioned against disclosing it to others, and that the number of persons to be informed of further developments in connection therewith be limited to an irreducible minimum.

Arthur A. Varala joined the team in May 1936 and developed a 200-MHz system, allowing a smaller antenna necessary for shipboard use. The system also used a duplexer, developed by Young and Page, for switching a single antenna between transmit and receive modes. This ingenious device, which connected the transmitter and disconnected the receiver and vice versa for very short periods of time, later became common to all radar equipments.

In April 1937, Admiral William D. Leahy, Chief of Naval Operations, witnessed a demonstration of the equipment at the NRL and became convinced of its capabilities. That same month, the first seaborne testing was conducted. The equipment was temporarily installed on the USS *Leary*, an old destroyer, with a Yagi antenna mounted on a gun barrel for sweeping the field of view. Planes at ranges up to 18 miles were located, with the distance limited by the low power available from the transmitter tube. The technology was fully disclosed to the Bell Telephone Laboratories (BTL) in July. This was the first industry brought in on the development.

Testing Aboard the USS Leary

In 1938, the Chief of Naval Operations directed that radio detection equipment be placed in the fleet for operational purposes. The NRL further improved its 200-MHz system and designated it XAF. To obtain higher power, Page developed the ring-oscillator circuit, allowing several tubes (in even numbers) to function in parallel. This delivered 15-kW, 5-μs pulses to a 20-by 23-foot, stacked-dipole "bedspring" antenna. The receiver used a single-line target display (later called an "A" Scope). Laboratory tests were successful in detecting planes at ranges up to 100 miles.

The NRL had succeeded in developing a radio detection and ranging system that would shortly revolutionize naval warfare. Despite this, they had difficulty in obtaining sufficient funds to rapidly pursue this and similar projects. In 1939, the Bureau of Engineering requested and obtained only $25,000, exclusive of engineering salaries, for internal electronics research at the NRL.

The XAF was installed on the battleship USS *New York* for sea trials starting in January 1939, and became the first operational radio detection and ranging set in the U.S. fleet. During a 3-month period, it routinely detected ships at 10 miles and aircraft at 48 miles. It was used for navigational purposes, for spotting shot falls, and, amazingly, for tracking in-flight projectiles. The equipment also proved to be very reliable under sea conditions; only two breakdowns occurred during the test period. In his report on the tests, Rear Adm. Alfred W. Johnson, Commander of the Atlantic Squadron, stated,

XAF on USS New York

> The XAF equipment is one of the most important military developments since the advent of radio itself. The development of the equipment is such as to make it now a permanent installation in cruisers and carriers.

The basic principles of radio detection and ranging equipment had been disclosed to RCA in March 1938, and the firm was given a contract for the development of a 400-MHz experimental set. The preliminary equipment, designated CXZ, although not satisfactorily developed, was tested on the battleship USS *Texas* during December 1938.

XAF Transmitter and Receiver

After testing of both the XAF and the CXZ, the Navy decided that the XAF would be more reliable. In May 1939, the principles of the XAF were disclosed to RCA and Western Electric, with competitive bids for the further development and production of the system requested from the two firms. The contract was awarded to RCA. Five of these search systems, designated the CXAM, were delivered in May 1940. One was placed aboard the USS *California,* a battleship that was sunk in the Japanese attack on Pearl Harbor. Others were on the aircraft carrier USS *Yorktown* and cruisers USS *Chicago*, USS *Northhampton,* and USS *Pensacola.*

Cruiser with FA Radar

Based on success at the BTL from internally funded work on higher-frequency technology, the Navy issued a contract to Western Electric for a 500-MHz shipboard fire-control radar designated CXAS (later FA), and deliveries started in June 1940. Even with an output of only 2 kW, the FA suffered badly from a short life of the transmitter tubes.

The acronym RADAR was coined from Radio Detection And Ranging. The origin is attributed to two U.S. naval officers, E. F. Furth and S. P. Tucker. In November 1940, the Chief of Naval Operations directed the use of the word as non-classified for reference to the then-secret project. The acronym quickly became the name "radar." For some time, this name was not publicly used and, even in official documents, "direction-finding" was often substituted for "detection."

SK Antenna

The CXAM was further refined into the SK system and produced by General Electric starting in 1941. Improvements included a 15- by 15-foot, rotating antenna (called the "flying bedspring") and, late in 1941, a Plan Position Indicator (PPI), a map-like radar screen developed at the NRL. With a 330-kW output, it could detect aircraft up to 150 miles. The SK remained the early-warning radar for aircraft carriers, battleships, and cruisers throughout the war.

SK Controls with PPI

Derivatives of the SK included the SC, produced by General Electric, with a 15- by 5-foot antenna and primarily for destroyers, and the SA, built by RCA, with an even smaller antenna for destroyer escorts and other smaller vessels. All of these systems were also used on Allies' ships. An airborne version, the ASA, intended for the Navy's R4D (C-47) aircraft, was developed, but the antenna was too large for practical mounting.

Aircraft safety drove the development of radar altimeters. Such systems needed very little transmitted power (the earth is a *large* target). A pulse radar operating at about 500 MHz was investigated by the NRL for this application. The project was turned over to the Army's SCL and in early 1940 the equipment was put into production by RCA as the SCR-518. These units ultimately gave accurate measurements to a relative altitude of 40,000 feet.

SD Radar Mast

The NRL also developed the SD, an aircraft detection radar for use by surfaced submarines. Operating at 114 MHz and with 100-kW output, it used a periscope-type antenna. The system allowed detection, but not bearing, of approaching aircraft; bearing, however, was relatively unimportant since crash-diving was the normal defense. Production of this radar was started by RCA in 1941.

One of the first applications of Operations Research in the U.S. Navy was in determining the best operating technique for the SD system. This was a trades study, comparing probabilities of target detection versus detection by the enemy of the submarine through intercept of the radar signal. It was decided to use the radar in most circumstances.

As radar came into being as a detection device, there was no means of identifying the targets as to friendly or enemy, particularly those airborne and under poor visibility conditions. In 1937, the NRL addressed this problem and developed the first recognition technique, later called identification friend or foe (IFF). Designated the XAE, this consisted of a unit on the aircraft that, when in the vicinity of friendly ships, transmitted a series of coded signals. On shipboard, a Yagi antenna on a hand-held mount was aimed at the aircraft and, upon receiving and decoding the signal, would flash a light, signifying the approach of a friendly aircraft. From that start, the NRL took the lead in what was then called recognition systems. Included were the ABD and ABE systems, incorporated in the airborne radars.

Antennas on NRL Building 1, 1940

As later described, with the advent of microwave radar, research on these new systems was primarily done elsewhere, but the NRL continued the improvement of high-frequency radars and other electronic equipment.

Army Radar

Upon becoming director of the Signal Corps Laboratories (SCL) at Fort Monmouth, New Jersey, in 1930, Dr. (Major, later Colonel) William R. Blair initiated low-level research in target-detection by infrared and radio. He patented a device for pulse-echo direction-finding and ranging; however, no operating device was made. Starting in 1932, progress at the Naval Research Laboratory on radio interference for target-detection was passed on to the Army, but it does not appear that any of this information was used by Blair. The SCL's first definitive efforts in this field started in 1934 when the Chief of the Army Signal Corps, after seeing a microwave demonstration by RCA, suggested that radio-echo techniques be investigated. The SCL called this technique radio position-finding (RPF).

Emphasis was placed on assessing capabilities of the existing microwave tubes. Tests were made with the German-built Hollmann tube, giving 5-watt output at 50-cm (600 MHz). The only available sample of this tube was X-rayed and duplicated at the SCL. Ranges were determined for 9-cm (3 GHz) magnetron equipment brought to the laboratories by Irving Wolff and Ernest Linder of RCA. Laboratory experiments were also carried out with RCA acorn tubes, producing wavelengths as short as 45 cm (660 MHz) in a Hartley circuit.

During 1934 and 1935, tests of microwave RPF equipment resulted in Doppler-shifted signals being obtained, initially at only a few hundred feet distance and later over several miles. These tests involved a bi-static arrangement, with the transmitter at one end of the line of transmission and the receiver at the other, and the reflecting target passing through or near the path. The state of the development was summarized by Blair in his 1935 annual report on the project:

> To date the distances at which reflected signals can be detected with radio-optical equipment are not great enough to be of value. However, with improvements in the radiated power of transmitter and sensitivity of the receiver, this method of position finding may well reach a state of usefulness.

In an earlier report, Blair had noted that the SCL might investigate another technique: "Projecting an interrupted sequence of trains of oscillations against the target and attempting to detect the echoes during the interstices between the projections." This suggestion for using a pulsed transmission possibly came from his knowledge of the pulsed transmitter supplied by the NRL to Breit and Tuve of the Carnegie Institution for determining the height of the ionosphere.

To house the activities of the SCL Squier Hall was constructed in 1935. The facility was named to honor Major General (Dr.) George O. Squier, founder of the SCL and Chief Signal Officer during WWI.

Squier Hall – Fort Monmouth

In 1936, a modest project in pulsed microwave transmission was started by William D. Hershberger, SCL's chief engineer at that time. After discussions at the NRL, where he had earlier worked, Hershberger changed from microwaves to a 200-MHz system with lobe-switching for the antennas. It was primarily intended for aiming searchlights associated with anti-aircraft guns. A first demonstration of the concept was made on May 26, 1937. The observers included the

Secretary of War, Harry Woodring. The next day, orders were given for full development of the system.

Before the end of 1938, a demonstration of a full radar system was given. The system used 16 transmitter tubes in a ring-oscillator circuit (developed at the NRL) and a complicated mechanical assembly of three antennas. Designated SCR (Signal Corps Radio) 268 ("Radio" for security purposes), production of this gun-laying system was started by Western Electric in 1939, and it entered service the next year. Later, the PPI was added and the system was designated SCR-516, a low-altitude early-warning radar.

SCR-268 – Army's First Radar

The demonstration in 1938 led to an urgent request from the Army Air Corps for a simpler, early-warning system. Operation at 100 MHz, was selected, allowing the use of new types of transmitter tubes in a conventional circuit. Other simplifications centered on the antenna, including elimination of lobe-switching and the addition of a duplexer developed at the SCL. The project was led by Dr. (Major, later Colonel) Harold A. Zahl. Good funding and a high priority was received; thus, development was quickly completed

SCR-270 Console

This design had two configurations – the SCR-270, a mobile set, and the SCR-271, a fixed-site version. Westinghouse put both into production, with deliveries starting in mid-1940. These systems had a range up to 240 miles. The SCR-270 remained the Army's standard early-warning radar throughout the war.

The Army deployed five SCR-270 sets around the island of Oahu in Hawaii. On the morning of December 7, 1941, one of these radars was being operated by Privates Joseph Lockard and George Elliot. At 7:02, a cluster of blips appeared on the screen at a range of 136 miles due north. They checked the movement for 18 minutes, first thinking there was something wrong with the radar, then passed the observation on to the Aircraft Warning System at Fort Shafter. The Lieutenant on duty dismissed it as "nothing unusual," believing the detection to be from a flight of U.S. B-17 bombers known to be approaching from the mainland. The alarm went unheeded. Lockard and Elliot continued tracking until

7:39 when the planes were only 20 miles away. Sixteen minutes later, the Japanese hit Pearl Harbor.

The Aircraft Warning System was America's first attempt to have a central radar information center similar to the one used very effectively by the British. It was being established jointly under the Army and Navy by Major Kenneth P. Bergquist and Lieutenant Commander William E. G. Taylor. Unfortunately, the incoming U.S. planes had no recognition equipment (IFF), so the unexpected Japanese aircraft were taken as "friendly."

Mobile SCR-270

Taking over an earlier project of the NRL, the SCL developed the SCR-518 pulsed-radar altimeter for the Army Air Corps. Operating at 518 MHz, this system was produced by RCA starting in 1940. The final system weighed less than 30 pounds and was accurate to about 42,000 feet above ground. The Signal Corps was also involved in an early version of a portable, radar-based instrument landing system, eventually designated the SCS-51.

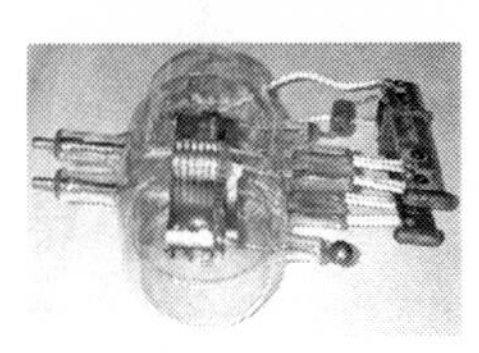

VT-158 Tube

Although the SCL initiated its radar research using microwaves, it never returned to this wavelength region during wartime activities. SCL did, however, push the frequencies higher, primarily through Zahl's development of the VT-158, a vacuum tube generating 240-kW pulse-power at up to 600 MHz.

RDF IN GREAT BRITIAN

In early 1934, as war clouds loomed over Great Britain, a large-scale Air Defense exercise was held, including mock raids on London and other potential targets. Although the routes of bombers were known in advance, over half of them reached their destinations without being detected. This led Prime Minister Stanley Baldwin to make the pessimistic statement, "The bomber will always get through."

The likelihood of air raids and the threat of invasion by air and sea drove a major effort in applying science and technology to defense. In November 1934, the Air Ministry established the Committee for Scientific Survey for Air Defense. Called the "Tizard Committee" after its chairman, Sir Henry Tizard, this group had a profound influence on technical developments in Great Britain.

There had been reports in the press that the controversial scientist, Dr. Nikola Tesla, had invented a death ray that could "bring down squadrons of airplanes 250 miles away." In early January, 1935, Robert Alexander Watson Watt, superintendent of the Radio Research Station in Slough of the National Physical Laboratory, was asked to comment on the potential of directed radio beams to be used as such a weapon. His scientific assistant, Arnold F. "Skip" Wilkins, made simple calculations concerning energy level necessary to kill or disable a human in an aircraft, and the corresponding generation and propagation of this energy. Wilkins' results showed that such a weapon was not feasible with foreseeable radio transmitter powers, but he noted that energy reflected from the plane might be sufficient for detection at a distance. A memorandum on this was sent to the Tizard Committee in late January.

Robert A. Watson Watt

Arnold F. Wilkins

Watson Watt had earlier performed scientific research in radio probes of the ionosphere and in directional antennas for locating lightning strikes. Thus, his thoughts turned to a different defense application of radio. After Wilkins made an analytical study, a secret memorandum entitled "Detection of Aircraft by Radio Methods"was sent by Watson Watt to the Tizard Committee on February 27, 1935. The memorandum also cited pulsed methods to measure the range.

The Tizard Committee reviewed the memorandum with great enthusiasm, but Watson Watt was asked to perform a simple demonstration before scarce funds were committed. For this, an attempt would be made to collect reflections from an aircraft flying along the beam of a powerful BBC transmitter operating at 6 MHz in Daventry. Wilkins set up a receiving van in Weedon, several miles from the transmitter. It used two spaced antenna arrays phased to reject the direct signal but pass any reflected signal.

Wilkins at Daventry Demonstration

On February 26, 1935, a Heyford bomber was flown up and down the transmitted beam. Watson Watt, Wilkins, and others watched a cathode-ray tube displaying the detected level of the reflected signal. The results were immediate and conclusive. The reflected signal was detected up to distances of over eight miles, confirming the calculations made by Wilkins. The experimental results were similar to observations made by Hyland at the NRL in June 1930 and confirmed in tests by Young the following November.

The Daventry demonstration so impressed Watson Watt that he declared, "Britain has become an island again!" The Air Ministry immediately classified the project as highly secret, and initial research funds were allocated. Dr. Edwards G. "Taffy" Bowen was added to the team in April, primarily to develop the transmitter. A preliminary system was designed, based on pulsed transmission as used by Breit and Tuve in 1925 for probing the ionosphere. The transmitter operated at 6 MHz and had a pulse width of 25 μs.

Edward G. Bowen

In May 1935, the activity was moved from Slough to an abandoned airfield near Orfordness, Suffolk, on the North Sea coast some 75 miles northeast of London. An antenna was strung between two 75-foot masts. After unsuccessful tests at a lower power level, the transmitter was eventually increased to 200 kW and on June 27, 1935, a clear echo was received from an aircraft at 17 miles distance.

Over the next several months, the frequency was changed to 23 MHz and by the end of the year a tracking distance of 60 miles was demonstrated to the Tizard Committee. The original memorandum from Watson Watt had noted that a working system would need to provide height and bearing information as well as distance. For this, Wilkins use a method that he had earlier developed at Slough for determining the incoming direction of long-distance radio signals and added crossed dipoles as the receiving antenna.

Like Page, Taylor, and Young in America, Watson Watt, Wilkins, and Bowen are generally credited with initiating radar development in Great Britain. Here it might be noted that some sources credit the "original discovery" of radio detection and ranging (radar) to Watson Watt; as previously noted, however, the NRL team demonstrated a system in December 1934. Others incorrectly credit this discovery to Drs. Wilhelm T. Runge and Hans E. Hollmann at Telefunken in Germany, but their first demonstration was in May 1935.

Initial RFD Systems

Bawdsey Research Station

The cover name "radio direction-finding" (RDF) was given to the technology. In September 1935, the Tizard Committee directed that a full RDF network be developed, with overlapping coverage from individual facilities. For this effort, the Bawdsey Manor near Orfordness was taken over in early 1936 and converted to the Bawdsey Research Station. Watson Watt was named as the superintendent and Bowen as the technical leader. The military personnel called the scientists and engineers at the station "boffins," and this name stuck for RDF participants throughout the war.

The resulting RDF design incorporated pulsed transmission for distance measurement and crossed receiver antennas for determining direction. A 25-Hz pulse-repetition rate (PRR) was selected, allowing different stations to be synchronized using the 50-Hz electrical power grid. The pulse length was 20 μs. Operation was initially 350 kW at 22 MHz, using fixed transmitting antennas strung between four (later three), 360-foot steel towers. Multiple receiving antennas on four 240-foot wooden towers in a rhombic configuration allowed approximations of azimuth and elevation. The operation eventually changed to 750 kW on four selectable frequencies between 22 and 55 MHz

Chain Home Transmitting (left) and Receiving Towers

Called the Chain Home (CH) RDF system, its first station was accepted by the Royal Air Force in May 1937. The analysis methods now known as Operations Research ("Operational Research" in Great Britain) were used to determine the best ways to implement and apply this new technology. The CH system could detect aircraft at 15,000 to 20,000 feet altitude and a range of about 100 miles, but with a significant measurement uncertainty. Two major deficiencies of CH were its inability to detect low-flying targets (limited to look-angles above about 2 degrees) and its inability to follow the attackers inland from their coastal stations.

Watson Watt took the position of Director of Communications Research in the Air Ministry in 1938, and Albert P. Rowe succeeded him as the Bawdsey superintendent. Rowe had previously been the secretary to the Tizard Committee and had represented the Committee in the very important Daventry demonstration in 1935. Although not technicall strong, he brought significant organization to the previously university-like atmosphere at Bawdsey. He instigated a weekly meeting (the "Sunday Soviets") of all the personnel, leading to highly productive exchanges between the technical staff and the military representatives.

Chain Home Control Center

Arnold Wilkins was responsible for construction of the CH stations. By mid-1939, the system had 18 stations in service along Great Britain's eastern and southern coasts. Target sightings were relayed by telephone to a central command and control center. Locations of enemy and defense aircraft were shown on a plotting board. Radio links were used to vector RAF pilots to the general vicinity of approaching aircraft. Although very imprecise in its positioning information, the CH system is given credit for making the difference in the outcome of the Battle of Britain – Germany's daylight bombing of London and other cities during 1940.

To compensate for the imprecision of the ground systems and also to allow nighttime engagements, Bowen and his team at Bawdsey developed a 195-MHz Airborne Interceptor (AI) search system for the RAF fighter aircraft. A second type airborne search system, the 176-MHz Air-to-Surface Vessel (ASV) for maritime patrol aircraft, was developed simultaneously. These systems used a common electronic design and the same type transmitting antenna (forward-looking and broad-beamed), but with different receiving antenna configurations – four for the AI and two for the ASV with a motor-driven switch for sequencing them to a single receiver. Although bulky and with complex antennas, the AI and ASV equipment set the stage for the vital airborne systems to come. Both began testing in September 1937 and became operational in 1940.

Army RDF Development

In 1938, the British Army established its own research laboratory at Bawdsey. With Drs. E. T. Paris and A. B. Wood as the technical leaders, this group was charged with developing gun-laying (GL) systems for anti-aircraft guns and coastal defense (CD) systems for directing coastal artillery. The Bawdsey group included W. A. S. Butement and P. E. Pollard. The two of them had, in fact, documented their concept for a pulsed, radio-detection system in January 1931, four years before the Watson Wattradum, but it was never accepted.

The GL effort led to a mobile system operating between 55 and 100 MHz, but, using small antennas, it did not provide good accuracy. Used for aiming searchlights, it was put into service in 1939. Operating between 180 and 210 MHz, the CD system used rotating, broadside dipole arrays with lobe-switching. With a narrow beam, it could detect with good accuracy aircraft flying as low as 500 feet and at a distance of 25 miles. In August 1939, upon Watson Watt's recommendation, the CD equipment was adopted to augment the CH system. Known as Chain Home Low (CHL), one was placed at most CH stations. Later, because of jamming in the 200-MHz band, the frequency was changed to 500-600 MHz.

CHL on Mast

When Great Britain declared war on Germany in September 1939, the Bawdsey operations were evacuated. After a brief move to Scotland, in May 1940 the Air Ministry group settled at Swanage, near London, and was renamed the Telecommunications Research Establishment (TRE). The Army group at Bawdsey, renamed the Air Defense Research and Development Establishment (ADRDE), moved to Christchurch. Both the TRE and the ADRDE contributed many major developments throughout the war.

GCI – Type 7

With use of the CHL, it was realized that positional control of the interceptor aircraft, was possible. This led to another system called Ground Controlled Intercept (GCI). Led by Pollard, the GCI was one of the first developments at TRE. It used a version of the CHL with the addition of a height-finder unit and, later, a Plan Position Indicator (PPI) from America. Configurations ranged from large, fixed systems to small, highly mobile sets. Put into service in early 1941, these systems permitted use of new tactics that increased enemy aircraft losses from 0.5 percent to about 7 percent.

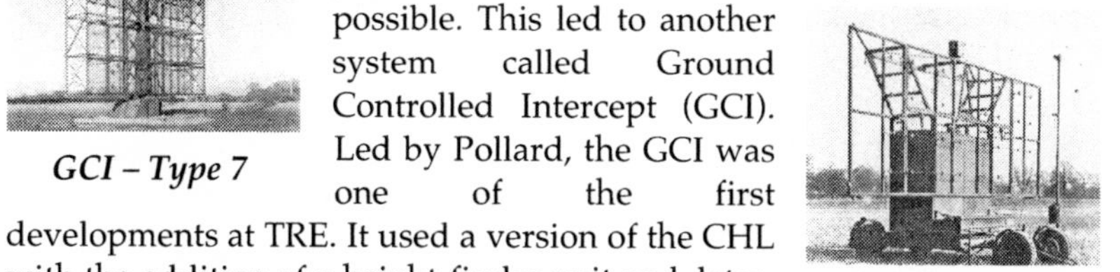

GCI – Type 8

Navy RDF Development

In mid-1935, shortly after the first RDF system was demonstrated, the Royal Navy received permission to conduct its own RDF research at His Majesty's Signal School in Portsmouth, later changed in name to the Admiralty Signals Establishment (ASE). As at the NRL and the SCL in America, its initial work was in microwaves, but soon changed to VHF. After a year with little success, the ASE then shifted to more practical frequencies. Alfred W. Ross led the development of an air-warning system designated Type 79. This operated at 40 MHz with separate, stationary transmitting and receiving antennas. Prototypes of this system were successfully tested in late 1938 on battleship H.M.S. *Rodney* and cruiser H.M.S. *Sheffield*, the first vessels in the Royal Navy with radar.

Type 79 Antennas

This initial system was improved with a gun-control set as Type 279 and placed into production, with the first deployment in early 1940. Type 279 was for large vessels (battleships, cruisers, and carriers), and had two large, rotable antennas that moved in synchronization. For smaller vessels, the Type 286 was an adaptation of the 200-MHz ASV, but it had fixed antennas and was considered almost useless.

Also at the ASE, a 600-MHz project was led by Dr. J. F. Coales. Sea trials of a 25-kW system designated Type 281 unit took place in 1939. This used a rotating pair of Yagi-array antennas and incorporated lobe-switching. Over the next year, three excellent gun-laying systems evolved: Type 284 for anti-armament, and Type 282 and Type 285 for anti-aircraft, short-range and long-range, respectively.

Type 281-Series Receiver

THE TIZARD MISSION AND JOINT DEVELOPMENTS

A state of war with Germany was declared by Great Britain on September 3, 1939. By mid-1940, it was recognized by the British leaders that survival in the war depended on access to the technical resources of the United States. In early September, a British technical mission was sent to the United States, empowered personally by Prime Minister Winston Churchill to disclose everything concerning secret weapons being developed for defense of their island.

Commonly called the "Tizard Mission" after its organizer, Sir Henry Tizard, its objective was for the two countries to share information about radar, sonar, and other critical technologies. Dr. Edward G. Bowen represented radar, and the other technologies were represented by the renowned nuclear physicist, John D. Cockcroft. Senior officers from each of the British military services were also included.

Having operated under the tightest levels of secrecy, both the United States and Great Britain believed that they had perfected the only working radio detection and ranging system. Each side was surprised by the progress made by the other. They were astounded to learn that the CLH/CD and CXAM radars had very similar characteristics. After several days of information exchange, Tizard sprang his trump card – the resonant-cavity magnetron, their solution to overcoming the microwave barrier. This device – small enough for Tizard to hold in his hand – could generate power of near 1,000 times that of existing microwave tubes.

Original Resonant-Cavity Magnetron (without magnet)

It is appropriate to mention the great importance of the resonant-cavity magnetron. This device made microwave radar possible – a system small enough to be carried on every type of military aircraft. Albert P. Rowe, secretary of the original Tizard Committee and superintendent of the TRE at the time of the Mission to America, later made the following comment:

> Few in a position to judge would hesitate to name the cavity magnetron as having had a more decisive effect on the outcome of the war than any other single scientific device evolved during the war. It was of far more importance than the atomic bomb.

First developed in 1920 by Dr. Albert W. Hull at General Electric in America, the smooth-bore magnetron was originally intended as a substitute for the triode but showed early promise of generating high power at microwave frequencies. Researchers throughout the world attempted to bring this promise to fruition, primarily through segmenting the anode and later with multiple cavities, but with only modest success. At Birmingham University, Harry A. H. Boot and John T. Randall were examining the klystron for possible improvements. In Boot's words, "We concentrated our thoughts on how we could combine the advantages of the klystron with what we believed to be the more favourable geometry of the magnetron."

With little more than a general concept, they built a device with the anode made from a solid copper block and containing six resonant cavities. This anode formed a part of the external envelope and, being "outside" the vacuum region, was easily cooled and could be placed very close to the surrounding magnet. The magnetic field caused the electrons to sweep past the cavities and generate resonate electromagnetic oscillations, similar to the effect of air sweeping across the pipes of an organ. Although essentially unanalyzed prior to being built, it produced about 400 watts of continuous power at near 3 GHz when first tested on February 21, 1940.

As noted, the invention by Boot and Randall was not the first cavity magnetron in the world; similar devices had previously been developed in a number of other countries, but none of these led to microwave radars at that time. Many of these devices were not only reported in the open literature, but magnetrons from American and German laboratories also received U.S. patents. It is not known if the researchers at Birmingham University were aware of these earlier developments but, in either case, Boot and Randall certainly provided the breakthrough that made the resonant-cavity magnetron into a practical generator of high microwave power.

Improvements were quickly made on the Boot and Randall device, particularly by Dr. Eric C. S. Megaw of General Electric (Great Britain). In mid-1940, a cavity magnetron weighing only 6 pounds and generating a peak power up to 50 kW at 9.6 cm (3.2 GHz) was released as the E1189. This was the device brought to America by the Tizard Mission.

At the TRE, a microwave development activity was set up under Dr. Philip I. Dee, and it received an E1189 resonant-cavity magnetron for application experiments. A breadboard pulsed system was put together by Dr. A. C. Bernard Lovell, and by August 1940, it demonstrated an ability to detect aircraft, ships, and surfaced submarines. A sea-worthy design was quickly made by the ASE and put into production for the Royal Navy as Type 271, the first high-power microwave radar.

The Rad Lab and Joint Activities

In the United States, the National Defense Research Committee (NDRC) had been formed in June 1940 to oversee scientific research for war. Dr. Vanover Bush, who had led the formation of the NDRC, was appointed Director. Bush, an electrical engineer, was then President of the Carnegie Institution in Washington. His comments upon appointment were as follows:

> We were agreed that the war was bound to break out into an intense struggle, that America was sure to get into it in one way or another sooner or later, that it would be a highly technical struggle, that we were by no means prepared in this regard, and finally and most importantly, that the military system as it existed, would never fully produce the new instrumentalities which we would certainly need.

Alfred L. Loomis, the wealthy "amateur" scientist of Tuxedo Park, chaired the NDRC Microwave Committee. Following the revelation of the resonant-cavity magnetron by the Tizard Mission, it was agreed that major development of microwave radar would be a joint activity, centered in a laboratory in the United States. There was debate, however, as to where this activity should be located. Loomis wanted the central laboratory to be at the Carnegie Institution, but Bush, saying that Carnegie did not have the necessary administrative strength, wanted MIT. Dr. Frank Jewett, president of the BTL, felt that his commercial organization was best suited for this undertaking. Bush, as director of the NDRC, prevailed, and MIT was chosen. Attempting to disguise its purpose, the name "Radiation Laboratory" (Rad Lab) was selected for the facility – the same name as a major nuclear research facility at Berkeley on the West Coast.

MIT freed up 10,000 square feet of space, the NDRC provided a half-million dollars for the initial year, and the Rad Lab opened in October 1940 with 48 employees. Dr. Lee A. DuBridge, a physicist who had been a Dean at the University of Rochester, was named as the Rad Lab director. The associate director was Dr. Isidor Isaac (I. I.) Rabi, a renowned physicist from Columbia University. A majority of the researchers initially at the Rad Lab were also physicists; most early work related to microwaves had been done in experimental physics laboratories at universities.

E. G. Bowen (seated), L. A. DuBridge (center), and I. I. Rabi with E1189 Magnetron

As war came closer for America, the Rad Lab expanded, then mushroomed after December 7, 1941. Space was taken in a number of MIT locations; eventually, Building 20 was built as the central facility.

Researchers in Great Britain and at the BTL undertook further perfection of the resonant-cavity magnetron, and Western Electric put the improved device into production. In February 1941, using one of the

MIT Building 20 – the Rad Lab

new tubes, the Rad Lab obtained echoes from its first microwave radar. This was a laboratory version of an airborne interception (AI) system. In addition to the high-power magnetron, effective microwave radars required four other major ultra-high frequency components: a local oscillator, a mixing diode, transmission lines, and a duplexer. Solutions were shared between the two nations, some during the visit by the Tizard Mission.

The klystron, already modified by Robert W. Sutton at His Majesty's Signal School to become the reflex klystron, was used for the local oscillator. For the mixer, Dr. Herbert W. B Skinner, a renowned solid-state physicist from Bristol University who headed detector research the TRE, developed a wax-encapsulated silicon diode. Initially, the Rad Lab used a vacuum-tube diode (actually a grounded-grid triode), but soon learned that Skinner's solid-state device performed better. Fabrication of crystal diodes with uniform characteristics was very difficult and remained a "black art" throughout the war.

Waveguides, developed independently in 1936 by Dr. Wilmer L. Barrow at MIT and Dr. George C. Southworth at the BTL, were ideal as output and input transmissions lines for microwave radars. The NRL's duplexer, developed earlier for lower-frequency radars, was unsatisfactory for these frequencies. The solution, called the T-R switch, came in March 1941 from Dr. Arthur H. Cooke of the Clarendon Laboratories at Oxford University.

The Tizard Mission also provided information on the British ASV 176-MHz search radar. With the addition of a duplexer, it was adopted by the Navy as the ASE for use on large patrol aircraft. Initially installed on PBY Catalinas, this became the first U.S. aircraft to carry radar in operational service. A Yagi antenna was fitted under each wing, skewed 7.5 degrees from the centerline; lobe-switching was incorporated.

ASE Yagis on PBY

The Ase was too large for the TBF Avenger torpedo bomber, so the NRL reduced the size and increased the frequency to 515 MHz. Like the ASE, it used a Yagi antenna fitted under each wing. First designated the XAT, then the ASB, this went into production by RCA just two days after Pearl Harbor. It was adopted by the Army Air Corps as the SCR-521. The last of the major non-magnetron radars, it was the most common airborne system of the Allies, with over 25,000 sets produced. Installed almost universally on Navy carrier-based aircraft, it was known as the "Workhorse of Naval Aviation."

ASB Antennas on TBF

Another initial project of the Rad Lab was development of LORAN, an acronym for Long-Range Navigation. This did not involve microwaves, but was a meter-wave system. In October 1940 at the second meeting of the Tizard Mission, a British-developed navigation system named Gee was described. The system had deficiencies in range and accuracy. Alfred Loomis, one of the hosts for the meeting, immediately had concepts for corrections, and his ideas were turned over to the Rad Lab for investigation.

A Combined Research Group (CRG) was established with participants from Great Britain, Canada, and the Navy, Army, and Marines of the United States. Among other activities, this Group was charged with developing a common Identification Friend or Foe (IFF) system used with Allied radars to sort out friendly from enemy targets. At that time, the U.S. had ABD and ABE basic "recognition" systems, and the British had the Mark II and Mark III "parrot" systems. Primary operation of the CRG was on the NRL campus in a guarded compound.

Initial Microwave Radars

After the Rad Lab's first successful demonstrations with a laboratory version of the AI radar, final development of the system was delayed. Attention was turned to 10-cm (3-GHz) radars: a mobile fire-control system for the Army and the SG (surface) and ASG (airborne) search systems for the Navy. The Army system would eventually become the SCR-584, certainly the most recognized of all radars from the Laboratory. It included a highly complex analog computer developed by the BTL and a precisely controlled antenna mount designed by General Motors. Although tested in early 1942, delivery of systems did not occur until the beginning of 1944. Some, 1,500 SCR-584 systems were produced by General Electric and Westinghouse.

With the NRL and the BTL participating in the final design, both the SG and ASG 10-cm (3-GHz) systems were put into production in mid-1941. These were the first fielded radars using the NRL-developed Plan Position Indicator (PPI), a device that greatly improved the usefulness.

SG Antenna

The SG surface-search system produced 50-kW, 1.3-2.0-μs pulses and had cut-parabolic antenna with a gyro-stabilized mount. It could detect large ships at 15 miles and a submarine periscope as far as 5 miles, and allowed firing "blind" at targets beyond the horizon in any weather conditions. It was also used for low-level air search. Manufactured by Raytheon, about 1,000 of these sets were installed on destroyers and larger ships throughout the war.

The small, highly mobile Patrol Torpedo (PT) Boat was introduced in 1942. For this, Raytheon modified the SG to a more compact version, the SO. These sets gave the PT Boats a great advantage, particularly for night operations.

The ASG, also designated AN/APS-2 and commonly called "George," was manufactured by Philco. This 10-cm (3-GHz) air-search radar was superior in performance to the ASB but was bulky and thus limited to large patrol aircraft and blimps. About 5,000 ASGs were built and used effectively against enemy submarines.

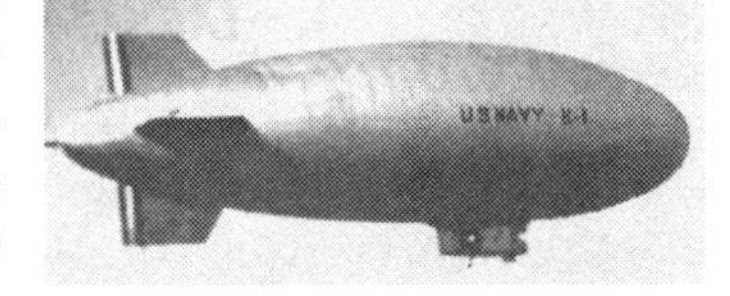

Blimp with ASG Under Gondola

At the BTL, Dr. Mervin J. Kelly instigated projects for developing powerful magnetrons operating at higher and lower frequencies, as well as at higher power. By December 1940, the BTL had built the first American-designed resonant-cavity magnetron, actually a UHF device operating at 750-MHz (40-cm). The BTL had also continued work on a successor to the FA (formerly CXAS). The 750-MHz magnetrons were put into production by Western Electric and used in two new 40-kW fire-control systems: the FC (Mark 3), for use against surface targets, and the FD (Mark 4), for directing anti-aircraft weapons. Deliveries of these systems started in the fall of 1941. Ultimately, about 125 FCs and 375 FDs were produced.

FD Antenna

Chapter 4

THE MAINTENANCE CRISIS AND ITS SOLUTION

In September 1939, when the Second World War started in Europe, the U.S. Navy had about 125,000 officers and enlisted personnel and a fleet of near 400 commissioned ships. The Marine Corps had 19,000 personnel and the Army 190,000. President Franklin Roosevelt declared the Country to be neutral; nevertheless, within a few months he authorized construction of about 100 more warships and suggested that industry make plans for eventually producing 50,000 aircraft per year for the military.

In the middle of 1940, Congress authorized the Navy to further expand the fleet by 70 percent, including 7 new battleships and 12 aircraft carriers, and to have 15,000 aircraft. The Navy and Marines initiated recruitment campaigns to double their personnel. The first peacetime draft in the Nation's history started in September 1940, and reservists were called to active duty in June 1941. At that time, the Navy had 284,427 personnel and the Marines had 54,359.

On December 7, 1941, Japan attacked the United States, and the next day Congress declared war on that nation. In his address to America, President Franklin D. Roosevelt described December 7 as "a date which will live in infamy."

Attack on Pearl Harbor

War was also declared on Germany and Italy on December 11, and the U.S. Navy was authorized an increase in strength to 500,000 personnel.

At that time, there were about 2,000 naval vessels and, although orders had been issued in 1939 for radars to be placed on all capital ships, only 79 sets of all types had been installed.

Signing War Act

ISSUES IN MAINTENANCE TRAINING

When the U.S. Navy placed the first radio telegraph equipment into service in 1902, the Secretary of the Navy was given a report that included the following:

> It would be necessary to employ at each station a competent person to act as operator and instructor, who should be an educated electrician, skilled in the care and adjustment of delicate electrical apparatus.

The report also suggested that a special rating be created for radio operators.

For the next 20 years, radio equipment was operated by men rated as Yeomen and Electricians, the latter being responsible for maintenance. The rating of Radioman finally came into being in 1922, but maintenance remained primarily under Electricians. On-the-job training, followed by an advancement-in-rate examination, was normally used by seamen who aspired to be Electricians or Radiomen – this was called "striking" for the rating – and they might be sent to one of the formal schools. The basic schools for Electricians centered on the operation and maintenance of electrical power systems, while those for Radiomen were primarily for gaining proficiency in Morse code and communication operations. These schools offered a few courses covering radio equipment maintenance, but most of the personnel gained such knowledge on the job and from private study of maintenance manuals.

As described earlier, the Radio Materiel School (RMS) had shared the Bellevue campus with the Naval Research Laboratory (NRL) since 1924, serving as the Navy's only advanced training activity in radio maintenance for enlisted personnel. Here, a small number of highly experienced Electricians and Radiomen were provided six months of instruction in applied mathematics, basic electrical and radio theory, and advanced maintenance techniques. Admission was through passing a comprehensive examination, primarily on mathematics and its applications. Graduates of the RMS who exhibited special capabilities might receive appointments as Warrant Officers. The NRL-Bellevue school also had a Warrant Officers' Radio Engineering Course, graduates of which were usually promoted to Chief Radio Electrician, one of the highest classifications possible for persons without commissioning as officers.

Through the years, the curriculum of the RMS was continuously upgraded, ensuring that the graduates were competent in maintaining the latest equipment coming from the NRL's Radio and Sound Divisions. With the addition of new communication equipment and the advent of radar, there were significant changes in early 1940. It was recognized that the graduates would need to be better trained in the basic subjects and the overall curriculum must be more rigorous. The school was increased to eight months and divided into Primary and Secondary parts. The physical facility on the NRL campus was also expanded to accommodate simultaneous classes.

The three-month Primary started with a brief review of elementary topics, mainly those contained in "Preparation Required for Candidates," a document available for self-study by persons before standing the entrance examination. This was followed by intense overview of mathematics and physics topics and basic electronic theory. Chief Radio Electrician Nelson M. Cooke, an early graduate from both the RMS and the Radio Engineering Course, had returned to the school and was made responsible for developing this portion of the program. The slide rule was extensively used in Primary courses.

Five months in length, the Secondary part of the RMS centered on communication systems, with some time given to radio direction-finding and underwater sound (sonar) equipment and an introduction to the newly developed radar hardware. Instructors in the RMS were senior petty officers and warrant officers who had previously graduated from the school and had extensive, hands-on experience with the hardware. For the advanced topics in underwater sound and radar equipment, civilian specialists from the NRL were sometimes used as guest lecturers.

By any evaluation, the RMS in early 1941 was an outstanding training activity for electronics maintenance personnel. There were, however, a number of significant problems. With an increase in the number of classes, more students were needed. Admission to the RMS had traditionally been limited to experienced naval personnel who passed a rigorous examination and had an intense desire for the training. While the admission exam continued, it was necessary to accept students lacking Navy experience and, often, the drive to succeed in the training.

Radio Materiel School at the NRL, early 1941

The consequence was an increase in the failure rate at the RMS, which over the years had typically been around 30 percent. This, then, required an even greater number of students to be admitted as well as a further increase in instructors – already in great demand. As students with weaker qualifications came into the program, instructors with greater capabilities were needed for the Primary part – those who could better teach science and mathematics as well as fundamental electronics theory.

Another significant problem was the production rate of the RMS. For most of its existence, the school had conducted sequential classes of about 60 students each and averaging about 40 graduates, giving an output of only some 80 personnel per year. In 1940 when the length changed from six to eight months, the operation was expanded to three classes during the year. By mid-1941, a new class of 135 students was starting every four months. Assuming the success rate remained the same, this input would have given an overall graduation rate of about 275 per year.

The Bureau of Navigation (BuNav), then responsible for naval personnel, was projecting near-term needs for maintenance technicians that would require a graduation rate in the thousands – far greater than the RMS production. What were the fundamental options for solving this problem? Could the basic process for training electronic maintenance men be improved?

Since 1924, the Navy had depended upon one school – the RMS at the NRL-Bellevue – to produce well-qualified technicians capable of handling maintenance on all types of electronic equipment. The only exception at that time was a special underwater-sound school operated at New London. Would it be more efficient to train large numbers of technicians at a lower level for maintaining the simpler radio communications equipment, and, for a much smaller number of the best students, send them on to specialized schools for advanced theory and communications, radar, and sonar hardware? This was, fundamentally, how training in the Army Signal Corps and in the British Commonwealth nations was being conducted.

Perhaps because of the historical success of the RMS and, at least to some extent, the tradition in the Navy of resisting change, no real consideration was given to reforming the training process. Men for electronic maintenance would be selected from the best available personnel, would continue to receive the best possible training, would be qualified in all types of equipment, and would remain the elite among enlisted ranks.

With the RMS at Bellevue already reaching capacity, a new, greatly expanded school was urgently needed. Deciding that this should be on the West Coast, Treasure Island, site of the 1939-40 World's Fair in San Francisco Bay, was eventually selected. On October 30, 1941, the BuNav directed that a school at Treasure Island be established, with an authorized complement of 800 students and a curriculum patterned after that of the existing RMS.

The RMS at NRL-Bellevue and its descendant at Treasure Island would center on shipboard and shore-based electronic equipment. At that time, there was no plan for an equivalent school in the Navy for airborne electronics, which was then limited to communications equipment. After America and Great Britain agreed to share radar secrets, a few Navy officers and enlisted men had been trained on airborne radar at Royal Air Force School #31 recently opened in Clinton, Ontario. By mid-1941, the Bureau of Aeronautics determined that they needed their own special school for airborne equipment. Plans were made for an Aviation Radio Materiel School (ARMS) to be established in late 1941 on the campus of the Naval Academy in Annapolis, Maryland. The initial instructors and officers were sent to the Canadian school for training.

The five-month ARMS would be devoted to radar, with students drawn from personnel already knowledgeable in communications; thus, the planned instructional program had essentially no background courses in science and mathematics and little introductory electronics. The output would be about 500 trained personnel per year. Even before the ARMS opened, however, it was realized that a much larger, more comprehensive school would be needed. This, of course, would also require more students with advanced qualifications, as well as more instructors.

With the expectation that amateur radio operators (Hams) would be a good source for potential RMS students, in mid-1941 an aggressive recruitment campaign was initiated with the assistance of the American Radio Relay League (ARRL). Qualified Hams could volunteer for the Navy V-6 Reserve and enter with Petty Officer Third- or Second-Class Radioman ratings. In addition to announcements in the ARRL publication, *QST*, Hams received personal letters from district recruiting officers.

This recruiting campaign had good results. Within a few months, a significant number of Hams volunteered into the Navy and, although rated as petty officers, were sent to Boot Camp. Since they were enlisted to attend the RMS, the customary admission examination was not taken until they started the school. This had two extreme consequences. On the lower end, not all Hams had the necessary ability in mathematics; thus, some quickly failed in Primary and had their ratings removed. On the other end, a number clearly showed that they had the knowledge to by-pass portions of the program, some even going directly to being instructors.

As the prospects of war loomed closer, the training officials at the BuNav were faced with many major issues concerning electronics maintenance. There was the critical need for supplying many more technicians for the most complex electronic equipment of the day. This naturally generated the need for additional training facilities and many instructors with superior qualifications. And the curriculums of the schools, particularly at the introductory level, required additional upgrading. Certainly not the least of these issues was in new recruiting methods for attracting large numbers of highly capable students to fill the classrooms.

All of these training issues were unresolved when World War II was brought to America.

SOLVING THE TRAINING ISSUES

Upon hearing of the attack on Pearl Harbor, William C. (Bill) Eddy, director of an experimental television station in Chicago, took a night train to Washington, D.C., intent on returning to active duty as a Navy officer. On December 8, with an unlit pipe clenched in his teeth, he searched through the chaos in the Navy Department, trying to find a receptive ear. Eddy described this in a 1956 interview with *The News-Dispatch* (Michigan City, Indiana). Eventually he found a senior officer in the Training Division of the Bureau of Navigation who had heard of Eddy's television accomplishments. When the officer told him that highly skilled personnel were desperately needed to maintain a new type of high-frequency radio equipment, Eddy responded, "You'll need lots of radar men. We can train 'em. We've got room, equipment, skilled personnel – you can have it all!"

The officer was dumfounded. The existence of radar was not public knowledge, and what would this television expert know about training maintenance technicians for naval electronics? In no time, however, he was convinced of Eddy's potential value. Eddy pointed out that television had much in common with radar, and he not only was an authority in television but also had Navy experience in underwater sound detection, holding patents in this equipment. He had earlier started an electronics school for the Navy and had done research in high-frequency transmission. As to knowing of radar, only days before Pearl Harbor, Navy recruiters had sent letters to radio amateurs, likely including some then working for Eddy, that contained the following statement:

> The Navy has been conducting considerable research work in the development of the ultra high frequencies. The results of this research are just being applied in the development of what the Navy calls 'RADAR'.

When Eddy made his brash offer, he was evidently thinking in terms of a small number of technicians – hundreds at most – that he might train at his Chicago facility. Many thousands, however, would actually be needed. A small *ad hoc* group at the Bureau of Navigation was already immersed in searching for solutions for the training issues. Eddy's drive and experience was quickly recognized as potentially beneficial to this effort, and he was invited to join the group. This included Lieutenant Commander Wallace J. Miller, Lieutenant Commander Sidney R. Stock, and Warrant Officer Nelson M. Cooke. A number of others also participated, including Lieutenant William F. Grogan and training specialists from the BuNav. The key members were Eddy, Miller, Stock, and Cooke, and to them is mainly due the credit for designing the Electronics Training Program, likely the most intense and highly successful training activity for enlisted personnel in the history of the U.S. Navy. They were well qualified for the task.

The Founding Team

The key members of the *ad hoc* team that planned the new Electronics Training Program were ideally suited for this endeavor. Although highly diverse in their backgrounds, their collective expertise was exactly what was needed. Between them, they understood the characteristics of students who would be successful, they knew the limitations of existing training programs, and they fully appreciated the magnitude and urgency of the task being faced.

William Crawford (Bill) Eddy was born in 1902 and grew up in Saratoga Springs, New York, where his father was a successful businessman and four-term mayor. In boyhood, Eddy was described as a technical prodigy with a Huck Finn spirit and enterprising bent. To channel his capabilities, he was sent off to complete his formative years in a private military school. In 1922, Eddy received an appointment to the U.S. Naval Academy. Although a relatively good student, his gadgeteering and entrepreneurial nature kept him in continual trouble. He finished in the Class of 1926 with the maximum allowable demerits.

William C. Eddy

Eddy's first assignment as an Ensign was on the light cruiser USS *Cincinnati*, initially being sent to Nicaragua to "fight the banana wars," then dispatched to China to protect American interests and "show the flag" along the Yangtze River. In 1928, he requested and received a transfer to the submarine service. At six-foot six, he was almost too tall for this assignment, but soon demonstrated his worth by developing a significant improvement in tracking techniques. Sent in early 1930 to the Submarine Base at New London, Connecticut, he was promoted to Lieutenant (jg) and qualified as a submarine commander, the youngest in the Navy at that time. An accomplished artist, Eddy designed the Submarine Warfare Insignia (commonly called "Dolphins") that was adopted by the Navy and is still in use today.

Remaining at New London, Eddy set up an electronics course for officers and had his own laboratory for conducting research in underwater sound gear and radio transmission from a submerged position; his research resulted in four secret patents. An increasing loss of hearing, however, forced him into disability retirement from the Navy at the close of 1934.

Joining the newly formed laboratory of inventor Philo Farnsworth in Philadelphia, Eddy worked as an engineer on the team that perfected all-electronic television. He later applied his technical talents to early television broadcasting with RCA in New York. While with RCA, he was awarded 43 patents and received wide recognition for his technical accomplishments. Invited by the Balaban and Katz theater chain to initiate electronic television broadcasting in Chicago, Eddy opened W9XBK as an experimental station in April 1941, operating from facilities in Chicago's Loop district. Eddy's ever-present pipe was actually a part of a miniature hearing aid that he had developed, with sound transferred to his ears through his clenched teeth and jaw.

Wallace J. Miller

Wallace Joseph Miller was the Officer-in-Charge of the Radio Materiel School at NRL-Bellevue. For the past two months, he had been planning the replication of this school in the San Francisco area, finally recommending to the Bureau of Navigation that it should be on Treasure Island. Born in 1903 and raised on a farm in Spencer County, Kentucky, Miller briefly attended Ohio Northern University before receiving an appointment to the U.S. Naval Academy, where he graduated as a classmate of Bill Eddy in 1926.

Miller was assigned as an Ensign to the cruiser USS *Cleveland* and, like Eddy, participated in "protecting the rights and interests" of America in Central America during the Nicaraguan revolution. A member of the landing force at Peurta Cabeza,

Miller personally delivered the ultimatum from the United States to the President of the Liberal Faction, an action that effectively terminated the revolution.

For the next several years, Miller attended primary and secondary flight training and served as an aviation officer on the USS *Arkansas*, the USS *Nevada*, and the USS *Neches*. Between 1933 and 1936, he studied radio engineering at the Naval Postgraduate School at Annapolis, then at the University of California in Berkeley, where he earned the Master of Science degree. Following this, he served as the Radio Officer on the USS *Omaha* and the USS *Arizona*, and then on the staff of Commander Flotilla One. Lieutenant Commander Miller was assigned as the Officer-in-Charge of the RMS in February 1940, and led the significant upgrading of this training activity.

Sidney R. Stock

Sidney Richard Stock was born at Fish Haven, Idaho, in 1895, the son of a Mormon pioneer who had emigrated from South Africa. He started at Utah State Agricultural College, majoring in mechanical arts, but interrupted his studies in early 1918 to join the Army Air Corps. After completing Aviation Ground School and training as an airborne observer, he was discharged at the close of WWI. Returning to Utah State, he divided his time between finishing his academic studies and teaching automobile mechanics in a veterans' rehabilitation program at the college.

Upon earning his B.S. degree in 1922, Stock was invited to join the Utah State faculty. With a strong interest in aviation and being a Ham radio operator (W6HUT), he developed courses in both areas, eventually forming the Aviation and Radio Technology Department in 1934. The electronics program gained wide recognition, being cited by NBC in 1939 as having one of the strongest radio engineering curriculums in the Nation. The aviation side was also highly successful, providing more students in the pre-war Army and Navy cadet flight training programs than any other western college.

As the Bureau of Aeronautics was planning a training program in aviation electronics maintenance, Stock's accomplishments came to their attention. A Navy representative visited Stock in Utah, described the secret radar activity, and asked him to head up the urgently needed training effort. At the first of October 1941, Stock was directly commissioned as a Lieutenant Commander by the Bureau of Aeronautics and assigned to implement its program for airborne electronics maintenance. He first visited a school in Clinton, Ontario, for orientation and training in British airborne radar. Following this, he became Officer-in-Charge of the Aviation Radio Materiel School that was being initiated in facilities at the U.S. Naval Academy, Annapolis, Maryland.

Nelson M. Cooke

Nelson Magor Cooke was born at Davis City, Iowa, in 1903, and enlisted in the U.S. Navy in 1920 as an Apprentice Seaman. His innate technical capabilities allowed him to become rated as an Electrician Petty Officer in minimum time. He attended the Radio Materiel School at the NRL in 1928. His performance there was such that he received an appointment as a Radio Electrician (Warrant Officer) at age 25, one of the youngest persons ever holding such a position during peacetime.

Between 1928 and 1934, Cooke served on the seaplane tender USS *Wright*, then was sent to the Warrant Officer Radio Engineering school at NRL-Bellevue. Following this, he was promoted to Chief Radio Electrician and again served on the *Wright* and then the aircraft carrier USS *Saragota.* He returned to NRL-Bellevue in 1938 to serve as the Senior Instructor in the Radio Engineering program and a leader in planning the upgraded curriculum for the Radio Materiel School.

An outstanding technical writer, Cooke prepared the mathematics lecture notes used in the Radio Engineering program. He also designed a special slide rule for electrical and radio calculations that was manufactured by Keuffel & Esser, and wrote an instructional manual for its use. Both his notes and the slide rule were adopted for the new Electronics Training Program. Cooke's lecture notes were later published as a classic textbook, *Mathematics for Electricians and Radiomen* (McGraw-Hill, 1942), possibly the most used textbook ever written on applied mathematics through trigonometry and vector algebra.

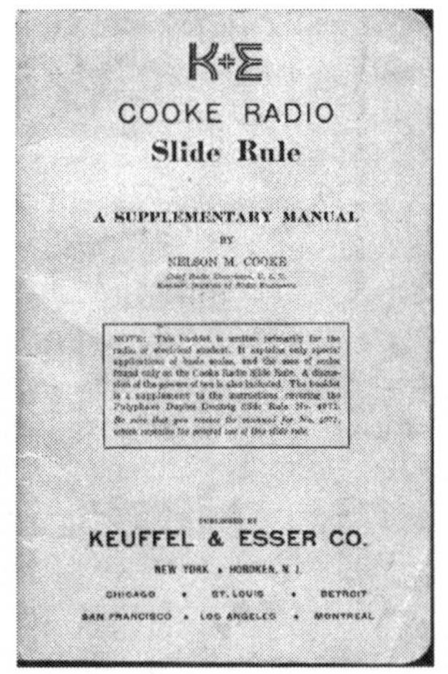

The Considerations

World War II had started, and thousands of technicians would soon be required for maintaining the great variety of advanced electronic equipment being added to the Navy. But there was no concise plan for obtaining these urgently needed specialists. Sheer numbers of men was not a problem – over 200,000 volunteers and draftees would be pouring in to fill the Navy's newly authorized strength. Essentially none of these, however, would be able to directly fill the maintenance needs – all such personnel would have to be trained.

Some of the broad considerations in planning a new electronics maintenance-training program would have included the following:

- The equipment being installed, particularly the radar, was far more complex and required a higher level of understanding than anything entrusted to Navy enlisted personnel in the past. Training courses would need to be very comprehensive.
- Even if the personnel became proficient on equipment as it existed at the time of training, new types of systems would become available faster than maintenance personnel could receive refresher courses. The training would need to emphasize basic theory.
- The equipment, especially the radars, was new and relatively unproven. Considering its importance to modern warfare, immediate repairs involving innovation were necessary, and preventive maintenance was very important.
- Technicians for equipment aboard vessels at sea would need to be highly self-reliant, functioning without access to extensive repair parts and substitute units, and without assistance from civilian representatives of the supplying industries.

The ideal solution would have been to establish additional maintenance schools and populate them with volunteers and draftees who had previously studied electrical engineering in college, or had extensive civilian experience in advanced electronics. But the Navy did not have facilities and experienced instructors for new schools, and the time lost while they were being acquired would be unacceptable. As to students with the desired engineering backgrounds, such persons were in short supply and, if they met the Navy's stringent physical requirements, would likely be drawn into Officers Candidate School and become communications officers, not maintenance technicians.

The *ad hoc* planning group visited the existing RMS at the NRL as well as the ARMS being developed at the Naval Academy and addressed a myriad of interrelated questions:

- What improvements could be made to the RMS structure?
- Where might appropriate existing facilities be found?
- How could qualified instructors be obtained?
- What type of training would be best for future technicians?
- What knowledge should technicians have after training?
- What might be the characteristics of appropriate trainees?
- How could persons with these characteristics be obtained?
- What minimum implementation schedule would be possible?

As to restructuring the RMS, it was readily agreed that the existing two parts should be made into distinct training activities: Primary School and Secondary School. If the NRL and Treasure Island operations were solely Secondary Schools, this would open facilities and the instructional staff to approximately double their output. A number of new Primary Schools – serving as filters and suppliers for the Secondary Schools – would be required. Facilities for these, not requiring laboratories for elaborate naval hardware, might be more readily available. Nelson Cooke, who was then responsible for the primary part of the NRL-Bellevue RMS, generally agreed, but insisted that his operation remain at the NRL, at least for some time. Sidney Stock, still developing the ARMS, agreed that this operation, whether at Annapolis or another location, could also be served by the proposed Primary Schools.

Determining possible locations for the new Primary Schools was surprisingly simple: they could be developed and operated under contracts with existing colleges and universities. This would also solve the problems of finding qualified instructors and in developing an improved Primary School curriculum. When the draft was implemented in the fall of 1940, it had a major effect on the population of male students on American campuses. By Fall Term of 1941, most engineering departments experienced difficulties in filling classes and were searching for ways that they could participate in the war-preparation effort. They would certainly welcome an opportunity to participate in this new training program.

There were two highly related precedents for college-operated preparatory programs of a similar nature. When the Canadians started the radar school at Clinton, they turned to academia for a preparatory program. Initial training in electronics was provided by colleges and universities across Canada. Richard Stock was familiar with the Canadian program, having visited the school at Clinton. In the United States, MIT had recently initiated the Radar School, set up by the Electrical Engineering Department for Army and Navy officers. For persons who did not have the background for the Radar School, a Pre-Radar School was conducted by Harvard University.

The combined Primary and Secondary schools, with a total duration of eight months (the length of the existing RMS), would need to convert persons with possibly no background in the field to specialists who could assume the responsibility for maintaining a wide variety of highly complex electronic equipment. In one sense, this would be more difficult than turning out electrical engineers in this relatively brief period – the graduate would not only have to understand the theory but also have some exposure to the varied types of naval electronic hardware. Then, what type of training would be appropriate?

The existing RMS provided a good guide. For years, this activity had taken students who had a minimum of formal training and turned out some of the best technicians in America. There were two significant differences, however, in the old RMS and the presently needed program. First, the incoming students would not have the experience or on-the-job training of earlier students. Thus, there must be much more emphasis on training starting at a lower level. Second, the emerging equipment was far more complex than that of the past, so the training would need to be broader and encompass more theory. As earlier noted, since electrical engineers were not available for maintenance, the Electronics Training Program would need to produce technicians close to this level, and in less than one-fourth the time of a standard college curriculum.

With appropriate selection of students, the college-operated Primary Schools might crowd the essential mathematics, physics, and electronic circuit theory into the available three months. Extensive examinations would be needed to ensure that only well-prepared persons would be graduated. The basic preparation for these persons would continue into the Secondary School, with classroom work dominating at the start, then building to mainly laboratory activities involving actual hardware. The keys to success would be "appropriate selection of students" and "extensive examinations."

What then should be the characteristics of potential students, and how could persons with these characteristics be brought into the program? The traditional RMS had been highly selective in admitting students, requiring experience as well as passing a difficult qualifying examination. Those admitted were highly motivated and above average in intelligence. Students in the new program would certainly need these characteristics, but, particularly with the short time for training, willingness and ability to devote an enormous effort to the study would be equally important. While applicable experience or prior study in electronics would be desirable, it could not be a requirement. Also, the existing admission examination that assumed extensive preparation would need to be significantly changed. Whatever the entrance criteria, extensive additional filtering would be required after trainees started the program.

In an effort to place persons directly into the RMS, a special activity in recruiting amateur radio operators (Hams) had earlier been instigated. This allowed Hams to enter the Navy with petty officer ratings (mainly second class) as Radiomen. While this might have been beneficial in quickly filling new classes in the RMS, it could not continue indefinitely. Not only would the supply of such persons be quickly exhausted, but it might have a negative effect in recruiting otherwise highly qualified personnel who did not have this credential. The *ad hoc* group would recommend to the BuNav that some level of advanced rating be given to all persons allowed into the program, regardless of their Ham experience. It was also recommended that a new classification be developed, giving special distinction to persons qualifying to maintain electronics.

Concerning the Secondary School curriculum, the *ad hoc* group believed that the existing one in the RMS was generally adequate in scope but should be improved in content, assuming that better prepared students would be available from new Primary Schools. The sections on various topics should give more attention to theory and general aspects, rather than particular hardware elements. Technicians, often acting alone on assignments, would be expected to maintain a wide variety of electronic equipment; thus, study of theory would be of much greater value than lengthy laboratory work on specific equipment.

Time was critical, and the *ad hoc* group felt that the new training program should start essentially immediately. Since it would likely take several months to get contracts in place and set up college-based Primary Schools, a developmental prototype school might be started at once. For this, Bill Eddy went back to his original offer: "We can train 'em. We've got room, equipment, skilled personnel – you can have it all!" His reference was to the experimental television station in Chicago. He reiterated his offer, saying that his staff of 15 television engineers could be made available to both develop a curriculum and teach the first classes in the station's facilities. Although at risk, he would start renovations and preparation of instructional materials, having the facility ready for opening within a month. Eddy also volunteered to develop the new classification examination. An interesting part of Eddy's offer was that his employing firm, Balaban & Katz Theaters, would cover the cost of this prototype school.

The Recommendations

Within a few days, the *ad hoc* group had arrived at an overall plan that might solve the electronics maintenance-training crisis. Their recommendations to the BuNav Training Division included the following major elements:

- The training program [hereinafter called the Electronics Training Program (ETP), a name coined by the author in the absence of any official name given by the Navy] should, like the existing Radio Materiel School (RMS), be divided into two sequential parts: Primary Schools of three-months duration, and Secondary Schools five to six months in length.
- A new classification examination should be developed for admission to the ETP. The current practice of giving petty-officer ratings to persons holding radio amateur licenses would continue; others would enter with seaman first class (S1/c) ratings.
- To gain access to highly qualified instructors and new laboratory facilities, the Primary Schools should mainly be conducted under contracts by colleges and universities, using their electrical engineering faculties and laboratory facilities.
- The Primary School should include a highly condensed coverage of the necessary physics, mathematics, and electrical engineering topics normally in the main courses of a college curriculum. A very high failure rate of students should be expected.
- The new Primary curriculum should be tested through a special school at the facilities of experimental television station W9XBK in Chicago. This school, operated without cost to the U.S. Navy, would use the station's engineering staff as instructors.
- A new petty officer rating of Radio Technician should be opened, with all persons in the ETP being changed to this rating at their next promotion. Persons entering as S1/c would be promoted to this rating upon completion of the Primary School.
- The existing RMS at NRL-Belleview and the soon-to-open RMS at Treasure Island should become the Secondary or Advanced Schools, devoted to fleet and shore hardware. The newly opened Airborne RMS at Annapolis would be equivalent to a Secondary School for flight hardware.
- The Secondary School curriculum should be an upgraded version of that existing at the RMS, with additional theory and other advanced topics, and with laboratories including various types of communication, radar, and other hardware.

- An intensive recruiting campaign should be conducted, indicating that the program was the equivalent of "two years of college in less than a year" and particularly attracting new high school graduates with aspirations of a technical career.

Richard Stock, with experience in college programs of this general type and also having seen colleges used in the Canadian radar-training program, was a major proponent of this plan for Primary Schools. In addition to providing students for the NRL-Bellevue and Treasure Island Secondary Schools, it could also serve as the main source for ARMS students at Annapolis. The Primary Schools would give all of the needed review and introductory theory, drawn from the existing beginning three-months of the RMS, as well as selected topics in electrical engineering programs.

The instruction in Primary Schools would be accelerated to an extreme – involving 60 to 70 hours per week and rapidly weeding out those students not having the needed capabilities and drive. The colleges and universities would be given general guidance as to the material to be covered, but the details would be left to their discretion. The curriculum would be general in nature – not directed toward a particular Secondary School – and include only unclassified information. Each participating institution would assign a senior faculty member to serve as the instructional coordinator.

Nelson Cooke, who was then responsible for the Primary part of the NRL-Bellevue RMS, and Wallace Miller strongly supported the content and intensity of the proposed Primary School curriculum, but felt that this activity at the RMS should continue for the immediate future, remaining the same in size and then possibly phasing out later. This was included in the recommended plan. It was also recommended that the Treasure Island activity initially include a Primary School, but then this be discontinued once the stand-alone Primary Schools became fully operational.

Bill Eddy continued his quest in Washington for being immediately returned to active duty. Unsuccessful in this, he went back to Chicago to tell Balaban and Katz, owners of W9XBK, about what he had offered the Navy – he had made this offer, then telephoned for permission! The theater chain readily blessed the offer, agreeing to provide the instructors, facilities, and laboratory equipment without charge. Although the Navy had not officially notified Eddy of acceptance of his offer, work was started in remodeling the facility. Assisted by members of the television staff, preparation began on instructional materials; this was necessary if he was to meet the promise of "opening within a month."

Acceptance and Implementation

The Training Division of the BuNav accepted all of the *ad hoc* group's recommendations, including Eddy's offer to immediately start a "no-cost" developmental-prototype Primary School in Chicago. Primary Schools were officially designated Elementary Electricity (a imisnomer and Radio Materiel (EE&RM). Some Navy records show the "EE" as meaning Elementary Electricical and still others show it as Electrical Engineering. The Secondary activity, conducted at NRL-Bellevue and Treasure Island, was designated Advanced Radio Materiel School.

The Aviation Radio Materiel School at Annapolis remained under the Bureau of Aeronautics. After the planned move to Ward Island, Texas, it would serve as another Secondary School and be fed by graduates of the newly formed Primary Schools.

The college-based Primary Schools would be official Naval Training Stations but under the full academic control of the contracting institutions. While the Navy would provide the basic curriculum and the examination requirements, each school would determine the details of instruction and be responsible for assigning the faculty. Such an arrangement was unique in naval training. The institutions would provide classrooms and laboratories for the instruction, as well as housing and food services for the students. The Navy would assign to each Station an Officer-in-Charge and a small cadre of enlisted personnel for administrative support, and some minimal level of military and physical training.

The Secondary/Advanced Schools would continue to be full naval activities, conducted in Navy facilities using Navy personnel as instructors. The classroom and laboratory instruction would include classified materials, and appropriate security must be maintained. Because of this, all persons admitted to the training program would need to be U.S. citizens, either native-born or through naturalization.

New petty-officer rating of Radio Technician (RT) and Aviation Radio Technician (ART) were formed for persons completing the ETP. Other personnel already holding a petty-officer rating of Radioman or Electrician could receive the RT rating upon passing a classification examination. Volunteers or draftees who passed the new admission examination would enter the Navy as Seaman First Class, could advance to Petty Officer Third Class after completing Primary School, and could be promoted to Petty Officer Second or First Class upon graduation from Secondary School.

On January 2, 1942, Captain Louis E. Denfeld was assigned as Assistant to the Chief of the BuNav with the responsibility for training the rapidly expanding personnel. A 1912 graduate of the Naval Academy, he had just been awarded the Legion of Merit for "exceptionally meritorious conduct as Chief of Staff of a Task Force, Atlantic Fleet" leading up to and during the first days of the war. Denfeld was a person who could quickly assess information and take action. As soon as he came aboard, he reviewed the recommended plans for the Electronics Training Program and gave his approval on January 7.

Louis E. Denfeld

Denfeld's approval was telephoned to Eddy, and final preparations for the school in Chicago were made. The Primary School at 190 North State Street opened five days later on January 12, 1942, inaugurating the Electronics Training Program. Ensign Eugene S. Pulliam was assigned as the acting Officer-in-Charge, awaiting Lieutenant Eddy's return to active duty. Pulliam, previously a journalist in Indianapolis, supported Eddy in a variety of functions for the next two years. A press release from the Navy Department dated January 13, 1942, and printed in the *Chicago Daily Tribune,* began as follows:

> The U.S. Navy is developing high-frequency radio as a new weapon for the nation's armed forces. This was revealed today when the Navy Department announced it opened a Navy primary school for training in high-frequency radio in Chicago yesterday.

Thus, the training program that would play a vital role in World War II and have a major effect on post-war electronics was publicly announced. This was also one of the first public disclosures of a new weapon, "high-frequency radio," which was actually radar.

Eddy had requested that he be returned to active duty and appointed Officer-in-Charge of the Chicago training school. Because of his deafness and other physical defects, however, this did not occur until August 1942; Frank Knox, then Secretary of the Navy, personally directed his reactivation. He reentered with the Naval rank of Lieutenant, but, through quick promotions, reached Lieutenant Commander, Commander, and finally Captain in slightly over two years.

Also in August 1942, Nelson Cooke was commissioned as a Lieutenant and served as the Officer-in-Charge of the NRL-Bellevue Primary School, followed in May 1943 as the Executive Office of all the

NRL-Bellevue school operations. He received a promotion to Lieutenant Commander in 1945 and was responsible for the school until it closed in early 1947.

Wallace Miller was promoted to the rank of Commander and continued to head the NRL-Bellevue school until May 1943. As an officer in the Regular Navy, Miller had obligatory fleet duty and was assigned as Commander of the destroyer USS *Caperton*. He then served with high distinction in command of several different vessels for the remainder of the war.

Sidney Stock continued with the Aviation Radio Materiel School at Annapolis and planned the expansion and transfer of the activity. In July 1942, the transferred school opened at Ward Island, near Corpus Christi, Texas. The operation was eventually designated Naval Air Technical Training Center (NATTC), Ward Island. Shortly after Ward Island opened, Stock returned to Utah State to serve as the Officer-in-Charge of the Navy's Primary School that had started there the previous March. In January 1944, he transferred to a similar position at the University of Houston, where he was promoted to full Commander.

As the training program got underway, Lieutenant William Grogan, one of the other supporting participants in the initial planning, was promoted to Lieutenant Commander and transferred to Grove City College in Pennsylvania to start another of the Primary Schools. After opening the school in March 1942, Grogan remained as the Officer-in-Charge.

In May 1942, the Bureau of Navigation was converted to the Bureau of Naval Personnel. Captain Denfeld was appointed Assistant Chief of Personnel and served in that position until March 1945. The award for a Distinguished Service Medal presented to him upon leaving that post notes the following:

> Undertaking this exacting assignment during a period of crisis in the history of the Navy, [he] achieved distinctive success in advancing the difficult and complex program of rapid expansion in personnel necessary for the manning of ships and shore bases in widespread areas of operation.

Having reviewed and approved the Electronics Training Program as one of his first actions in the new assignment, Denfeld retained a personal interest in the program and participated in subsequent reviews and other activities.

Main Facilities of the U.S. Navy in Washington, D.C. during WWII

Above: The Naval Research Laboratory

Left: The Washington Navy Yard

Chapter 5

THE ELECTRONICS TRAINING PROGRAM

The Electronics Training Program (ETP) was developed to prepare Naval enlisted men for maintaining highly complex electronic equipment. This name, the ETP, was coined by the author; insofar as is known, the Navy never gave the program an official name, but simply referred to it as the Radio Materiel School (RMS), the name for a school existing since 1924 at the Naval Research Laboratory.

Training in the program was initially planned to be in two parts. The first would be a three-month Primary School, officially designated Elementary Electricity (or Electrical) and Radio Materiel (EE&RM), similar to the initial portion of the existing RMS at NRL-Bellevue, but emphasizing topics drawn from then-existing college engineering curricula.

The second part, a five-month Secondary School, officially designated Advanced Radio Materiel School, would be based on the upper portion of the existing RMS, upgraded with more theory appropriate for personnel with a stronger background. For airborne electronics, a Secondary School was planned as an expansion and relocation of the Aviation Radio Materiel School – then located in Annapolis at the U.S. Naval Academy – and supplied with students who had prepared at the Primary Schools.

At the same time that the ETP was being developed, the RMS at NRL-Bellevue was being replicated at Treasure Island in San Francisco. When it opened in January 1942, the RMS at Treasure Island was organized into Primary and Secondary Schools, but the former was discontinued in March when the college-based Primary Schools opened. The RMS at Bellevue, however, continued to have both Primary and Secondary Schools until May 1943.

GENERAL DESCRIPTION

A major distinction of the Navy's ETP was the emphasis on fundamental theory. In justifying this, Captain Harry F. Breckel, Commander of the ETP at Treasure Island, stated in a report prepared when he vacated this command: "The entire curriculum is based on the premise that if a man is a master of fundamentals, he will be, in a short while, an excellent maintenance technician."

In his report, Breckel provided the following excellent, concise description of the program:

> The mission is to train and qualify men in the fundamentals of electronics theory and the practical applications thereof so that graduates will be able to perform the duty of installing, servicing, and repairing all types of Naval radio, radar, underwater sound gear, and other related electronic equipment.
>
> In training a man for these duties, it is considered of greatest importance that he understand and be thoroughly grounded in electronic fundamentals rather than specific equipment. He may come in contact with one or more of 80 different models of communication transmitters and 50 models of communication receivers; more than 30 models of sonar or echo-ranging equipment; and more than 25 separate and distinct models of radar equipment.
>
> The radio technician is expected to also be familiar with many other devices, such as radio-frequency direction finders, power equipment, rotating machinery, selsens, syncro-systems, etc. Technicians must be able to use more than 100 types of special test equipment.

Ideally, persons who had studied electrical engineering or who had extensive experience in this field would be best suited for such training. But they were not readily available, certainly not in the numbers required. The training program, therefore, needed to include the essential topics to provide a basic background in electrical engineering – covering the essential topics in at least the first part of a standard college curriculum in this field.

Just as the first two years of an electrical engineering program provided the foundation for advanced courses in the last two years, the three-month Primary School in the ETP would prepare Navy students for the advanced topics of the five-month Secondary School – particularly the theory behind advanced communication and radar systems. Here a brief description of a "standard college curriculum in electrical engineering" of that era is in order.

America's first engineering school – the U.S. Military Academy – was opened at West Point, New York, in 1802. This was only a short time after 1794, when France established the École Polytechnique, the world's first such school. The words 'ingenuity' and 'engineering' in English and 'ingéniosité' and 'ingénierie' in French are linked to the same Latin word-root, and the verb 'to engineer' means 'to be ingenious.'

Prior to WWII, electrical engineering was a highly practical curriculum in many American colleges – often referred to as "handbook engineering." There was a strong distinction between physics and engineering. In fact, the study of electrical engineering had evolved out of physics departments to provide the applied aspects of electricity and magnetism. While higher mathematics – mainly calculus and differential equations – were included in engineering curricula, it was primarily so professors could show the derivation of equations ("formulas") for practical applications. These equations were then used to generate tables and graphs – tools for the practicing engineer. When reduced to their elements, the equations seldom exceeded exponential relationships, and these were most often applied using the engineering slide rule – a formidable device to the uninitiated.

The importance of handbooks in that era is readily illustrated in the *Radio Engineers' Handbook*, by Terman (McGraw-Hill, 1943). This was likely the most popular book for radio engineers ever published, and, in preliminary form, had for years been the "bible" in Stanford's undergraduate curriculum. Dr. Frederick E. Terman was Professor and Head of Stanford's Electrical Engineering Department. At the time his handbook was first published, Terman was Director of the Navy's Radio Research Laboratory at Harvard University.

F. E. Terman

Many people practicing or teaching engineering did not have a college degree in that field, or even a degree at all. Some examples have previously been mentioned. Reginald Fessenden, the father of AM radio, professor at Purdue, and holder of many important patents was not college educated. Leo Young, co-inventor of radar and an electronics leader at the NRL, never took a college course. One of the ETP planners, Nelson Cooke, obtained all of his formal education in the schools at the Naval Research Laboratory.

ORGANIZATION AND HISTORY

Based on the existing curriculum of the Radio Materiel School at the NRL-Bellevue, the planners of the ETP believed that all of the necessary basic theory could be absorbed by intelligent and motivated students in the three-month period of a Primary School, especially if the instruction was by experienced faculty members in colleges and universities. The success, or failure, of the Primary Schools would be shown in students' performance in the Secondary Schools. It was imperative, however, that this performance be excellent – graduates of the Secondary Schools must have exceptional capabilities and they were urgently needed.

For success in the Primary Schools, a critical balance would be necessary between the ability of instructors to present the required material within the allocated time, and the capability of students to understand and retain this material. Prior study or experience in radio would appear to be desirable for incoming students, but would this be sufficient for success in the intensive training?

Typical Recruiting Advertisement

The Naval Recruiting Offices were already very active in signing up radio amateurs, offering them entry as Radioman Third- or Second-Class Petty Officers, and it had been agreed that other persons passing a special entrance examination would be admitted to the program as Seaman First Class. An intense recruiting effort was initiated. Physical requirements were relaxed, and persons between 17 and 50 years old were eligible. Special enticements included a new rating for graduates, Radio Technician (RT). Recruitment advertisements and literature from the Navy indicated that the program had a value of $5,000. Later, as the program matured, this value was increased to $12,000.

Initial Organization

The March 23, 1942, issue of *Time Magazine* included a brief article with "Loop Sailors" as a heading and opening with the following

> There go those gobs again. This was getting monotonous – Chicagoans thought – kind of funny, too. Twice a day the sailors filed smartly out of the Naval Armory, 100 of them, headed for the Loop. (Hup, two, three four! Column right, hutch!) The Navy must have gone nuts. There weren't any ships in the Loop.

The article continued to say that Chicago newspapers discovered that they were the first class in "aircraft detecting devices," studying in the facilities of television station W9XBK in the State-Lake Theater building.

State-Lake Theater at 190 North State Street

Bill Eddy had offered to establish – without cost to the Navy – a Primary School in Chicago to test out the curriculum on students. Eddy's prototype Primary School at 190 North State Street opened on January 12, 1942. The 15 instructors were mainly from the engineering staff of experimental television station W9XBK. Most of the trainees in the initial class were Ham radio operators, and the others had something in their backgrounds indicating that they qualified for the school. A preliminary version of an admissions examination, prepared under Eddy's direction, was given to the incoming students, with results compared with their performance and used for test revisions.

The detailed curriculum, developed in a just-in-time manner, was tested on the first class. Two major problems were immediately apparent. Many of the students lacked the background or aptitude for the highly accelerated training. Also, the scheduled three months was too short to cover the material that Eddy considered essential, as well as a necessary "refresher" in elementary topics. Nevertheless, all of the students in the initial class completed the studies.

In February 1942, six institutions were selected to give the college-based Primary School: Bliss Electrical School (Maryland), Grove City College (Pennsylvania), Oklahoma A&M College, Texas A&M College, University of Houston, and Utah State College of Agriculture. Programs at these schools started in March, each with a class of 100 students who were either Hams or had passed the now-implemented Eddy Test. The initial success of these schools, however, was not comparable with that of Eddy's prototype school; there were many failures in the first classes.

Successive classes in the prototype school also experienced an unacceptably high failure rate. After reviewing the situation, it was realized that a significant change was needed in the Primary Schools. Several options were considered, including decreasing performance expectations, reducing the coverage and intensity of the curriculum, lengthening the period of the Primary School, and making the entrance requirements tougher. The first two options were considered to be unacceptable. After reviewing the performance in various portions of the curriculum and with some experimentation at the prototype school, a solution combining the latter two options was adopted.

By far the most failures were associated with the first month of the program, where there was an intensive review of fundamental topics and an introduction to the slide rule. Students who made it through this portion would usually perform satisfactorily in the remaining two months. This situation was found not only in the new schools but also in the continuing Primary Schols at NRL-Bellevue and Treasure Island.

Upon recommendations from Eddy, the Bureau of Naval Personnel (BuPers, formed from BuNav) directed changes for the ETP. The overall schedule was increased by four weeks, with the additional time placed in a new school called Pre-Radio. All new students in the program would first attend this school. Conducted at a staggering pace, Pre-Radio would serve as a filter for Primary School. In late 1942, Eddy opened a temporary Pre-Radio School in the Chicago Naval Reserve Armory, and from April through June 1943, a Pre-Radio School was also conducted at Treasure Island.

With the Eddy Test serving as admission and the intensive review and introductory topics given in Pre-Radio, only exceptionally capable students would pass on to Primary School. The full three months of Primary School could then be devoted to the remainder of the curriculum, greatly increasing its efficiency. The revised Primary School curriculum was implemented in late 1942 when Pre-Radio started. Graduation certificates from one Primary School (Texas A&M) shows 850 school hours, indicating over 70 hours per week in a combination of classroom and laboratory work and required outside study time; beyond this, time was also needed for physical training and military activities.

Starting in June 1943, Pre-Radio Schools were established at four facilities in and around Chicago. It had been found that students who scored well on the Eddy Test did about as well as Hams in the overall program. Based on this, entry to the program was changed to be solely on the Eddy Test, and the Petty Officer rating for entering Hams was eventually eliminated.

To meet the need for instructors, Eddy implemented a Teacher Training School at the State-Lake facility in Chicago. Here, in a three-week activity, Navy and Marine graduates of the program were given special preparation for serving as instructors in the new Pre-Radio Schools and the Prototype Primary School, as well as for augmenting the civilian instructional staffs in the college-based Primary Schools.

In September 1943, the Pre-Radio, Primary, and Teacher Training activities directly under Eddy were officially designated the Radio Chicago Naval Training Schools. Throughout the remainder of the war, Eddy served as the Commanding Officer of Radio Chicago, and is credited with a large part of the overall success of the ETP. An accomplished artist, Eddy personally designed the logo for this operation.

Eddy's Logo

In mid-1943, the Bureau of Naval Personnel directed the various schools to change to a weekly rather than monthly schedule, allowing a more even scheduling of classes between schools. This resulted in the three levels of schools operating with 4-, 12-, and 20-week durations. Ward Island had changed to a 24-week schedule the previous April, and the other Secondary Schools later increased to this duration.

Revised Organization

When the ETP was initiated in 1942, the anticipated strength in the Navy was about 640,000 men and 142,000 in the Marines. In 1943, the numbers more than doubled to 1,750,000 and 309,000, and the need for radio technicians went up accordingly. This increase by a factor of about 2.6 had saturated the various ETP schools, even with the introduction of Pre-Radio. The overall strength projection for 1944 then went to 3,000,000 and 475,000, and another significant increase was anticipated for 1945.

On December 15, 1943, a conference was held in Washington to discuss the Electronics Training Program. Included were representatives of the Commander in Chief of Naval Operations, Bureaus of Ordnance and of Ships, and Procurement and Assignments Divisions of the Bureau of Naval Personnel. Radio Chicago, the Secondary Schools at NRL-Bellevue and Treasure Island, and the Naval Training Center at Chicago's Navy Pier were also represented. The major topics were ways to meet the increased need for ETP graduates and what changes were needed in the curriculum.

The college-based Primary Schools and the Treasure Island Secondary School had already expanded significantly since their start. The NRL-Bellevue operation still had both Primary and Secondary Schools, but these had always been filled. Tentative plans were already underway for taking over the full facilities of a small college in Arkansas and moving the NRL-Bellevue Primary School there, thus freeing facilities for enlarging the Secondary School. This facility at NRL-Bellevue, however, was much smaller than that at Treasure Island, and this change would not add significantly to the program.

Another Secondary School (a fourth when including NATTC Ward Island), was urgently needed. Like the others, it should be in a highly secure facility located in a city that already had a major Navy presence. With existing schools on the East, West, and Gulf Coasts, a location in the Mid-West would have geographical and political advantages. A large facility would be needed, and there was no time for extensive construction. Navy Pier in Chicago was an obvious choice. The Navy had taken over this facility in August 1941 and converted it to a training center for aircraft mechanics and similar skills that could accommodate up to 10,000 personnel. It was believed that the existing schools at Navy Pier could be relocated and the facility made ready for a Secondary School within a few months. The critical path in the development would be in establishing the faculty.

Extensive increases were also necessary for the Primary Schools. While the college-based schools had provided a solution to the initial crisis, particularly in using the engineering faculties and facilities to initiate the program, the conference attendees felt that the expansion should now be through Navy-operated Primary Schools. Thanks to the colleges, the curriculum was now well established. After two years of operation, the ETP had produced many electronic specialists who would be qualified as faculty members. A pool of such persons was available in naval and marine men who had served at sea or in overseas locations after graduating and were now ready to be rotated back to the States for new assignments.

The conference attendees agreed to recommend that new Primary Schools be established as soon as practical at the Gulfport Naval Training Center – a collection of service schools that had opened in 1942 – and at Great Lakes Naval Training Center – the historical location for Navy technical service schools. In both locations, the Primary School would displace existing schools and use their housing and classrooms.

A more readily available facility was brought to the conference attendees' attention. Since February 1942, the Navy had operated a 1,000-man Pre-Flight School at the huge Del Monte Hotel in Monterey,

California, but this school was now being closed. The buildings had previously been converted for use as a Navy operation; thus, a large Primary School might be opened there in a few weeks. This would also give a Primary School on the West Coast, and it would be relatively close to the Secondary School at Treasure Island.

As to possible revisions of the curriculum, the attendees felt that the schools were producing well-trained technicians for the great variety of electronic equipment in the Navy. It was noted, however, that there was increased use of sonar in the fleet and the ETP graduates might benefit from more advanced instruction in this area. To that date, there had been minimal attention given to sonar, much less than that allocated to radar and communications. Since there was no feeling that any existing instruction should be reduced, adding advanced sonar would require lengthening the Secondary School.

The conference attendees made their recommendations to the Bureau of Naval Personnel. Based on these recommendations, as well as those of his staff, Captain Louis Denfeld, Director of Training, ordered that the following be implemented:

- A Navy-operated Primary School would be immediately opened at the Del Monte Hotel in Monterey, California.
- Navy-operated Primary Schools would be opened as soon as practical at the Naval Training Centers at Great Lakes, Illinois, and Gulfport, Mississippi.
- The Primary School at NRL-Bellevue would be immediately moved to the College of the Ozarks in Arkansas and the Secondary School enlarged in the vacated space.
- The facilities at Navy Pier would be converted for opening a new and enlarged Secondary School as soon as practical, with Treasure Island assisting in this implementation.
- To acquire additional qualified Secondary School instructors, Treasure Island would immediately initiate a formal teacher-training activity.
- The Secondary Schools at NRL-Bellevue, Treasure Island, and Navy Pier would be increased by four weeks, with advanced sonar added to the curriculum. This would make all four Secondary Schools 24-weeks in duration.
- Radio Chicago and most college-based programs would continue, but programs at Texas A&M and Utah State would be closed.

All of these actions were implemented by mid-1944. For the next year, the ETP was relatively stable, reaching a high point in number of students in early 1945. The remaining college-based Primary Schools were phased out by mid-1945. For reasons that are unclear, the Navy then opened Pre-Radio and Primary Schools at the existing Naval Training Center in the Rouge complex of Ford Motors at Dearborn, Michigan. This school existed for less than a year.

The following descriptions of the ETP elements are based on operations after the program somewhat stabilized with the addition of Pre-Radio in early 1943.

THE EDDY TEST

In an article in *The Mathematics Teacher* (December 1943), Nelson Cooke described the essential standards for selection of ETP trainees:

> The candidate for radio technician training must first indicate an interest in this type of work. Secondly, he should be a highly intelligent individual with mechanical aptitude and a good background in mathematics. In addition, he must be able to get along with others and be physically capable and willing to work a 15-hour day during his training – and like it.

In January 1942, Eddy and his staff in Chicago developed a classification examination, popularly called the "Eddy Test," to identify potential students with the desired intellectual and other characteristics. Initially, this test was required only of persons who did not hold an amateur or commercial radio license, but in a short while was required of everyone wanting to enter the program. It was also given to persons starting Boot Camp who scored well on the Otis Higher Examination, a type of intelligence test often given men upon entering the Navy. There was a hard, universally used pass-fail criteria for the Eddy Test, and a second chance was never allowed. Because of the importance of this test in selecting trainees, the content was tightly controlled – no publicly available copies seem to have survived.

Eddy later described the test as having questions with multiple-choice answers, with each of the answers giving some indication of the test-taker's knowledge, creativity, reasoning ability, and general aptitude. Thus, the answers were weighted – not simply right or wrong – and speed certainly affected the results. Persons taking the test remember it as containing questions involving mathematics, physics, elementary radio, and shop practices, as well as psychological indicators.

There are statements from Eddy and others to the effect that the test was not to determine native intelligence. In his paper, however, Cooke noted that persons who passed the Eddy test and then successfully finished Pre-Radio scored an average of 75.8 on the Otis Higher

Examination. This roughly corresponds to around 133 on the Stanford-Binet intelligence-quotient scale. If correct, this would place the personnel in ETP Primary School in the top two percent of IQ in the Nation. Nelson went on to justify this highly selective process:

> Many men capable of becoming good radio technicians under ordinary circumstances must be eliminated as such because they are incapable of absorbing the training in the time allotted. This means that we must have a very high type of man as an applicant in order that time will not be wasted by a high rate of attrition.

Men who passed the Eddy Test before entering the Navy have differing recollections as to how they were informed. Some believe that the tests were graded "on the spot" at the recruiting office. However, since a second chance on the test was not allowed, it would appear that a central administration must have been involved. In any case, considering the importance placed on this exam, they must have been closely controlled. It is known that Eddy's staff in Chicago graded some, if not all, of the tests. Many men received a form letter from Eddy stating the following:

Grading Eddy Tests at the State-Lake Facility

> In your Radio Technician classification examination [Eddy Test], graded at this school, you attained a mark that is considered passing. May I congratulate you on your qualification for this highly specialized training in the Navy.

There is no information available on the pass-fail ratio for individuals taking the Eddy Test. Persons entering Pre-Radio, however, were often told that they represented a small fraction of the original test-takers. Those admitted to the program came from highly diverse backgrounds. (See Appendix III on Representative Students.) There were recent graduates from big-city, science-oriented high schools, as well as clever men from rural communities who had learned much through experience and self-study. Many were college students, even graduates, who volunteered for the Navy as an alternative to the draft, but some had little more than an elementary-school education. Prior knowledge of radio ranged from none to holding a commercial operator's license. Most were young, often needing their parents' consent before joining, but a few were "older" – in their 30s and 40s. Like Eddy himself, they were often "tinkerers." All had a good aptitude for mathematics and a thirst for technical knowledge.

Those joining the Navy after passing the Eddy Test were immediately given Seaman First Class ratings, and some with advanced education or experience received ratings as Petty Officers. When sent to Boot Camp, they were assigned to special RT (Radio Technician) companies, composed entirely of persons destined for Pre-Radio School. It was common in these companies for persons to receive late-arriving commissions as Ensigns or Warrant Officers and be sent off to special schools or assignments. The length of Boot Camp varied, depending on the demand for technicians, sometimes being only two or three weeks but usually about eight to ten weeks. Throughout the war years, success in the Eddy Test served as the passport to the best technical program available in the Armed Services. In high schools and colleges, the test took its place alongside those for the Navy's V-5 and V-12 officer-preparation programs and the short-lived Army Specialized Training Program (ASTP).

PRE-RADIO SCHOOL

Pre-Radio was added as the third type of school in the ETP in late 1942. After passing the Eddy Test, persons must then successfully complete Pre-Radio. This school would give special preparation as well as provide further weeding. The intent was to increase the graduation rate in the costly and very demanding Primary and Secondary Schools.

A temporary Pre-Radio School was set up at the Chicago Naval Reserve Armory. With the urgent need of getting trainees into the program, Boot Camp for persons passing the Eddy Test was cut to a minimum; thus, the training at the Armory was six weeks in length, with the time divided between Pre-Radio and completion of Boot requirements. For the permanent Pre-Radio Schools, Eddy convinced Chicago's mayor and the Board of Education to lease three of their facilities to the Navy. In June 1943, Wilbur Wright Junior College was taken over, followed by Theodore Herzl Junior College and Hugh Manley High School. A Pre-Radio School was also established at the Naval Reserve Armory in Michigan City, Indiana, close to Eddy's residence.

The four-week Pre-Radio curriculum centered on a lightning-speed review of high-school mathematics (through second-level algebra) and basic physics (mainly electricity and magnetism). Each entering student was immediately issued a Cooke slide rule and these were used from the start; a large demonstration slide rule was in each classroom. Lack of speed and accuracy in using the slide rule was often the downfall of many otherwise capable students.

Slide Rule Class

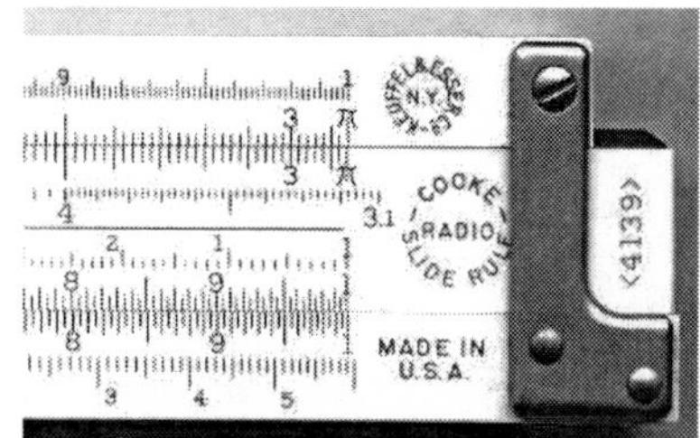

Special Radio Slide Rule Designed by Nelson Cooke

On weekdays, lecture and laboratory classes ran from 7:30 a.m. to 5:00 p.m., followed by organized problem sessions to 9:00 p.m. Note-taking during lectures was mandatory, and homework was assigned daily.

Eddy directly coordinated the preparation of much of the instructional materials. He was particularly involved with training to frame practical problems in mathematical terms. The notes from mathematics instruction at the Pre-Radio Schools were published in a book: *Wartime Refresher in Fundamental Mathematics*, by Eddy *et al* (Prentice-Hall, 1942). Divided into four major sections (corresponding to the four weeks of Pre-Radio), it started with arithmetic and went through simultaneous equations, with a substantial part devoted to "word" problems.

About half the classroom time during Pre-Radio was spent on mathematics and the related slide rule. Second to this was time allocated to physics, chemistry, and electrical components. The remaining time was given to direct-current circuits and introductory alternating-current theory. Laboratories involved soldering techniques, use of hand tools, basic machine shop practices, and electrical instruments.

Electrical Lecture

The intensity was more than some otherwise capable students could take – there were many voluntary dropouts. Pre-Radio graduation rate was likely less than 50 percent. Typically, the successful students had done well in high school, many had a year or more of college, and some were college graduates. A few, however, did not have good educational backgrounds, but found their talents after being immersed in the program.

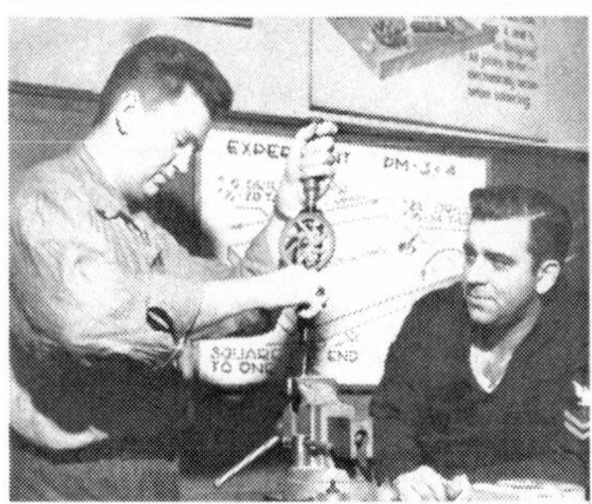

Shop Practice

Whatever their backgrounds, essentially everyone in Pre-Radio found it difficult. Homer E. Jackson, a student at Herzl in 1944, noted: "I already had a degree in mechanical engineering, but found the pace of Pre-Radio almost too much." Exams were held Saturday mornings, and a fraction of each class disappeared at the start of the next week. Repeating was not allowed, but all persons receiving failing grades on only one or two subjects were interviewed, during which it was decided to either let them continue or leave the program.

PRIMARY (EE&RM) SCHOOL

The three-month Primary or EE&RM School was unlike any other training for enlisted personnel in the armed forces. In this brief period – an academic quarter – students would be taken through major sections of courses normally given in the first two years of a college electrical engineering curriculum. As the ETP was being initiated, Primary School also included all of the review and preparatory materials later given in Pre-Radio.

The first Primary School under the newly organized ETP was privately funded by the Balaban & Katz theater chain and operated under Bill Eddy's direct tutelage at B&K's State-Lake facilities in downtown Chicago. The initial class started there on January 12, 1942. Primary Schools were also initially conducted at NRL-Bellevue and Treasure Island as part of their Radio Materiel School. In June 1942, Treasure Island discontinued its Primary School, and at the end of 1943, the Primary School at NRL-Bellevue was moved to the College of the Ozarks in Arkansas.

Eddy's Primary School served as the prototype of all such schools in the program. It was initially concentrated in 8,500 square feet of space adjacent to the W9XBK television studios in the State-Late Theater building in Chicago's Loop District. It was officially called the 190 North State Street Primary School. The initial instructional staff was 15 engineers from the television station, led by the station's Chief Engineer, Arch H. Brolly. Upon receiving approval to proceed from Captain Denfeld, the staff completed preparations for opening.

No classified materials were involved in the Primary School. Security was not a problem, so the television station continued to operate with the school in its midst. To fill in for the W9XBK engineers while they were involved as school instructors, Brolly quickly recruited and trained a number of women for the station's technical staff, in the process gaining much publicity for "freeing men for war work." Both Brolly and Eddy continued with "moonlighting" activities for W9BX, eventually gaining its conversion in 1943 to WKBK, the first commercial television station in Chicago.

Arch H. Brolly and Some Wartime Television Engineers

While awaiting Eddy's return to active duty, the Navy assigned Ensign Eugene S. Pulliam as the acting Officer-in-Charge of the Chicago Primary School. Several enlisted personnel were also assigned in supporting functions. Although initially intended as a temporary assignment, Pulliam quickly became indispensable to the school and remained in various functions for the next three years.

Eugene S. Pulliam

Students in the first classes were a mixture of seasoned Radiomen and Electricians and recruits without Navy experience, all personally interviewed and selected by Eddy. In the rush to fill the first class, a number of fresh recruits were admitted prior to even being issued uniforms. For several of the first classes, many of the volunteers bypassed Boot Camp and were sent directly to school.

The original plan was for Primary Schools to be at colleges, making extensive use of existing faculty members and laboratories. In early February 1942, representatives of institutions from throughout the Nation met with Navy officials in New Orleans to learn about the program and were invited to submit proposals for participating. Proposals and their evaluation, including on-site visits, took place with unbelievable speed. Contracts to successful bidders were issued from the BuNav over Captain Denfeld's signature on February 19.

Schools selected were Bliss Electrical School (Maryland), Grove City College (Pennsylvania), Oklahoma A&M College, Texas A&M College,

Utah State College of Agriculture, and the University of Houston. They began training at varied dates during March.

One of the first college-based schools to open was at Oklahoma A&M. A summary of an article in the *Oklahoma A&M College Magazine* (July 1943) illustrates the speed of the program's initiation:

- Notification of selection by the Navy was on February 19. In the remainder of the month, instructors were chosen from the regular faculty and, using the basic curriculum, they quickly developed the instructional details.
- Textbooks, slide rules, and laboratory supplies were ordered with high priority. Classrooms, laboratories, and dormitory space were prepared. The military personnel – two officers and three petty officers – arrived on March 1.
- On Sunday morning March 2, the first class of 100 students arrived. That same day they were assigned dormitory rooms, given an orientation, issued instructional materials, and told to begin studying. Classes started Monday morning.

All of these schools had excellent academic reputations and offered programs in electrical engineering. Most were experienced at working in special training for the Government. To handle administrative and personnel matters, a small staff of Naval officers and petty officers was assigned to each school. Included was a School Commander or Officer-in-Charge who had overall responsibility for the school; this was often a directly commissioned faculty member.

Each college selected one of its own faculty members to be responsible for the instructional program. Initially, all instructors at the college-based schools were regular faculty members and, as needed, hired civilian specialists. As more classes were added, Navy and Marine personnel, all graduates of the overall program, augmented the civilian faculty

The basic curriculum was included in the contract from the Navy, but the details were mainly left to the individual schools. When the college-based schools initially opened, there was no Pre-Radio and the review materials, including slide rule, were crammed into the first part of Primary School. As previously described, there was insufficient time for all that needed to be covered, and this led to an unacceptable failure rate. Effective January 1943, the four-week Pre-Radio activity was added, serving both as preparation for Secondary School and as additional filtering of students. With more elementary materials removed, the Secondary curriculum became much stronger and was very effective in preparing students for Secondary School.

While Primary was just as rigorous as Pre-Radio, most students now had a good foundation review and were accustomed to intensive studies. Weekday classes ran eight hours, with a required study period in the evenings. A minimum amount of physical training and drill were also involved. Saturday mornings were for examinations. The graduation rate was likely between 60 and 80 percent.

Homework Every Night

Textbooks were used at the discretion of each school, but most schools issued *Industrial Electricity I & II,* by Dawes (McGraw-Hill, 1938), *Fundamentals of Radio,* by Terman (McGraw-Hill, 1938), *Mathematics for Electricians and Radiomen,* by Cooke (McGraw-Hill, 1942), and *Radio Physics Course,* by Ghirardi (Radio & Technical Publishing, 1933). If not issued, copies of these books were available in study areas, as were electrical handbooks and books of mathematical tables. The *ARRL Handbook* and *The Radio Handbook* (from Editors & Publishers) were popular additions to students' personal libraries. Each student was issued an engineering slide rule (usually a log-log duplex decitrig model), and, in a case strapped to his belt, became a part of the uniform of the day

Much of the instruction was from specially prepared notes. A faculty member from Utah State commented, "These were like taking every tenth page from lecture notes for a regular course on that subject." It was up to the student to fill in portions that he did not understand. Most schools already had classroom instructional aids suitable for the program. A common item was an RCA Dynamic Demonstrator, a superheterodyne receiver with components mounted on a large circuit board with access terminals for connecting test instruments.

The curriculum included a continuation of Pre-Radio math, covering logarithms, trigonometry, complex numbers, vector algebra, exponential functions, and graphical representation. Some basic concepts from differential and integral -calculus were introduced, but mainly to allow an understanding of certain time-related electrical functions and were normally interwoven into related lectures. The log-log duplex engineering slide rule was used for all numerical calculations, but students were exercised in mentally obtaining "order-of-magnitude" estimates.

Typical Lecture / Demonstration Classroom

Electrical and electronics lectures dominated the classroom time and centered on d-c and a-c circuit theory, electrical motors and generators, power components and supplies, vacuum tube characteristics, radio-frequency components, amplifier and oscillator circuits, frequency response and filter calculations, modulation and detection theory, general receiver and transmitter circuits, and introductory transmission line and antenna theory. This was all taught from a generalized standpoint; there were essentially no references to specific Naval systems or applications.

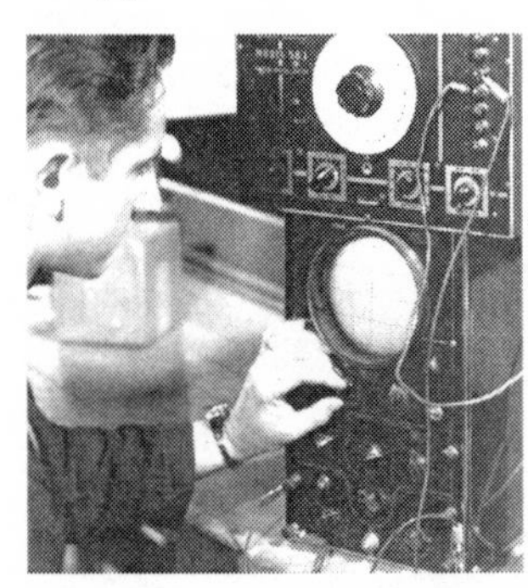

Electronics Laboratory

All theory was accompanied by laboratory exercises, including the use of electronic test instruments and operation of electrical power devices. Electronics laboratories included standard volt-ohmmeters, vacuum-tube voltmeters, audio and radio-frequency signal generators, oscilloscopes, and mutual-conductance tube testers. Each student constructed several electronic devices, including a full superheterodyne radio, and was required to demonstrate a competency in signal tracing and circuit repair.

While most of the laboratory work was electronics related, each school also had an electrical power laboratory in which basic motors, generators, and their controls were operated and tested.

Power Devices Laboratory

In 1944, with the demand for technicians continuing to increase, the Navy opened four new Primary Schools: one at Great Lakes Naval Training Center; another at a new Naval Training Center at Gulfport, Mississippi; a third at Del Monte Hotel in Monterey, California; and a fourth in the facilities of the College of the Ozarks at Clarksville, Arkansas. By this time, a reservoir of trained and experienced military personnel was available as instructors, particularly those who had served overseas and were cycled back to the States. Instructors at these new schools, therefore, were all Navy and Marine personnel.

Constructing a Radio

SECONDARY (ADVANCED) SCHOOL

In 1941, the Navy had taken over existing municipal facilities at Treasure Island in San Francisco and Navy Pier in Chicago. The 300-acre Treasure Island in San Francisco Bay, site of the 1939-40 Golden Gate World's Fair, and the 3000-foot-long Navy Pier, extending into Lake Michigan, were converted into broad-based Naval Training Centers.

Golden Gate / Treasure Island

Pre-War Navy Pier

Already offering the Radio Materiel School, NRL-Bellevue, in January 1942, converted to an eight-month combined Primary and Secondary Schools, of three- and five-months duration respectively. This was replicated at Treasure Island, with this school opening February 2, 1942. The Bureau of Navigation was changed to the Bureau of Naval Personnel (BuPers) in May 1942, and these Secondary Schools (sometimes called Advanced RMS), as well as the Primary Schools, came under this Bureau. Another Secondary School at Navy Pier began operations on June 5, 1944.

The Bureau of Aeronautics opened the Aviation Radio Materiel School (ARMS) in January 1942. With Sidney Stock as Officer-in-Charge, this school was in a special facility built on the campus of the Naval Academy at Annapolis. Recognizing that a much larger and more secure school was needed, the Navy acquired undeveloped Ward Island in Corpus Christi Bay in March 1942 for the ARMS home. Construction of the new facilities started in May. Personnel and equipment of the ARMS were transferred from Annapolis and the facility was commissioned on July 1, 1942.

The Naval Air Technical Command was formed in September, and in January 1943, Ward Island was designated as a Naval Air Technical Training Center (NATTC). Although under a different Bureau, NATTC Ward Island was generally considered as another Secondary School under the overall Electronics Training Program. Ward Island students were normally graduates from the Primary Schools.

The Primary School at Treasure Island was discontinued in June 1942, and at the end of 1943, the NRL-Bellevue Primary School was transferred to the College of the Ozarks, allowing Treasure Island and NRL-Bellevue to concentrate on the Secondary School. The curriculum for the advanced portion of the original Radio Materiel School, with the instruction expanded and upgraded for highly selected students, formed the basis of the curriculum at NRL-Bellevue and Treasure Island. From its start, Navy Pier offered only the Secondary School, closely emulating the advanced curriculum then existing at Treasure Island.

The schools at NRL-Bellevue, Treasure Island, and Navy Pier were devoted to electronic systems used by surface vessels and shore bases. Some of the graduates went on to New London, Connecticut, for special courses on submarine electronics. The school at Ward Island was devoted to airborne electronic systems. While Pre-Radio and Primary Schools involved only unclassified information, the Secondary Schools included classified instruction and the laboratories had secret radio, radar, and other electronic equipment. The physical locations for the four advanced schools afforded excellent security.

Students in the Secondary Schools came from the U.S. Navy, Marines, and Coast Guard. Ward Island also had students from the Royal Air Force and the Canadian Royal Air Force, as well as some Australian and Brazilian servicemen. While mainly intended for enlisted personnel, officers were sometimes included in certain blocks of instruction. All of these schools had special refresher and advanced courses for both officers and enlisted men. Class sizes varied between locations and at different time, but about 120 students seemed to be average. Classroom lectures were usually limited to 30 students, and specific laboratory set-ups normally served no more than four persons at a time.

Although continuing with the rigor of Primary, the Secondary Schools had a less intense pace and a greater emphasis on laboratory and hardware work. Instructors were Navy and Marine senior enlisted personnel and sometimes warrant officers. Many instructors were college educated, some with graduate degrees, but these persons usually could not meet physical requirements for officer's candidacy. Instructors were also selected from recent Secondary School graduates with superior grades.

There were eight hours of instruction on weekdays, and examinations were held on Saturday mornings. The instruction averaged about half classroom and half laboratory, but this varied by the subject. The instructional materials were specially prepared notes on subjects and Navy manuals for equipment, but essentially all of this was classified and could not be taken from the classroom buildings. Students were required to take notes, but these were treated as classified; they were retained in the instructional facilities and later shipped to duty stations if requested. All classrooms and laboratories were in secure compounds at each facility.

While homework was usually assigned for theory topics, it was not taken up or graded – it was for the student's benefit. Most students, however, quickly found that after-hours study, including homework if assigned, was necessary for satisfactory grades. For this, they often referred to standard radio engineering texts, as well as new publications on subjects such microwaves. All of the schools had areas outside the secure compound designated for additional study. Special study sessions were sometimes conducted in the evenings in regular classrooms, usually for students making lower grades.

Most students had books of their own, and each of the schools had a library, well stocked with math, physics, radio, and other technical books. *Practical Radio Communications* by Nilson and Hornung and the *ARRL Handbook* were popular for outside study on the communications portions of the curriculum. Upon its publication, Terman's *Radio Engineers' Handbook* (McGraw-Hill, 1943) was purchased and used by many students. In April 1944, War Department Technical Manual 11-467, *Radar System Fundamentals,* was released but not issued in the program. This TM, the first unclassified treatment of radar, was often purchased from the Ship's Service Stores by students, although the level was generally considered somewhat below that of other materials.

The curriculum at the Secondary Schools continued where Primary Schools left off, giving a mix of theory and hands-on hardware exercises. Initial theory topics included advanced receivers, transmitters, transmission lines, antennas, and wave propagation. Consideration was next given to transients in passive networks and to waveform-modifying electronic circuits. General elements of full radar systems were then examined. Target characteristics and IFF principles were also covered.

This led to examining limitations of conventional tubes and components. A considerable time was then spent on new components, including waveguides, mixing crystals, "acorn" and "lighthouse" tubes, reflex klystrons, and, finally, the resonant-cavity magnetron. Attention was also given to directional antenna arrays and reflectors and the electro-mechanical devices associated with their pointing, including drive motors, syncro motors and generators, and servomechanisms.

In parallel with the classroom studies, much time was devoted to practical work on hardware. This involved wiring diagrams and circuit tracing, performance measurement and tuning, operating techniques, practical trouble-shooting, and safety procedures. Specific systems varied between the Secondary Schools and were under constant change as new systems were released.

At the surface schools, the equipment studied included a variety of low-, medium-, and high-frequency radio communications systems, high-frequency and microwave search radars, fire-control and anti-aircraft radars including height-finding units, IFF units, direction-finding receivers, LORAN and other radio-navigation systems, sonar apparatus, and special topics such as combat information centers.

At Ward Island, the equipment was essentially all airborne, suitable for both carrier-based aircraft and larger patrol and rescue aircraft. Included were aircraft receivers and transmitters, direction-finding and homing radios, high-frequency and microwave search radars, radar altimeters, IFF units, LORAN systems, and a variety of special topics such as introductions to radio-controlled aircraft, the Norden bombsight, and Magnetic Anomaly Detection (MAD) systems.

Laboratories at Advanced Schools

About midway through Secondary School, attendees with satisfactory grades and who had not been promoted upon completing Primary School were promoted to Radio Technician (RT) Third Class or, for those at Ward Island, Aviation Radio Technician (ART) Third Class. At graduation, those with top grades received a promotion to Second Class and some outstanding students were made First Class. In 1945, the ratings were changed to Electronic Technician Mate (ETM) and Aviation

Electronic Technician Mate (AETM). In addition to the regular students in the total program, some (mainly senior Radiomen) were admitted directly to Secondary School. Their reclassification to RT or ART (later ETM or AETM) depended on satisfactory grades in this school. Ratings and promotions for students from other services or countries were handled by their cognizant organizations.

The graduation rate from these advanced schools was high. Also, unlike in the two preceding schools, persons with unsatisfactory performance in a particular block of instruction were usually allowed to repeat. Records for Ward Island indicate that the graduation rate over the first two years of operation was about 93 percent, considering attrition for all reasons. The disposition of inept students was in accord with BuPers directives and varied through time. In general, however, students not completing the program but already holding an RT or ART (earlier Radioman) petty officer rating were reduced to Seaman First Class and transferred to a receiving station for further assignment.

Saturday-Morning Inspection

Less the wrong impression is conveyed, the Secondary Schools were strictly military. There were official "uniforms of the day." Navy enlisted students marched, often with a band playing, when moving as groups. An objective at each station was that students would have one hour of physical exercise or military drill each weekday; while the exercise was generally on a regular schedule, the drill was sometimes concentrated in widely spaced review parades.

Chapter 6

THE SCHOOLS, FACILITIES, AND LOCATIONS

William C. (Bill) Eddy, Lieutenant (jg) USN Retired, had arrived in Washington, D.C., on December 8, 1942, hoping to be returned to active duty. There he had found his way into an *ad hoc* group that was urgently seeking a solution to a crisis in Naval electronic maintenance training.

Thousands of technicians were needed for the new electronic marvels on ships, submarines, aircraft, and shore bases. The Radio Materiel School (RMS) on the campus of the Naval Research Laboratory (NRL) was the only existing source for such men, but its production was far too small and its graduates lacked the knowledge of theory that would be needed for understanding the myriad of expected equipment. Eddy was brash and self confident to the extreme. Speaking for his experimental television station in Chicago, he had said, "We can train 'em. We've got room, equipment, skilled personnel – you can have it all!"

The next week he returned to Chicago, disappointed at still being a civilian, but facing three formidable tasks. First, he had to explain to his employer, Balaban & Katz, why he had offered to supply facilities and personnel for Naval training – and he had promised this to be without charge. Second, he had said that he would have a school available to receive students within a month. The third task – the result of which would ultimately carry his name – was to develop a test that would identify appropriate trainees.

Balaban & Katz Theater Corporation had more than 100 movie houses in the Chicago area, and in 1940 entered television broadcasting with experimental station W9XBK. Eddy had been hired away from fledgling NBC Television in New York to build and operate the facility, the first all-electronic television station in Chicago. W9XBK occupied the top floor of the State-Lake Theater building and was not yet authorized for commercial operations. Balaban & Katz were more than willing to allow part of the facility to be converted into a Navy school and to provide the station's engineering staff to operate the school, with their corporation covering the costs.

The Bureau of Navigation – at that time responsible for training within the Navy – had not officially approved the overall plan for the Electronics Training Program (ETP) put forth by the *ad hoc* committee, but, because of the importance of time, Eddy had been told that there was essentially no risk in proceeding with developing the school. This was conceived to be a prototype of schools that would eventually be operated by colleges and universities, and, as such, would only be needed until the permanent schools were well established.

For developing the admission test, Eddy had the existing exam that had been used by the RMS for many years. The RMS exam was a comprehensive assessment of extensive preparation by experienced Radiomen and Electricians. But to identify thousands of men from varied civilian backgrounds for a highly technical, very intense training program, a much different examination was required. Such an examination needed to assess the individual's basic abilities in mathematics and physics, but, more important, it should identify those who had the fundamental drive and capabilities to devote themselves to nearly a year of unbelievably hard study.

The facilities were made ready, beginning instructional materials were prepared, and on January 12, 1942, an initial class of 100 students started the first class of the Primary School of the new Electronics Training Program. A draft of the admission test was ready, but since the Training Division of the BuNav had not approved it, the test was not used for the first class. Within a short time, however, passing the Eddy Test became the main criteria for admission to the most unusual training program of World War II.

In the following March, college-based Primary Schools were started, using the basic curriculum as it was being developed by Eddy and his staff in Chicago. Before long, however, it was realized that additional screening was needed for students entering Primary School and too much elementary material was in the curriculum. Consequently, a number of Pre-Radio Schools were started in the Chicago area. Eddy also initiated a Teacher Training School at the State-Lake facility.

In this same period, a new school was started at Treasure Island, replicating the existing RMS at Bellevue, and the advanced portions of both of these became Secondary Schools in the ETP. In addition, the Bureau of Aeronautics had started an Aviation RMS on the campus of the Naval Academy, then moved it to Texas to become the Naval Air Technical Training Center (NATTC) Ward Island, functioning as a third Secondary School. As the war progressed and many more technicians were needed, a number of Navy-operated Primary Schools were started, and a fourth Secondary School was established at Navy Pier in Chicago.

Each of these schools and their leaders played a vital role in solving the electronics maintenance crisis of World War II.

RADIO CHICAGO

In the ETP plan, colleges and universities would operate Primary Schools, but time was of the essence, and the program could not wait for starting these schools. In addition, there needed to be a "prototype" for the college-based schools, developing the basic curriculum and testing it on the students. Thus, after Bill Eddy returned to Chicago from Washington, he immediately set about establishing such a school. As earlier offered to the Navy, this was in the facilities of experimental television station W9XBK on the 12th floor of the State-Lake Building in Chicago's Loop district.

Eddy had come back from his quest in Washington with no official status in the Navy, but, while still an employee of Balaban & Katz, he started and directed the prototype-developmental Primary School. In a 1985 interview with *The News-Dispatch* (Michigan City, Indiana), Eddy described his subsequent return to Naval duty:

> Apparently in my civilian role, I stepped on a few toes and invariably got called to Washington. Secretary of the Navy Frank Knox asked why I was out of uniform? When I pointed out that I had retired with 11 physical defects, my medical records were pushed aside and in one fell swoop I was back in uniform.

Eddy returned to duty as a Lieutenant on August 17, 1942. A year later he was promoted to Lieutenant Commander and in September 1943 became a Commander.

Bill Eddy in 1943

It was found that a significant portion of the students in Primary Schools were not sufficiently prepared or did not otherwise have the requisite characteristics for this rigorous training. While the Eddy Test was excellent for initial screening, further filtering in an intensive review school was needed. Thus, beginning January 1943, a four-week Pre-Radio School became the required preparatory activity for all Primary Schools. Eventually, four Pre-Radio Schools were operated in and around Chicago and, to obtain instructors, a Teacher Training School was started.

Although the State-Lake Primary School was originally intended to be temporary – a prototype – the Navy realized the importance of this operation and the school was enlarged and made permanent. In recognition of the continuing contributions being made to the overall ETP by Eddy and his staff, the operation at the State-Lake Building, the Teacher Training School, and the Pre-Radio Schools were officially designated the Radio Chicago Naval Training Center in September 1943 under Commander Eddy. The 10th floor of the State-Lake building was also taken over and a four-floor warehouse at 65 West Lake Street was converted for use by Radio Chicago. As the operation enlarged, Lieutenant John W. Proffer, a "mustang" Radioman with 28 years of continuous naval service, was assigned as Radio Chicago's Executive Officer.

John W. Proffer

Eddy was promoted to Captain on November 5, 1944, and continued as the Commanding Officer of Radio Chicago until this was deactivated on September 1, 1945. On September 12, Eddy was placed on terminal leave. He was awarded the Legion of Merit "for exceptionally meritorious conduct in the performance of outstanding services to the Government of the United States as Commanding Officer of Naval Training Schools." The citation included the following:

> Captain Eddy pioneered in nationwide recruiting and developed selection tests which greatly facilitated the procurement of men for radio technician training. Exercising sound judgment and superb professional ability, he consistently supervised the improvement of teaching methods by developing training aids, text material, and specialized instructor training . . . . By his keen foresight, unwavering zeal, and meticulous attention to detail in discharging all duties pertaining to the recruiting, selection, and training of radio technicians, Captain Eddy contributed materially to the successful prosecution of the war against the enemy.

190 North State Street Facility

The Balaban & Katz Theater Corporation had been very successful in Chicago, and they acquired the State-Lake Theater in 1938. The theater was named for its location on the corner of State and Lake Streets, across from the State and Lake "El" Station on the north-east corner of the

Loop. Electronic television was emerging in America, and Balaban & Katz wanted to have the first such station in Chicago. In 1940, they hired Bill Eddy away from RCA's embryonic NBC Television in New York to build and operate the station.

Eddy was already well known in this field, having assisted the inventor of all-electronic television, Philo T. Farnsworth, and then leading the studio efforts for NBC. The FCC granted a license in April 1941 for experimental station W9XBK. Operating at 60-66 MHz, the studio and transmitter were located on the top floor of the State-Lake Building with the transmitting antenna on a tower atop the building. (In 1943, W9XBK become Chicago's first commercial station, WBKB.)

With the financial support of Balaban & Katz, Eddy and the staff of W9XBK set up the first Primary School in 8,500 square feet of space adjacent to the television facilities. This was officially designated by the Navy as the 190 North State Street EE&RM School, but was commonly called the State-Lake Primary School. The school opened on January 12, 1942, only five days after the program was officially approved.

190 North State Street

The Navy did not have sufficient candidates immediately available for the first class, so Eddy and his staff personally recruited students, some directly entering the Navy by attending the school while still wearing civilian clothes. Students were initially housed in the Navy Reserve Armory, a large facility on the Chicago River at the foot of Randolph Street just north of the downtown area, and a few blocks from the State-Lake Building. An article in the *Chicago Daily Tribune* on March 8, 1942, described the typical day:

> At 5:30 a.m., the students rise in the Naval Armory, dress, shave, and eat breakfast, and by 7 o'clock are in class in the Balaban & Katz television studio. They study or do class work until 11, march to the Armory to eat, and are back in class by 12:15. More classes or study follow until 5. Then the students return to the Armory for another meal, to be followed by study until 8:30. Lights are out by 9, and everyone's asleep by 9:30.

Initially, the school had two classrooms and two laboratories. The faculty was the 15-man engineering staff of W9XBK. Several of these men were commissioned into the Navy to prepare instructional materials. Others remained as civilians, serving in classrooms and laboratories under Archibald H. (Arch) Brolly, the Chief Instructor. Brolly was an electrical engineering graduate of the University of California at Berkley, had earned a master's degree in electronics from Harvard, and then pursued further graduate study at MIT. He had been the Chief Engineer for Farnsworth Television Laboratories when Eddy worked there, and had been brought to Chicago for a similar position with W9XBK.

Arch H. Brolly

The curriculum was patterned on that being given in the first three months at the NRL-Bellevue RMS, intended to allow successful graduates to smoothly transition into the advanced training at NRL-Bellevue, the new RMS at Treasure Island, or the new Aviation RMS when it opened at Ward Island. The electronic test equipment in the laboratories started with that of W9XBK but was soon augmented by Navy-furnished instruments and specially built test sets. The staff developed various types of breadboard experiments for d-c and a-c circuits, as well as electronic devices, the latter culminating in a complete superheterodyne receiver.

The operation at State-Lake was started as a prototype Primary School, developing and testing a curriculum that would be the basis of college-based schools. As such, the BuNav considered the school as temporary. In May 1942, BuNav was reformed as the Bureau of Naval Personnel (BuPers), and the Electronics Training Program came under their Training Division. Louis Denfeld, the official who had initially signed off on the program and now a Rear Admiral, took personal cognizance of this activity. With the realization that the overall curriculum and characteristics of available students would likely be continuously changing, Denfeld directed that the 190 North State Street Primary School be made permanent, allowing the needed development and coordination to be at a single location.

As multiple classes started, additional space was needed. The Navy leased the 10th floor of the State-Lake building for additional offices and classrooms. Later, a Butler Building was set up for more classrooms in a parking lot adjacent to the main building. When the housing capacity of the Naval Reserve Armory was exceeded, the students were moved to the Navy Pier, and were bussed daily between there and the State-Street Building.

Another important function at the State-Lake facility was the handling of the classification exams – the Eddy Tests. While these were given nationwide at recruiting stations, high schools, and colleges, the results were normally sent to the State-Lake facility for grading. Records were kept to ensure that persons were not allowed a second chance for passing. WAVES were primarily involved in the Classification Office, as they were in many other State-Lake administrative and instructional support activities.

Pre-Radio Schools

Before the end of 1942, it was realized that the three-month curriculum for Primary School was not in sufficient depth to fully prepare students for the upgraded Secondary Schools. In addition, it was found that a significant portion of the students in Primary Schools were not adequately prepared or did not otherwise have the requisite characteristics for this rigorous training. This included men who had passed the Eddy Test, as well as those admitted by holding amateur radio licenses. It was then that the Pre-Radio Schools were started, with the elementary and review material being covered by these new schools and replaced in the Primary curriculum by more time and greater depth in topics.

The four-week (actually three and one-half) Pre-Radio would cover the material previously included in the first portion of Primary School – a "crash" review of fundamental mathematics and physics as well as an introduction to the slide rule and basic shop practices – but the major purpose would be screening; quickly determining if potential students had the aptitude, capability, and drive to pursue the great rigor of the program to follow. The Navy could not afford to allow persons into Primary School, much less Secondary School, unless there was a good expectation of subsequent graduation.

Actually, the first schools of this general type, preceding the official Pre-Radio classification, were set up in late 1941 in Los Angeles, California, and Noroton Heights (Darien), Connecticut. The Los Angeles school was in the Naval Reserve Armory and the one in Noroton Heights was at a large Radio Operators School located in a facility formerly occupied by the Fitch Home for Soldiers. These schools gave the first month of the Primary School, but only operated briefly in early 1942 while Treasure Island was being set up.

In late 1942, BuPers made the decision to require all persons entering a Primary School to first successfully complete a Pre-Radio School, with passing the Eddy Test as an entrance prerequisite for Navy personnel. Bill Eddy and his staff at the State-Lake facility were assigned the task of establishing Pre-Radio within the framework of the ETP. From their operation of the prototype Primary School, they were intimately familiar with the material that needed to be covered, and they were already handling the grading of Eddy Tests.

A temporary Pre-Radio School was set up at the Chicago Naval Reserve Armory, the facility that had been earlier used for housing students for the State-Lake Primary School. It appears that during its few months of existence, this Pre-Radio School functioned as an off-site activity of the Primary School, and it is likely that Eddy's assistant, Lieutenant Eugene Pulliam, was in charge of the operation. One of the students, Eugene H. Fellers, remembers that they taught a lot of mathematics at this school: "I studied two years of high-school algebra, much of which I had never had before, in four weeks!"

Chicago Naval Reserve Armory

To fill the projected needs of Primary Schools, Pre-Radio Schools graduating up to 1,500 students each month would eventually be needed. It was highly desirable that the schools be self-contained, combining housing and instruction in one facility. Eddy personally persuaded Mayor Edward J. Kelly of Chicago to turn over three city schools for this activity. Wilbur Wright Junior College was taken over in June 1943, followed that fall by Hugh Manley High School, and a year later by Theodore Herzl Junior College. Crane High School in Chicago had originally been in Eddy's plans, but it apparently was never used for this purpose.

As previously noted, in late 1942, before the city facilities were negotiated, a temporary Pre-Radio School was set up in the Chicago Naval Reserve Armory. This was followed by a school opened at the Naval Reserve Armory in Michigan City, Indiana. In addition, from March to June 1943, a large Pre-Radio School was operated at Treasure Island. For the next two years, entry of most Navy and Marine personnel into the ETP was through Pre-Radio conducted at the Chicago city schools or in Michigan City.

Bunking at Pre-Radio

All three of the Chicago city schools were in single, three-story buildings that were readily suitable for classrooms and laboratories. The two gymnasiums in each school were filled with three-tier bunks, and students were not allowed in except at night on weekdays. Each person had a locker for school materials and items needed during the day.

The opening of a Pre-Radio School at Wright was described in an article in the July 11, 1943 issue of the *Chicago Daily Tribune*:

> By the end of this week, there will be 1,000 men 'aboard ship' in Wright Junior College, taken over last June 14 by the navy. Approximately 800 are enlisted seamen studying to be radio electricians [sic] and the rest are marines taking the same . . . four-week intensive course.

An Officer-in-Charge, usually a Navy Lieutenant, and an Instructional Officer were assigned to each school. A Marine Lieutenant was also assigned where there was a large number of Marine trainees. The instructional staff was all Navy and Marine enlisted men, often former high school teachers. Eddy made frequent visits to the Pre-Radio Schools, sitting in classes and talking with students and faculty members. He often brought a television receiver – the first ever seen by many of the students – showing the broadcasts from his station, W9XBK.

Schools of this nature – used for "filtering" – are always tough. Joseph A. Weygandt, a student at Manley, wrote his parents in April 1944:

> This joint is sure a sad place to live (except the chow). We are in a classroom 12 hours a day, in bed 8 hours, and the other 4 are split up in washing and dressing, standing in line, and eating. The 'rest' of the time we read books and play checkers.

Any physical training had to be outdoors, and this was only when the weather permitted. There was no entertainment, religious services, or other group activities at the schools, but, after examinations on Saturday mornings, students had a free weekend. Nearby churches, synagogues, and community organizations were outstanding in providing for the students, with at least one "serviceman" facility close to each school. Downtown Chicago, however, was the most popular destination. Local transportation was free, as were tickets to shows. Service personnel in Chicago were treated as honored guests.

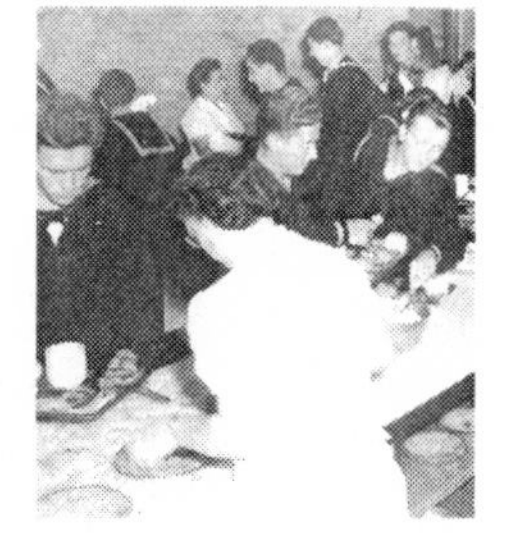

Pre-Radio Cafeteria

The food service was a major advantage at the Pre-Radio Schools. As a part of the lease from Chicago, the cafeterias continued to be privately operated. After Boot Camp, most of the students found the food and service to be outstanding.

In 1945, a Pre-Radio School was opened in association with a Primary School in the Naval Training Center at Ford Motor Company's

Rouge Plant, Dearborn, Michigan. The Dearborn schools operated for only a short time.

The Pre-Radio function was transferred to the Great Lakes Technical Training Center on September 1, 1945. The facilities in Chicago were returned to the City at that time. Some documents indicate that the combined Pre-Radio schools of Radio Chicago handled 68,000 students during their existence, including those persons who were unsuccessful.

Wilbur Wright Junior College was located at 3400 North Austin Avenue, about 12 miles northwest of downtown Chicago. During 1944-45. Lieutenant (jg) Edwin B. Nickerson was the Officer-in-Charge; before being commissioned, he had completed the ETP. Named to honor one of the Wright Brothers of aviation fame, the Junior College was originally formed in 1934 and remained in the same building until 1993. It continues as Wright College in another Chicago location.

Hugh Manley High School is located at 2935 West Polk Street, near Douglas Park about three miles west of the Loop. Lieutenant T. F. Kane, who had prior experience in Pacific battles, was the 1944-45 Officer-in Charge. Constructed in 1928, Manley was named to honor a school facility engineer who lost his life while preventing a catastrophe. In recent years, it has operated as the Hugh Manley Career Academy High School.

Theodore Herzl Junior College was located at 3711 West Douglas Boulevard in the North Lawndale industrial area, about a miles southwest of Manley High. The Officer-in-Charge was Lieutenant (jg) H. E. Crow, also a Radio Technician prior to being commissioned. The school was originally formed in 1911 as Crane Junior College, but closed in 1933 due to the depression. It reopened in 1934 as Theodore Herzl

Junior College, named for Binyamin Ze'ev (Theodore) Herzl, founder of the Jewish Zionist Movement. In 1954, the facility was converted to be used as an elementary school.

Michigan City Naval Reserve Armory was in Indiana, only a few miles around Lake Michigan from Chicago. The Armory, a white-stucco, Art Deco building, is located in Washington Park. Bill Eddy lived in Michigan City, and, for personal involvement, wanted a Pre-Radio School there. For establishing this school, Eddy used Lieutenant (jg) Eugene Pulliam, previously his assistant when starting the State-Lake facility. Later, Lieutenant (jg) Richard J. Mueller, another former Radio Technician, became the Officer-in-Charge. The Armory still stands and is fully used.

64 North State Street Facility

With the opening of Pre-Radio Schools and the enlargement of the State-Street Primary School, additional facilities were needed for Radio Chicago. In early 1944, the Navy leased a large warehouse at 64 North State Street, just two blocks from the State-Lake Building, and converted this to space for housing and a variety of support services. Some indication of the condition of this building is shown by the nickname it was given: the "dust bowl."

Included in the three floors plus basement of this facility were Primary School student barracks (moving from Navy Pier), a large study area with a library, officers' wardrooms, a civilian-operated food service center (chow hall), a ship's service center, recreation rooms, and a reception center for visitors on week-ends and special guests. Military and physical training activities – very difficult in the Chicago Loop area – were also

Alvino Rey and the Radio Chicago Orchestra

conducted from this facility. Well-known musician Alvino Rey (Alvin H. McBurney) was a Radio Technician stationed locally and led the Radio Chicago Orchestra.

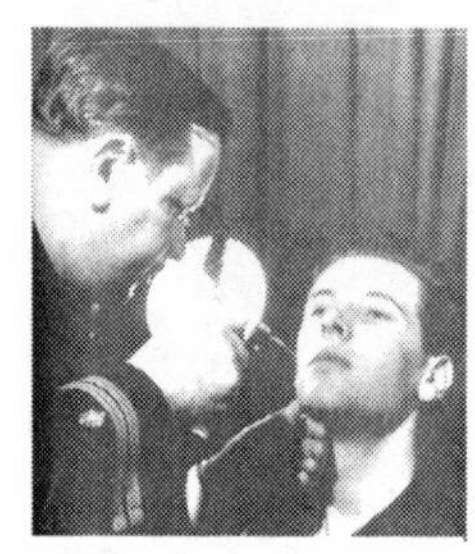
Medical Services

An area was set aside for consultations with chaplains and Red Cross representatives, both having regular visiting hours. The facility had a sickbay with a Navy physician (Lieutenant Commander N. B. Pavletic, who had earlier served in the fleet), a dental clinic, a pharmacy, and full-time manning of the sickbay by Pharmacist Mates. For more serious medical problems, the Navy had two floors of the Wesley Memorial Hospital, only six blocks away at 25 South Dearborn Street.

Certain service operations at 64 North State Street were used by all of Radio Chicago. There was a Construction and Repair Shop that maintained facilities and equipment at the State-Lake School as well as at the Chicago Pre-Radio Schools. Fleet men who had "earned time on the beach" primarily manned this. The Visual Aids Shop had highly skilled artisans who developed large-scale models of tools and radio items (such as cut-aways of vacuum tubes), prepared charts and classroom diagrams, and created artwork and displays for every need. Included in this Shop were photography and printing equipment, used in preparing students' manuals and text materials. The central Post Office for Radio Chicago was also in this building.

Visual Aids Shop

With the opening of the Pre-Radio Schools, there was a major need for new instructors. Eddy started a Teacher Training School that was eventually in the 64 North State Street facility. Here Navy and Marine personnel, all graduates of the program, were trained in teaching methods and the use of instructional aids. The instructors-to-be ranged from recent graduates of Secondary Schools to seasoned veterans of the fleet. All were specifically selected for this function and had characteristics indicating a potential as instructors.

PRIMARY SCHOOLS

Before 1942, the Radio Materiel School (RMS) at NRL-Bellevue had an eight-month integrated curriculum. This was duplicated at Treasure

Island when it opened. As the new Electronics Training Program got underway in January 1942, the Bureau of Navigation directed that the training at NRL-Bellevue and Treasure Island be divided into Primary and Secondary Schools. The Primary Schools at these facilities were later discontinued. The Aviation RMS at Annapolis activity had started with a five-month advanced program; thus, as it was transferred to Ward Island, it was designated a Secondary School.

In planning the ETP, the most attention was given to the Primary School. It was here that, in a three-month period, the thousands of men would need to be brought to a level suitable for entering the Secondary School – the advanced portion of the program. These entering students would have highly diverse backgrounds – many with prior college work, some with amateur and commercial radio experience, and a few who had shown special capabilities as a Navy Radioman or Electrician – but a significant portion would be starting "cold" in electronics. The Navy did not have a large pool of personnel qualified as instructors or existing facilities for the schools. Thus, considering the curriculum to be taught and the characteristics of the students, there was no question that the Primary Schools would best be mainly conducted by colleges.

The new Primary School portion of the ETP was officially called Elementary Electricity and Radio Materiel (EE&RM) School, but this deceptive misnomer was seldom used. The activity was initiated through a prototype-developmental school in downtown Chicago, planned and operated by Bill Eddy and privately funded at the start by Balaban & Katz. Although originally considered to be temporary, this soon became a permanent part of the ETP, eventually enlarging to become the central activity of Radio Chicago.

In early February 1942, the BuNav called a meeting in New Orleans, with attendance extended to all higher-education institutions with engineering programs and interested in providing a Primary School under the ETP. George W. Whiteside, the Dean of Engineering at Oklahoma A&M, attended for his college and noted, "Practically every engineering college in the United States was represented and was active in trying to get the assignment of one of these proposed schools." After the specifications were presented, the colleges that were still interested were asked to submit full proposals – due in one week!

There is no record of how many schools submitted proposals, but Bliss Electrical School, Grove City College, Oklahoma A&M College, Texas A&M College, University of Houston, and Utah State College were selected. It is noted that none of these schools had existing programs with the Navy, such as V-5 (Aviation Preparatory), V-12 (College Training), NROTC, or any of the other Naval Training Schools then being established at American colleges; therefore, this was possibly a criteria for selection. Certainly the criteria included costs, availability and qualifications of faculty, housing, and instructional facilities offered; but responsiveness must have been highly important. In early March 1942, Primary Schools were opened at the six selected colleges. These were in addition to Eddy's State-Lake school in Chicago and the combined Primary and Secondary Schools at NRL-Bellevue and Treasure Island.

Mixed Navy and Marine Students

For the next two years, the college-based Primary Schools were the main source of new trainees for the Secondary Schools. Initially, classes averaging 100 students started monthly at each school, some, but not all, mixing Navy and Marine personnel. About 500 graduated each month from the six college-based schools plus the "home" school at Radio Chicago.

Treasure Island discontinued its Primary School in June 1942 to make room for a larger Secondary School, thus increasing the need from the colleges. The incoming class size at the colleges increased to 110 and the period of instruction was changed from 3 months to 12 weeks; certain of the colleges then started new classes every other week. With these changes, the total output of Primary Schools increased to about 750 every four weeks (nominally a month).

As the war continued and the need for radio technicians further increased, the Chicago and college-operated Primary Schools could no longer supply the requirement for Secondary Schools. In the first half of 1944, four Navy-operated Primary Schools were opened. These were at the College of the Ozarks in Arkansas, the Del Monte Hotel in California, and at Gulfport and Great Lakes Naval Training Centers. In 1945, just as the war was ending, another Navy-operated Primary School with its own Pre-Radio School was opened at the River Rouge Training Center in Dearborn, Michigan.

College-Based Primary Schools

The six college-based Primary Schools, while academically under the institutions, were official Naval Training Schools. A Commanding Officer or Officer-in-Charge, usually a Lieutenant or Lieutenant Commander, was assigned to each school; these men were often drawn from the college faculty and placed on active duty. A complement of about 10 Navy and Marine enlisted men were also assigned at each school, including two or more Chief Petty Officers and/or Master Sergeants, several Yeomen and other administrative personnel, and physical training and military instructors.

Each institution provided a program administrator, a director of instruction, and necessary lecture and laboratory instructors. They also furnished dormitory and food services, classrooms, laboratories, study halls, health services, and physical training facilities. Later, when some Navy or Marine personnel were assigned to the institutions as instructors, they functioned under the director of instruction. In accordance with their contract with the Navy, the institutions provided their services and facilities under a flat rate per student-month.

Bliss Electrical School

Bliss Electrical School was the only privately owned, for-profit institution that participated in the program. Bliss, located in Takoma Park, Maryland, was started in 1893 by Louis D. Bliss and provided instruction only in the electrical engineering field. The regular curriculum centered on an eight-volume text by Dr. Bliss, *Theoretical and Practical Electrical Engineering*. Students started an experimental radio station 2XH in 1913; this was later converted to WBES.

The school gained an excellent reputation and many honors were bestowed on it, as well as on Dr. Bliss. Its graduation speakers included U.S. Secretary of State William Jennings Bryan and President Warren Harding. During World War I, the school not only trained Army personnel but also organized a company of 57 engineers from the 1917 graduating class. Led by a Reserve Officer from the faculty, this group served in France as an Army Searchlight Company

Bliss Electrical School – Main Hall

The Primary School at Bliss started in March 1942 and continued to July 1945, with a total of about 4,000 students served. During this period, the school was fully devoted to the Navy program – the regular school closed for the war's duration. The civilian faculty initially taught the courses, and was later augmented by Navy personnel. Dr. Bliss had overall responsibility for the school, and his son, Donald S. Bliss, served as the Director of Instruction. Lieutenant Barry F. Dayton initially represented the Navy as the Officer-in-Charge. Wayne S. Green, II, an ETP graduate and later well known as a writer and publisher in electronics, has commented: "Training at Bliss was superb, and gave me an understanding of electronics that has been the backbone of my life's work."

Although all of the college-based Primary Schools followed the same basic curriculum, each had its own characteristics, mainly based on their pre-Navy specializations. At Bliss – an institute dedicated to producing electrical engineers who could immediately make very practical contributions in industry – there were extensive machine shops, power-generation laboratories, and drafting rooms.. Thus, in addition to the emphasis on electronics, the trainees at Bliss received extra instruction in these facilities.

Drafting Class at Bliss

Students who attended Primary School at Bliss recall this was a very pleasant experience. They had excellent dormitory space (two to a room with plenty of study area) and dining facilities. This was not a regular college and did not have the usual auxiliary facilities, but it did have a tree-lined campus and academic-style buildings for classrooms and laboratories. With only Navy men attending, there was a minimum of off-duty entertainment on the campus, but Tacoma Park was directly adjacent to the District of Columbia, with the downtown area of Washington less than 10 miles distance.

A Bliss Class in 1943

Following the war, Bliss Electrical School trained about 2,000 persons under the G.I. Bill, but, with the aging of Dr. Bliss and the problems of operating a private school under government regulations, Bliss Electrical School graduated its last class in October 1950. It was then taken over by

Montgomery College and now operates as the Tacoma Campus of this State school.

Grove City College

Located in Grove City, Pennsylvania, about 60 miles due north of Pittsburgh, Grove City College is a Christian, Presbyterian-affiliated school founded in 1876 by Isaac C. Ketler. When the invitations to propose on a Primary School were released in January 1942, Grove City was already severely affected by the draft and saw this as potentially of great benefit to sustaining the college. They immediately responded and were quickly inspected and approved by the Navy.

Although Grove City did not yet offer a full electrical engineering degree, it did have many engineering courses and students could major in this area. Also, it had good offerings in physics and mathematics and was known for its academic excellence. The first class of 100 Navy trainees arrived on March 1. At that time, the college had an enrollment of some 900 regular students, less than half being males. By the fall of 1943, there were only 81 civilian men in the student body.

Hall of Science

Memorial Dormitory

Housing was provided in Memorial Dormitory and a portion of Lincoln Hall. Classrooms and laboratories were in the Hall of Science. Lieutenant Commander William F. Grogan, one of the *ad hoc* group that originated the ETP and who was previously involved in a special "pre-radio" type of activity at the Naval Radio School in Noroton Heights, Connecticut, was assigned as the Officer-in-Charge. Professor Russell P. Smith was named as Director of Instruction. A teaching staff of 14 instructors was assigned, most being from the faculty but also including several persons hired specifically for the

William F. Grogan

program. Navy personnel later augmented this staff. Each month a new class of 100 men arrived, until a complement of 300 trainees was reached. In 1943, this was increased to 330 men.

The city of Grove City had a population of only about 6,000 persons, but they made the trainees feel at home. There was a local USO club, and the city swimming pool was reserved on Sundays for Navy students and their guests. Grove City College was a resident school and the regular students – now mostly females – welcomed the trainees to campus activities.

Laboratory – Instructor Smock (standing)

The program at Grove City continued to April 1945. Library documents at the college show that there were 49 classes graduating 3,759 persons, indicating a graduation rate of about 71 percent.

Following the war, enrollment and programs at Grove City swelled, particularly with veterans seeking an excellent engineering curriculum. The School of Science and Engineering was formed, and the Department of Electrical Engineering included a number of the faculty from the Navy's electronics program. Among these was Dale O. Smock, who continued as a Professor of Electrical Engineering until he retired in 1980 and provided information for this book. Still a small school with fewer than 2,500 students, Grove City is consistently ranked by *U.S. News & World Report* as one of "America's Best Colleges."

Oklahoma A&M College

Founded in 1879 and located in Stillwater, Oklahoma A&M had started an electrical engineering curriculum early in the century. The college eagerly responded to the call for Electronics Training Program participants. It had an enrollment of 5,300 students in the Fall of 1941, but like similar institutions, the student body had been significantly reduced by the draft and enlistments, particularly in the Engineering School. George W. Whiteside, the Dean of Engineering, attended the Navy's meeting in New Orleans and came back with the recommendation that Oklahoma A&M make every effort to be selected as one of the ETP Primary Schools.

George W. Whiteside

Upon acceptance of their proposal by the Navy on February 19, 1942, everything was made ready to start classes 13 days later. Whiteside, a graduate of the Naval Academy, was activated as a Lieutenant Commander (later Commander) to serve as the Officer-in-Charge. The newest and best dormitory, Cordell Hall, was made available for housing and meals. All classes were held in the modern and well-equipped classrooms and laboratories of the Electrical Engineering Department. The regular EE faculty provided the instruction, with a few Navy personnel added later. Professor Emory B. Phillips, earlier on the faculty of the Naval Academy, was appointed the Instructional Director. Two of the instructors were commissioned as Lieutenants and sent to the MIT Radar School, then returned as part of the Navy staff.

The first class of 100 students started March 3, 1942. In this class was Bennett L. Basore, who had been an honors student at Oklahoma A&M the previous semester. For the first year, a new class began monthly, but this was then changed to two-week intervals and the size increased to about 110 students. Oklahoma A&M gave 15 semester hours academic credit to graduates, and the Sigma Tau honorary engineering society awarded a trophy to the top man in each class. During its three and one-third years of operation, this school served approximately 7,000 students, with about 82 percent graduating. Both Whiteside and Phillips remained throughout the life of the program.

Engineering Hall

The Primary School at Oklahoma A&M was generally considered the most desirable in the Electronics Training Program. The housing and meals were excellent, instruction was maintained in a collegiate atmosphere, and there was a very friendly campus environment. The author attended Primary School there and remembers the time as one of the most enjoyable periods of his life.

Cordell Hall Dormitory

The Student Center gave regular privileges to Navy personnel, including renting typewriters, bicycles, and similar items at very low cost. Cordell Hall was only a few hundred feet away from the open end of the stadium, and from windows in their rooms students could watch football games and other events. Just off the campus was a shopping center with a movie theatre, drug store, and bookstore, all frequented by the Navy trainees. During some of the same years as the Primary School, the Navy also had a Yeoman training program for WAVES (Women's Navy Reserves) and SPARS (Coast Guard Women's Reserves) on the campus.

WAVES at Theta Pond

The institution greatly expanded after the war and, in 1957, was renamed Oklahoma State University with campuses in four cities. Bennett Basore, from the first class, came back to complete his studies and, later, again returned to serve as Professor of Electrical Engineering and Head of the School of General Engineering. Robert B. Kamm, another student in one of the first Primary School classes, had brought his wife to Stillwater. They so loved the college and community that he later returned as a Professor and then President (1966-77).

Texas A&M College

Located in College Station, this school was opened in 1876 as Texas's first public institution of higher education. Electrical Engineering as a degree program dates from 1912. An amateur radio station (5XB) was started by the electrical engineering students in 1921, later leading to establishing WTAW (Watch The Aggies Win). Until after WWII, Texas A&M was an all-male military college and regular students were required to be in the Corps of Cadets, then a part of the Army's ROTC program. The Aggie Class of 1941 entered military service *en masse*. During the war, more officers were graduates of Texas A&M than any other military institution in the Nation, including the Army and Navy Military Academies.

Academic Building

In late 1941, the college had an enrollment of about 2,000 students, similar in size to the town of College Station. When the opportunity to participate in the Electronics Training Program arose, the administration and faculty gave full support. The Primary School at Texas A&M began in March 1942 with the arrival of

100 trainees and continued until August 1944.

The trainees were housed in Dormitory 6, with the cafeteria directly adjacent; these were in a group of new facilities built in 1939. The Navy administration was in the Academic Building, one of the most recognized images at the heart of the campus. Some of the classes were also in this building, but most of the classrooms and laboratories were in the Electrical Engineering Building, Bolton Hall. Regular faculty members were used for instruction, led by the Department Head, Dr. Frank C. Bolton.

Dormitory 6

Bolton Engineering Hall

During the time that the ETP operated at Texas A&M, the Navy trainees constituted a good portion of the campus population, and there was a natural competition between the "men in blue" and the Corps of Cadets. The Primary School, therefore, functioned in a very military fashion, including marching under a Chief Petty Officer when the company crossed the campus. The campus offered a minimum of leisure-time activities for the trainees, and Houston and Austin were about 100 miles away.

The railroad station was in Bryan, an adjacent town not much larger than College Station. One student remembers the welcome by local officials upon arriving at the station: "They were cordial, but the message was that military uniforms were not new to the area and the local girls were strictly off limits."

Dr. Bolton later became President of Texas A&M (1948-50). ROTC enrollment at the college was made optional in 1954, and women were first admitted as regular students in 1957 (they had long attended the Summer Session). In 1963, the institution was changed to Texas A&M University, with campuses added at several other locations, including one at the former NATTC Ward Island. Texas A&M University now has the fourth-largest main campus in the nation, with approximately 200 buildings on 5,200-acres. This campus is home to the George H. W. Bush Presidential Library and Museum.

The University of Houston

This school was founded in 1927 as Houston Junior College and renamed with university status in 1934. It initially operated as a unit of the Houston Independent School District, with Dr. Edison E. Oberholtzer simultaneously serving as President of the University and Superintendent of Schools.

In 1939, a new campus was started about three miles southeast of downtown Houston. Dr. Walter W. Kemmerer, Director of Curriculum, became the *de facto* president. Within a year, the enrollment was near 2,500, but, because of the draft and male enlistments, this was projected to significantly decrease in the following years. (It reached a low of around 1,100 in 1943.) This prompted a major effort to be selected as one of the ETP Primary Schools.

Dr. Kemmerer was the driving force in a successful proposal to the Navy, made jointly by the recently formed Engineering and Community Service Colleges. Still under construction, the campus was dominated by two buildings: the Roy Gustav Cullen Memorial and the Science Hall. Navy offices were in the Cullen building, and most of the engineering and physics class-rooms in the Science Building were adopted for use in the program. Laboratories were in the Manufacturers' Industrial ("Shop") Building, hastily completed for the Navy program.

Cullen Memorial Hall

No dormitories had yet been built on the campus, but a new facility, called the Recreational Building ("Rec") was under development. A portion of this building was completed in a few weeks under "emergency construction" to serve as the dormitory and cafeteria space for the 300 trainees (a new class of 100 students arrived each month). When completed, the Rec building also contained a gymnasium and other recreational facilities. The first class of Navy students started March 12, 1942 Dr. Kemmerer noted that he "watched them march in

Science Hall

with great satisfaction, and a sigh of relief" – completion of facilities had been a condition upon which the contract was awarded.

"Rec" Hall - Dormitory & Dining

The program was under the academic administration of Naason K. Dupre, Dean of the University. Instructors were from the engineering, physics, and mathematics faculties. Lieutenant H. E. Wood was the first Officer-in-Charge. During some of the period that the Primary School existed, there was a Navy V-5 pilot ground school operating on the campus and also housed in the Rec. This led to some minor conflicts between the officers-to-be and the highly select ETP enlisted men.

The new campus of the University of Houston had been constructed in a relatively isolated area. It had also been designed with the assumption that students would commute via city busses. Since there were no nearby shopping or entertainment facilities and the busses had only intermittent service at night and on weekends, the trainees mainly stayed around the Rec during off-hours. Weekends, however, were no problem to trainee Chester P. Wilkes; he was from Algo, a tiny town only a few miles from the University, and he usually hitch-hiked home after Saturday-morning inspection. Wilkes noted that "We could go anywhere on liberty that was no more than 50 miles away."

In January 1944, Lieutenant Commander Richard Stock transferred from Utah State to become the Officer-in-Charge of the University of Houston program. The Primary School operated through March 1945, providing instruction to 4,278 trainees. The School of Technology became a descendent of the ETP, occupying the Industrial Building and receiving the laboratory equipment when the Navy program closed.

The University of Houston separated from the school district in 1945, then became a state institution with three campuses in 1963. The rapid construction of the Rec building led to structural faults in a few years, and it had to be demolished in 1966.

Utah State College of Agriculture

Utah State was founded at Logan in 1888 as Utah's Land Grant College. A School of Engineering was created in 1912 and, following WWI, an Industrial Division within this School was formed. Led by

Sidney R. Stock, the Radio and Aviation Department of the Industrial Division received widespread recognition. This was eventually split, with the Radio and Electronics Department evolving. The student body at Utah State reached over 3,000 on the eve of World War II, but during the war dropped to less than 1,000. As the number of regular students decreased because of the draft, the college pursued various activities to maintain their faculty. In July 1940, a major program in National Defense Training was started, including an activity for the Army Air Corps at Hill Field.

Utah State Campus

Dean of Engineering George D. Clyde represented Utah State at the New Orleans meeting concerning the ETP in February 1942, and directed the proposal development. The preceding October, Sidney Stock had been commissioned a Lieutenant Commander and called to Washington to develop an aviation electronics school for the Navy and subsequently served on the *ad hoc* planning group for the ETP. Thus, it was natural that Utah State would be one of the colleges selected to conduct a Primary School when the program was initiated.

Upon acceptance of the proposal by the Navy, Ernest C. Jeppsen, Director of the Industrial Division, was assigned overall responsibility and Assistant Professor Larry S. Cole was named Program Director. Lieutenant Carlos J. Badger, a graduate of the Naval Academy and an attorney in Salt Lake City, was called to active duty and initially assigned as Officer-in-Charge.

The Smart Gymnasium was converted to house the trainees and the Commons cafeteria was used for the mess hall. The Old Main was fitted out for classrooms and laboratories, and study rooms were in the Engineering Building. Instructors were drawn from engineering, physics, and mathematics faculties, with Waldo G. Hodson as the Chief Instructor. The first class of 99 trainees started March 23, 1942. W. Arnold Finchum, a member of the first class, later returned to graduate from

Smart Gymnasium - Housing

Utah State and, still later, again returned as a Professor of Electrical Engineering.

Old Main - Classrooms

The Logan community rallied behind the Navy trainees at the school. A United Service Organization (USO) Club was set up by a citizen's committee, and local residents furnished this, as well as providing food and stage entertainment on certain nights and weekends. A service was established with the Navy administration for arranging student visits with local families.

In October 1942, Stock returned to Utah State as the Officer-in-Charge, remaining until January 1944, when he was transferred to the University of Houston. He was replaced by Lieutenant Commander Nelson H. Randall, who had previously headed the Radar Training School at MIT. The Primary School at Utah State continued until August 1944. In its two and one-half years of operation, it served some 2,750 trainees in 30 classes. Of these, about 2,400 graduated, giving a graduation rate of approximately 87 percent.

Sidney R. Stock

The Department of Radio and Electronics became the Department of Electrical Engineering in 1945, to a large extent due to the experience in conducting the training under the RTTS. The school was granted university status in 1971, becoming Utah State University.

Navy-Operated Primary Schools

America continued to build ships and aircraft, with a consequential demand for more technicians to maintain the electronics. The Secondary Schools at NRL-Bellevue and Treasure Island could no longer keep up with the demand. At the end of 1943, BuPers directed that a new Secondary School be established at Chicago's Navy Pier, and that the Secondary School at NRL-Bellevue be increased in capacity by transferring its Primary School to a new Navy-operated facility at the College of the Ozarks in Arkansas.

Three other Navy-operated Primary Schools would also be established, two at existing Naval Training Centers at Great Lakes, Illinois, and Gulfport, Mississippi, and the third at the Del Monte Hotel in Monterey, California. Compared with the existing college-operated schools, these three Primary Schools were very large – classes of about 250 – and conducted in a military manner. With the increased capacity of Navy-operated Primary Schools, the schools at Texas A&M and Utah State were closed.

Essentially at the closing of the war in 1945, another Navy-operated Primary School was opened in the Naval Service Training School at Ford Motor's River Rouge plant in Dearborn, Michigan. The reasons for opening this school, which also included a Pre-Radio unit, are unknown. This school actually trained very few technicians and was closed within a year.

College of the Ozarks

This small liberal-arts college, located in Clarksville, Arkansas, traces its roots back to 1835 when it was started as Crane Hill School by the Cumberland Presbyterian Church in a two-room, hewn-log building. It became Arkansas Cumberland College in 1891, then changed to College of the Ozarks in 1920, affiliated with the Presbyterian (USA) Church. It had initially proposed to participate in the Electronics Training Program in 1942, but, not having electrical engineering offerings, was not selected.

Never very large, as the school year started in the Fall of 1943, the College of the Ozarks was down to only about 150 students. To survive, the Board of Trustees offered its full facilities for use by some military activity. BuPers decided to move the Primary portion of the NRL-Bellevue school to another location, and the offer from the College of the Ozarks was accepted in early December 1943. In a frantic three-week effort, the student housing, classrooms, laboratories, library, and cafeteria were moved off the campus into the First Presbyterian Church and the adjoining manse; in January, college classes began in the new location.

In this same time period, the dormitories were converted to Navy-style barracks, a mess hall was set up, and classrooms and laboratories were made ready. The Munger Memorial Chapel was used as a general auditorium and administrative offices. The faculty, administrative staff, and classroom and laboratory equipment were moved from Washington over the holidays, and on January 5, 1944, the first class of 100 students began the College of the Ozarks Primary School. Thereafter, a new class started every other week, with an ultimate size of near 600 students. All of the instruction was by Navy and Marine personnel, mainly transferred from NRL-Bellevue.

Munger Chapel

Although the college and town tried to make the trainees feel at home, Clarksville, with a population of only a few thousand offered almost nothing to do on weekends. Little Rock, the nearest city of any size, was 125 miles away. Most trainees, nevertheless, found their stay enjoyable. "Navy life picked up dramatically when I arrived in Clarksville," remembered John Westkaemper, an early trainee, "The location was attractive, the duties weren't onerous, the people were pleasant, and the weather was nice." The program continued through May 1945, providing training for about 3,000 students.

Faculty and Staff, 1945

The only direct benefit to the college was the $1,100 per month rent paid by the Navy, an amount not even covering the cost of renovations. The minutes of the Board of Trustees from June 1, 1945, succinctly record that "the true benefit was the college's satisfaction in serving the country." In 1987, the college changed its name to the University of the Ozarks. Like Grove City College, it is ranked by *U.S. News & World Report* as one of "America's Best Colleges."

Del Monte Hotel

Located in Monterey, California, one of the most picturesque places in America, the Del Monte Hotel seems an unlikely place for an ETP

Primary School, but one existed there from February 1944 to June 1947. Opened in 1880, the Del Monte Hotel was billed as one of the finest luxury resorts in the world, hosting captains and kings. The main building burned twice – once in 1887 and again in 1924 – but was rebuilt in even greater splendor each time. By the early 1940s, however, business had greatly dwindled and Samuel F. B. Morse, the hotel's owner and grandnephew of the telegraph inventor, offered to lease it to the Navy.

Del Monte Hotel

The Navy took over the Del Monte hotel and prepared it for a Pre-Flight Training School. This school opened in February 1943 and soon had over 1,000 students, housed as many as eight to a room. In the single year of its existence, the school graduated over 4,000 cadets. It was best known for fielding a football team that ranked eighth in the final 1943 AP national poll. When the Pre-Flight School closed, the ETP Primary School moved in.

The first class of 200 students for the Primary School arrived at Del Monte at the end of February 1944. Thereafter, classes of similar size reported every two weeks. Classrooms were in temporary buildings constructed on the hotel grounds, while laboratories were in other temporary buildings across the highway from the hotel. The main building housed the administrative offices, a dispensary, a dentist office, the chaplain's office, the mess hall, and a large auditorium. The faculty, mainly transferred from Treasure Island, was all Navy and Marine personnel. Commander Paul K. Bryant, previously head of the Naval Training School at Dearborn, Michigan, was the initial Commanding Officer. Commander William M. Cashin followed Bryant in 1945, and then Commander Max Blackford was in charge until the school closed.

Administration & Services

The facility was listed as having quarters for 1,300 men in the East and West Wings – called "Saratoga" and "Lexington," respectively, from their Pre-Flight School designations. Although students were housed six to a room, each room did have a private bath. In the mess hall – operated by the hotel – food was served on large platters at each table. All 1,300 students were served at one seating. A swimming pool and other recreational facilities were on the grounds. There was excellent weekend entertainment in the auditorium, and for those who wanted to visit San Francisco or Los Angeles, a train station was adjacent to the hotel.

Bob Hope Entertaining

In its three and one-half years of operation, the Del Monte Primary School served about 6,000 students, making it the largest of all such schools in the Electronics Training Program. The facility was decommissioned in June 1947. A year later, Congress authorized purchasing of the 608-acre property and construction of a new home for the Naval Postgraduate School, then located in Annapolis. This new school opened in December 1951, incorporating the old Del Monte Hotel facilities.

Gulfport Naval Training Center, Mississippi

America's World War II defense plans called for a non-congested, deep-water port to serve the Caribbean area. Gulfport, Mississippi, met this need, and in April 1942, a 1,098-acre plot of land was acquired a mile northwest of the Port of Gulfport. An Advanced Base Depot opened on June 2, 1942, and served as the receiving station for a Naval Construction Battalion (Seabees). An Armed Guard School and Cooks & Bakers School were added in October.

Typical Instructional Building

In March 1944, the mission of the Center changed to a U.S. Naval Training Center, and provided courses for ratings as Carpenter's Mate, Electrician's Mate, Gunner's Mate, Quartermaster, and Radioman, as well as advanced schools in basic engineering, diesel mechanics, and recognition. An ETP Primary School at Gulfport was also opened at that time with Commander B. L. Stewart assigned as Officer-in-Charge. The instructional staff was all Navy and Marine personnel.

The overall Gulfport Training Center was very crowded, with a peak in all of the schools of near 25,000 personnel in late 1944 – much larger

Initial Quonset Complex

than the adjacent city. The instructional facilities were all temporary buildings. Initially, students were housed in Quonset huts. Toilet facilities were in a separate "head" building, connected by plank walkways raised over the ever-present dust, mud, or water. Later, standard two-floor barracks were added.

Typical New Barracks

There was little entertainment during the week, and essentially none was available on base during weekends. Biloxi, only a little larger than Gulfport, was a few miles to the east, but the nearest large city was New Orleans, about one and a half hours by bus or train to the west.

Because of these living conditions, Gulfport was considered to be the least desirable of the Primary Schools. Davis L. Brewer, a student in late 1945, has noted:

> After formation of the entire school at noon on Saturdays, those leaving the base were marched downtown to the bus and train stations. This was to prevent all that mass of sailors from dispersing into the residential neighborhoods of Gulfport.

At the end of the war in the Fall of 1945, the Gulfport Primary School slowed its program, and the last class graduated in March 1946. There is no record of the number of personnel trained during the existence of the school, but it is estimated at about 2,500. The Gulfport Naval Training Center was decommissioned in 1946, and the base became a storage facility for critical materials. Later it was reactivated as a Naval Construction Battalion Center.

Great Lakes Naval Training Center

Naval Station Great Lakes is located on the shore of Lake Michigan, 30 miles north of Chicago. In 1904, President Theodore Roosevelt signed the act authorizing the construction of the station and directed the Navy to make Great Lakes the biggest and the best Naval Training Station in the world. It received its first recruits in 1911, "turning civilians into seamen and seamen into sailors."

In late 1940, as the Navy increased its personnel in preparation for war, the Bureau of Navigation authorized the opening of the Great Lakes Service Schools. By early the next year, schools were in place for training Electrician's Mates, Fire Controlmen, Gunner's Mates, Torpedomen, Radiomen, Quartermasters, Signalmen, Yeomen, Storekeepers, Machinist's Mates, Boiler Tenders, Metalsmiths, and Carpenter's Mates. Schools for other specialties were added later.

Main Gate

The Service Schools reached a high of 11,000 in late 1943, and trained a total of 115,000 men throughout the war years. Including the Recruit Training Schools, during World War II Great Lakes provided training for about one million persons, making it the largest training base in the Navy.

Main Side – Great Lakes Training Center

In 1944, as enrollment at the existing Great Lakes Service Schools decreased, plans were made for opening an ETP Primary School at this center with a 1,200-man capacity, but the opening was delayed until early 1945. At that time, Captain K. E. Bond was the Commanding Officer of the Training School. Like other Navy-operated Primary Schools, the instructors were all servicemen, mainly Navy and Marine graduates from the Secondary School at NRL-Bellevue or Treasure Island. For the next year, most of the trainees took Boot Camp at Great Lakes, were sent to Chicago for Pre-Radio, returned to Great Lakes for Primary School, then returned to Navy Pier in Chicago for the final Secondary School. It is estimated that about 1,200 trainees at a time were in the program at Great Lakes for the next year.

Typical Training Center Building

When the Pre-Radio Schools in Chicago closed at the end of 1945, this activity was added at Great Lakes. Then in mid-1946, the Secondary School at Navy Pier was transferred to Great Lakes, becoming a full operation in the Electronics Training Program. At this time, the overall curriculum was reorganized, making it more appropriate for the regular, peacetime Navy. At the end of 1946, the Secondary School at NRL-Bellevue was also transferred to Great Lakes.

For a number of years, Great Lakes continued with Primary and Secondary Schools, with graduates of the Primary School split between Great Lakes and Treasure Island. The overall program for training electronic technicians for maintaining sea- and shore-based equipment was eventually changed to Class A, B, and C schools, shared between Great Lakes and Treasure Island. In more recent years, this training has been concentrated at Great Lakes. All Navy recruits now go through Boot Camp at Great Lakes, and the Training Center remains the largest in the Navy. It presently includes 1,153 buildings on 1,628 acres and uses 50 miles of roadway to provide access to the Center's facilities.

Dearborn/River Rouge Service School

Located in Dearborn, Michigan, a few miles south of Detroit at the confluence of the Rouge and Detroit Rivers, the River Rouge complex of Ford Motor Company was the world's most famous industrial plant. Here Henry Ford achieved self-sufficiency and vertical integration in automobile production – a continuous work flow from iron ore and other raw materials to finished automobiles. "By the mid-1920's," wrote historian David L. Lewis, "the Rouge was easily the greatest industrial domain in the world . . . without parallel in sheer mechanical efficiency."

Now a National Historic Landmark, this multiplex of over 90 buildings with more than 15 million square feet of floor area, was a mile-and-a-half wide and more than a mile long, crisscrossed by 100 miles of railroad tracks. It had a multi-station fire department, a modern police force, a fully staffed hospital, and a training operation that was likely larger than any public technical school. During WWII, the giant complex had 120,000 employees producing jeeps, amphibious vehicles, parts for tanks and tank engines, and aircraft engines used in fighter planes and medium bombers.

Ford River Rouge Plant

As the war approached, the Ford Motor Company offered its training capabilities to the Navy. In January 1941, the Naval Service Training School at the Rouge plant in Dearborn was opened. This used on-the-job formats, with the Navy personnel engaged in direct defense work and instruction provided by Ford personnel. Barracks and a mess hall to accommodate about 800 trainees were constructed just outside the Rouge complex, and a newly built recreational facility was used for physical exercise. By the time the school closed in May 1946, it had trained more than 22,000 men in electrical and machine-shop skills, repair of diesel and gasoline engines, and many related activities.

In early-1945, a Navy-operated ETP Primary School was added at the River Rouge training facility. To make this operation self-sufficient, a Pre-Radio School was included, the only such training activity at that time outside of Radio Chicago. The trainees were sent to Dearborn directly from Boot Camp at Great Lakes. Unlike the other training activities at the Rouge facility, the Pre-Radio and Primary Schools used only Navy instructors, mainly recent graduates from the Navy Pier Secondary School. Commander G. M. Holley was the Officer-In-Charge.

Leaving School at Dearborn

The entire Dearborn-River Rouge Service Training School closed in May 1946. Roger F. Harrington, a faculty member at the Secondary School, made this observation: "Essentially all of the time of their existence, the radio technician schools at Dearborn were either being set up or dismantled. Although assigned as an instructor, I never taught a class." It is likely that only a few hundred persons were graduated.

SECONDARY SCHOOLS

By 1940, the Navy's electronic systems had evolved to the extent that maintenance personnel needed to be knowledgeable in theory as well as hardware. New types, as well as models, of equipment were constantly being added. Training on each of the great variety of sets was impractical; thus, it was necessary for maintenance technicians to start with basic theory to understand their function and consequent maintenance.

Electrical engineers, already having a background in the basic theory, would have been ideal for advanced training in hardware maintenance, but such persons were not available – certainly not in the thousands that were projected to be needed during wartime. As an alternative, the Electronics Training Program came into being. As initially planned by the *ad hoc* group in late 1941, the program would provide the vital parts of the needed background through college-based Primary (or EE&RM) Schools, and Secondary (or Advanced) Schools) would continue with more theory, as well as a significant introduction to the hardware.

In 1942, the Secondary Schools for fleet and shore electronics were at NRL-Bellevue in the District of Columbia and at Treasure Island in California. For aircraft electronics, the Secondary School was at Ward Island in Texas. A fourth Secondary School opened in Chicago at the Navy Pier in 1944. The following is a discussion of the locations, facilities, and leaders of these schools.

NRL-Bellevue (District of Columbia)

In 1924, shortly after the Naval Research Laboratory was opened in the Bellevue section of the District of Columbia, space in one of the initial five buildings was allotted to the Radio Materiel School (RMS). Here instruction was provided on the products coming out of the NRL's Radio Division. Students in the RMS were senior Radiomen and Electricians. A student in the early days was future entertainer Arthur Godfrey, who, while serving in the Coast Guard, graduated in 1929. Through the 1930s, the NRL had operated with two classes per year of six-months duration and with about 30 students in each class.

Closely associated with the RMS was the Warrant Officers' Radio Engineering School. Started in 1927, this school allowed Radio Electricians (Warrant Officers) to devote a year to intensive study in advanced mathematics, physics, and radio theory. From the beginning, the class size had been set at six men (ten percent of the existing Radio Electricians) and was the normal preparation for becoming a Chief Radio Electrician, perhaps the highest technical rank for warrant officers.

Wallace J. Miller

With the advent of radar and other new equipment, and in the Navy's preparation for war, there were significant changes in the RMS. In February 1940, Lieutenant Commander Wallace J. Miller was assigned as the Officer-in-Charge. Miller had an excellent background for this position. A 1926 graduate of the Naval Academy, he had taken flight training and served in the fleet as an aviation officer. He then studied radio engineering for three years at the Naval Postgraduate School and the University of California in Berkeley, earning a master's degree from the latter in 1936. Before joining the RMS, he had served four years as the Radio Officer on several fleet assignments.

In early 1940, the RMS was increased to eight months in length, divided into a three-month Primary and five-month Secondary. Chief Radio Electrician Nelson M. Cooke, then Senior Instructor in the Radio Engineering Course, led the upgrading of the Primary portion of the RMS. Miller took personal responsibility for upgrading the Secondary portion.

A new class of 60 trainees started every four months. For this and projected expansion, Building 28 was built on the southwest corner of the NRL campus, providing sleeping, eating, and classroom space for over 200 students. The Warrant Officers' Radio Engineering School was included in this facility. In early 1941, the entering RMS class size was increased to 135 men. Two temporary facilities, Buildings 32 and 34, were constructed to accommodate entering Primary students. A significant number of these new students did not make it past the first month (what was later the main content of Pre-Radio), and perhaps as little as 50 percent eventually made it into Secondary School

Building 28 – RMS at the NRL, 1941

After the start of the war and the initiation of the expanded Electronics Training Program, a new class started every two months. In August 1942, Miller was promoted to the rank of Commander and became the Commanding Officer of the NRL-Bellevue schools. At this same time, Cooke was commissioned a Lieutenant (jg) and named Officer-in-Charge of the Primary School.

A move of the school to the southeast of the campus was begun. Building 45, a large temporary dormitory complex, was constructed in late 1942. This included a sick bay and several other special facilities. A chapel was built nearby. A number of laboratory buildings were also added during the next year

New Dormitory Complex

Miller, an officer in the regular Navy, was required to have periodic sea duty and thus was transferred in May 1943, and was replaced by Lieutenant Commander R. Cole. Cooke was then named Executive

Officer of the NRL-Bellevue schools. Both Cole and Cooke retained these positions until the operations closed in late 1946.

Nelson M. Cooke *R. Cole*

The initial preparatory component of the Primary School was discontinued in mid-1943, and all incoming students were Pre-Radio School graduates. The student complement was increased to approximately 1,200 men, mainly Navy but with about 15 percent Marines.

At the start of 1944, the entire Primary School was transferred to a Navy-operated school at the College of the Ozarks, thus making space for additional Secondary School trainees. With this, however, came the need for more laboratories. At the same time, the BuPers directed that the Secondary School be increased to six months (actually 24 weeks), with the additional time used in instruction on sonar equipment, an activity requiring still more space. A typical graduating class in 1944 was about 75 men.

Typical Laboratories

The laboratories were the primary facilities of the school. Building 34 had the general electronics laboratory; Building 31 with three wings served the special circuits laboratories; the communications laboratories were in Buildings 28 and 53; Building 28 also had the sonar laboratory; and the second floors of Buildings 34, 40, and 43 contained the radar laboratories. Other new facilities included Buildings 59, 61, 62, and 64. Much of the hardware studied, including all of the radar and sonar, was classified, and the related lecture areas and laboratories were within a secure compound.

Many other buildings were also being added to the NRL. "So much construction is in progress on the Station that NRL is beginning to look like a modern boom-town," read an article in the July 1944 issue of the *NRL Pilot,* the Laboratory's newsletter at the time. The NRL then operated with eight divisions, with the Radio Division still being the largest. The Navy had about 450 officers and enlisted men stationed at the NRL, and there were some 1,100 civilian professional and supporting personnel plus about 550 contractors. A number of the Navy men had previously been scientists and engineers at the Laboratory. Some of these had been directly commissioned as officers; others, unable to meet requirements for commissioning, had volunteered to escape the draft and returned as enlisted men to their earlier activities.

Naval Research Laboratory, 1944 – School in Upper Right Corner

The instructors were primarily Navy petty officers with some Marines, essentially all graduates of the ETP. Special topics were taught by warrant officers and, occasionally, by lecturers from the NRL. Students were encouraged to keep notebooks, but, when not in use, these were secured within the compound and later sent as classified documents to the graduates in their duty stations. Although officers often took some parts of the curriculum, there was no mixing of officers and enlisted men in the classes. Lecturers for these officers' courses were often enlisted men.

Starting in 1944, a new class entered the Secondary School every two weeks. In 1945, an optional four weeks was added to the period of instruction, mainly in sonar equipment. Thereafter, some students had 24 weeks while others, at the option of the Navy, had 28 weeks. When it peaked in 1945, there were nearly 2,400 enlisted men attending the NRL-Bellevue Secondary School. After the start of the war, the Radio Engineering School had increased the class size to 30 warrant officers, and special electronics courses for officers (primarily reservists) were conducted. Altogether, at peak there were about 200 officers in the various NRL-Bellevue instructional programs. During the war years, it is estimated that about 8,000 persons graduated from the NRL-Bellevue Secondary School.

The buildings used in the program, although mainly temporary and hastily constructed, were very adequate for the purpose. The barracks for trainees, Building 45, was divided into rooms, each accommodating four students with two double-deck bunks, lockers, and a large, well-lighted study table. As the enrollment peaked, there were six men assigned in most rooms. The Recreation Club (Building 52, adjacent to the barracks) opened in 1944 and contained a gymnasium. A drill field was located adjacent to these buildings and, in good weather, exercise sessions were often conducted there. This field was also used for the full parades that were sometimes conducted on Saturday mornings. Formal entertainment was absent on the NRL campus, but plenty was available only a short bus-ride away. The RMS did have an outstanding choir, often singing with the Navy and Army bands in Washington concerts.

After significantly decreasing in attendees after the war, the school at NRL-Bellevue closed at the end of 1946. This activity was combined with the operations at Navy Pier, which had been transferred in June, to form a new advanced school at the Great Lakes Naval Training Center.

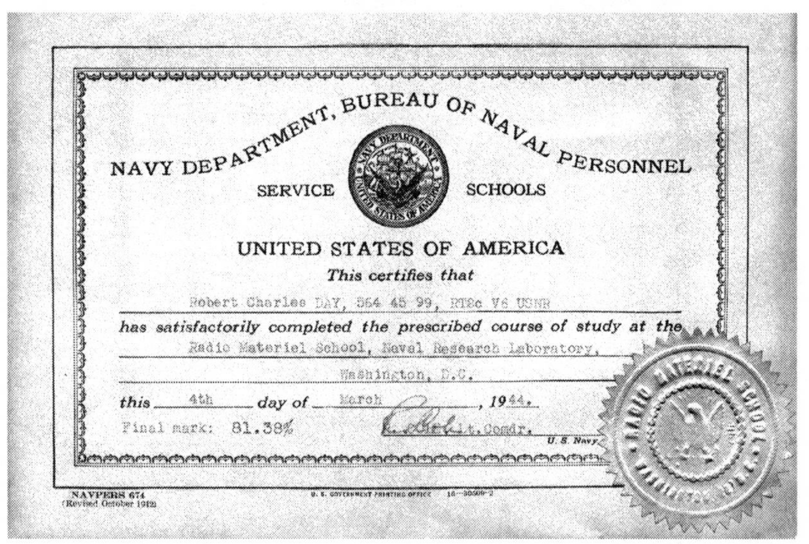

NAVY DEPARTMENT, BUREAU OF NAVAL PERSONNEL

SERVICE SCHOOLS

UNITED STATES OF AMERICA

This certifies that

Robert Charles DAY, 564 45 99, RT2c V6 USNR

has satisfactorily completed the prescribed course of study at the

Radio Materiel School, Naval Research Laboratory,

Washington, D.C.

this 4th *day of* March, 1944.

Final mark: 81.38%

Lt.Comdr.

U. S. Navy

NAVPERS 674
(Revised October 1942)

Treasure Island (San Francisco, California)

Yerba Buena Island, also called Goat Island, is a 300-acre parcel of rocky land along the causeway and bridge connecting San Francisco and Oakland, California. Yerba Buena (Spanish for "Good Herb") was first occupied in 1775, and a lighthouse on the island has operated since 1875. In the late 1930s, Treasure Island was constructed adjacent to Yerba Buena as the site of the 1939-1940 San Francisco World Fair, officially, the Golden Gate International Exposition. The only land access to Treasure Island is a short isthmus connecting it with Yerba Buena.

A Federal project of the Public Works Administration, the construction of Treasure Island began in February 1936 and was completed in January 1939. To build the 403-acre island, 29 million cubic yards of sand and gravel were dredged from the Bay and the Sacramento River delta. The name "Treasure Island" refers to the gold-laden fill soil that washed down from the Sierras into the Bay. Many buildings were constructed, with 20 nations having major pavilions. Along one side of the island there were two huge airplane hangars that served as the American Terminal for the Trans-Pacific Clipper Service of Pan American Airways. The Golden Gate Exposition ran from February through October 1939, then May through September 1940.

When the Exposition closed, the City and County of San Francisco started converting the island to an airport. In early 1941 the Navy interceded, leasing Treasure Island and a portion of Yerba Buena for potential use as a naval base and training operations. Most of the Exposition buildings had already been razed and little was done in new construction until the war came closer. In the last months of the year, however, there was a crash building program and a number of training schools were started. Later, a "declaration of taking and deposit" was filed, and the U.S. Navy seized Treasure Island for ownership.

Treasure Island as seen from atop Yerba Buena

The Bureau of Navigation directed the Officer-in-Charge of the Radio Materiel School at NRL-

Bellevue, Lieutenant Commander Wallace J. Miller, to examine Treasure Island and make recommendations on starting another Radio Materiel School there. Although Miller had some reservations, he recommended adapting an abandoned fairground building, known as Palace "H," to school use and constructing a highly secure radar training facility on Yerba Buena Island. His report led to the authorization on October 30, 1941, to establish such a school with a curriculum patterned on the one used at NRL-Bellevue.

Construction was immediately started on two school buildings (Palace "H" was not initially used), several barracks, and a mess hall on Treasure Island, and a radar building atop the 400-foot rocky hill on Yerba Buena. By mid-January, the instructional buildings were nearing completion, and the administrative staff was assembled. Lieutenant Commander Harry F. Breckel was assigned as the initial Commanding Officer, and Lieutenant Commander Ray H. Parker was the Executive Officer. The Instruction Department was headed by Lieutenant Paul G. Fritschel. Barracks were not ready, so the students were temporarily housed in one of the large buildings left from the Exposition.

Harry F. Breckel

Breckel was well suited to lead this training facility. An early amateur radio operator, he had joined the Navy as an enlisted radio operator in 1915 and was commissioned as a reserve Ensign three years later. Leaving active duty as a Lieutenant (jg) in 1919, he became an engineer with Precision Equipment Company in Cincinnati, where he built and operated experimental station 8XB. This became licensed as WMH, one of the first broadcasting stations in America. For 15 years, he developed broadcasting and high-frequency radio equipment and published many technical articles. Having remained in the reserves, Breckel was promoted to Lieutenant in 1925. In this capacity, he organized the Naval Communication Reserve in Ohio and Kentucky and participated with Naval Intelligence in several special projects. Between 1933 and 1935, he was on active Navy duty, attached to the Civilian Conservation Corps. In 1940, he was promoted to Lieutenant Commander and assigned to the Naval Reserve Radio School, Noroton Heights, Connecticut, becoming the Executive Officer in 1941. At Treasure Island, Breckel was promoted to Commander, then received the rank of Captain in early 1944.

Trainees at RMS Treasure Island were initially supplied from a personnel pool developed by the Twelfth Naval District, headquartered in San Francisco. This pool served as a filter for identifying potential students for the program. About 25 percent were radio Hams, entering the Navy with Petty Officer ratings as Radioman Second Class, and often had not even attended Boot Camp. To accelerate the anticipated training, some from this pool were sent to special one-month, abbreviated Boot Camps and preparatory courses that had been hurriedly set up at the Naval Reserve Armory in Los Angeles and at the Radio Operators School in Noroton Heights, Connecticut.

The Treasure Island school was commissioned on February 9, 1942, and an initial complement of 566 students began their training. All of the entering students were given a placement test to determine if they were qualified for the difficult training to come. They were then divided into groups of 30 to 40 and, based on the admission scores, started at various monthly levels. The initiation of instruction, however, faced two major problems: lack of instructors experienced in this type of school and lack of equipment for the laboratories. Drawing instructors from the first class readily solved the first problem. A significant number had either advanced experience in radio or teaching backgrounds, some with both. The placement test also identified potential instructors for various segments of the school, particularly for the early portions, where mathematics and physics were emphasized.

Administration and Classroom Buildings

The laboratory problem was more difficult – the equipment was the same as urgently needed in the fleet, and the school had a lower priority. To solve this, commercial test equipment and communications receivers were purchased, and "breadboard" transmitters were fabricated from purchased components. Many instructional aids were also built, including large demonstration boards for circuit tracing. The Ham radio operators were very helpful in these endeavors. Radar was presented in the last months, and by then some of these

Radar Laboratory on Yerba

systems and related test equipment had been obtained for the laboratories. A portion of the laboratory instruction took place on Yerba Buena, where the large radars were set atop the rocky mountain in a highly secure compound.

At the end of February, a new class of 150 students entered; the initial expectation was that 100 persons per month would complete the course. The required monthly output was almost immediately increased to 200, with the entering number building to 275 per month. Near the end of May, the Bureau of Navigation was reformed into the Bureau of Naval Personnel (BuPers) and ordered a review of the Treasure Island school. This was conducted by a Board of six officers, led by Lieutenant Commander Wallace Miller, Officer-in-Charge of the NRL-Bellevue school and the individual who had first recommended the Treasure Island school. This Board made very favorable comments but recommended a restructuring to concentrate on the advanced portions of instruction.

Marching Between Classes

In June 1942, BuPers directed that Treasure Island discontinue the first three months (actually 12 weeks) of training and draw on graduates of the newly formed Primary Schools in Chicago and various colleges. Thereafter, the program at Treasure Island was designated a Secondary or Advanced School, and the input was increased to 250 students per month. The first class of 87 students graduated at Treasure Island on June 26, 1942, several months ahead of schedule. This was possible because the placement test allowed them to start in the fifth month of the program.

Palace "H" as a Barracks

When Pre-Radio was added to the ETP, BuPers directed that the operation at Treasure Island temporarily add such a school. Palace "H," one of the original fairgrounds buildings, was used for this training as well as student housing. This opened in March 1943, then closed the following June when the Pre-Radio Schools in Chicago started turning out graduates.

A major conference on the Electronics Training Program was held in December 1943 at Washington.

This centered on how to meet the demand for even more qualified technicians, as well as examining topics in the Advanced Schools. More students were obtained for the Secondary Schools by doubling the number of trainees at certain Primary Schools, changing from a new class monthly to every two weeks. There was concern over instruction in sonar – a technology becoming of increased importance in the fleet. Treasure Island was directed to take the lead in this, and a special sonar instructional laboratory was constructed in early 1944. The curriculum was increased from 20 to 24 weeks. At this same time, Palace "H" was converted for additional radar, LORAN, and IFF laboratories, and large wings were added to the three original barracks.

Sonar Laboratory

When in full operation, the Treasure Island school consisted of the following facilities: For instruction there were Buildings 27, 28, 127, 155 (old Palace "H"), 224 (atop Yerba Buena), and 266 (the sonar laboratory). The barracks included Buildings 17, 18, 19, 20, 21, and 22, three of which had been greatly increased in size during the years by adding wings. A portion of Building 28 was also used for administrative offices, and Building 29 was used jointly for recreation and shop purposes.

Radar Instruction Facility

The demand for qualified technicians continued to increase, and more trainees were fed to Treasure Island from the Primary Schools, especially from the new Navy-operated schools at Gulfport, Great Lakes, Del Monte, and College of the Ozarks. In May 1944, BuPers directed that a new Secondary School be established at Navy Pier in Chicago. Treasure Island was called upon to build the instructional aids, as well as to provide a number of instructors for this school. This generated an immediate need for additional qualified instructors, so an Instructor Training Course was established at Treasure Island in June 1944. After two weeks in the classroom, the participants went into a four-week apprenticeship, teaching under the guidance of seasoned instructors.

From the beginning of the Treasure Island school, short courses of various lengths had been given to officers and enlisted personnel. Recognizing the increasing need for retraining, BuPers directed that a Special Courses Division be established in June 1945. Here, maintenance officers and enlisted men from Fleet Units were given a formal eight-week refresher course or attended one of the many short-term courses, particularly on the new types of radar and sonar.

Traffic on and off the Causeway

Treasure and Yerba Buena Islands, collectively called the Naval Station Treasure Island (NSTI), became the headquarters of the 12th Naval District. NSTI served as the "Gateway to the Pacific" throughout the war. By the end of hostilities in August 1945, the Treasure Island RMS had graduated about 10,000 electronics technicians. After the war, the Advanced School continued, mainly receiving students from the Primary School at Great Lakes Naval Training Center. The curriculum, however, was changed – reorganized to accommodate the needs of the peacetime Navy. The last class under the drastic wartime curriculum graduated in June 1946. Treasure Island remained under Navy ownership until 1996, when NSTI was decommissioned and the land opened to public control.

Ward Island (Corpus Christi, Texas)

Late in the summer of 1941, the Bureau of Aeronautics sent 25 officers and potential instructors to Clinton, Ontario, to attend the Training School #31 that Canada had just set up for the Royal Air Force. A facility was being built for a similar school on the campus of the Naval Academy at Annapolis, and the group was sent to Canada to prepare for its opening. Lieutenant Commander Sidney R. Stock was assigned as the Officer-in-Charge. The Aviation Radio Materiel School opened with its first class of 100 students on January 2, 1942. Even before it was opened, the Navy realized that a much larger and more secure facility would be needed. The Bureau of Aeronautics searched for a suitable location and selected Ward Island, an undeveloped area near the newly opened Naval Air Station at Corpus Christi (NASCC) in Texas. The NASCC had been authorized by Congress in 1938, but was not commissioned for pilot training until March 1941.

Ward Island was an excellent choice for an aviation electronics maintenance school. It could be made very secure and draw on the NASCC for needed support functions. The island was actually a large, stable sandbar in Corpus Christi Bay. It was named after John Ward, an investor who acquired the then-farmland in 1882 and developed it into a popular hunting and fishing site. With an irregular shape spanning approximately 1.2 by 0.6 miles, it covered some 240 acres in area, which, at its highest point, was only about 12 feet above sea level. With Corpus Christi Bay forming the northeast border, it extended to the southwest into the Oso Bay. A causeway road, Ocean Drive, ran from the Corpus Christi city limits – less than a mile to the northwest – to Ward Island and on to the existing NASCC some two miles to the southeast.

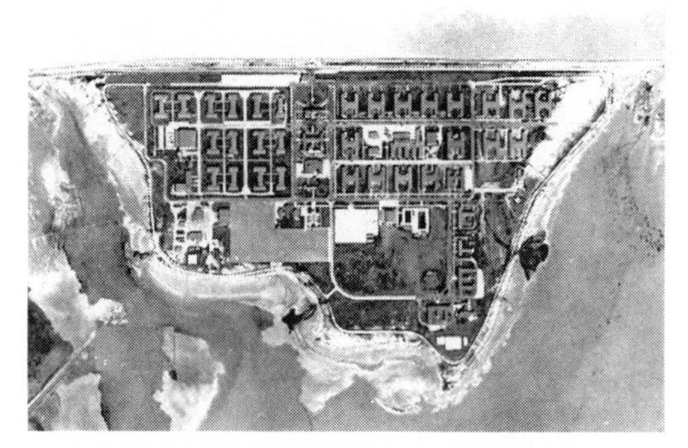

Aerial View of NATTC Ward Island

Ward Island was acquired by the Navy in early February 1942 and construction began in May. By the end of June, ten buildings had been completed or were near so; these were the Administration Building, five barracks, two instructional buildings, and two mess halls. Personnel and equipment of the Aviation Radio Materiel School were transferred from Annapolis, and the new training school was commissioned on July 1. Commander George K. Stoddard was the Commanding Officer, and Lieutenant Commander John W. Davidson was the Executive Officer.

George K. Stoddard

A 1909 graduate of the U.S. Naval Academy, Stoddard had been promoted to Lieutenant Commander by the end of WWI. During the next two decades, he mainly had assignments as a communications officer and retired with the rank of Commander in the late 1930s. In 1941, when the Bureau of Aeronautics decided to have its own radar maintenance school, Stoddard was called back to duty and assigned as the Officer-in-Charge of the U.S. Naval Detachment at the RAF radar school #31 in Clinton, Ontario. As the Aviation Radio Materiel School was being set up at Annapolis, Stoddard was searching for a permanent, larger facility. He was assigned to Ward Island in February 1942 to oversee the construction, and remained as the Commanding Officer when the facility opened.

Administration Building

There were 10 officers and 40 enlisted men on the initial instructional staff at Ward Island. The first class started on July 1, 1942; this included a number of students who had come with the move from Annapolis. Stock, who had been largely responsible for setting up the original curriculum and coordinating the transition to Ward Island, came to the school as its Superintendent of Training. He soon left, however, to head the Primary School at Utah State and was replaced by Lieutenant R. J. Beaudoin as Director of Instruction. In September, the school was transferred from the Bureau of Aeronautics to the Air Training Command and officially designated the Naval Air Technical Training Center (NATTC) Ward Island.

The curriculum was initially 20 weeks in length, devoted entirely to search radars, recognition (IFF) systems, radar altimeters, and related electronics. The first regular class graduated 106 students in late September 1942. This was followed in October by the second class with 152 graduates. Meanwhile, new students were arriving from the Primary Schools and other sources. Within a short time, a new group of 200 trainees convened every two weeks.

Under Four Flags

In the early period, many of the students were amateur radio operators who had enlisted in the Navy with petty-officer ratings. There were also students from the Marines and the Coast Guard, as well as the Royal Air Force, the Canadian Royal Air Force, and a few personnel from Australia and Brazil. The Marines were from the Division of Aviation, and the Ward Island Detachment was sufficient in number and importance to warrant a Lieutenant Colonel as the Commander (initially Charles C. Bradley). The Marines also received training in hand-to-hand combat and military tactics, arranged not to interfere with the regular electronics curriculum.

A course in airborne radar for supply officers was started in December 1942. This was three months long and covered many of the same topics as in the regular curriculum. The students were primarily graduates of the MIT special program in radio engineering. After a year, this course was discontinued, but was followed by a similar course for WAVE officers that continued until the end of the war. In addition to these, special refresher courses for both enlisted men and officers were occasionally given. Throughout 1943 and 1944, a highly classified instructional program on maintaining the new Magnetic Anomaly Detector (MAD) system was given for officers and enlisted men.

In March 1943, the regular curriculum was enlarged to include aircraft communications and the instructional period increased from 20 to 24 weeks. A year later, with more types of radar added to the curriculum, the period increased to 28 weeks. Also added at this time was a course in radar ordnance – bomb-release and gun-laying systems. In June 1944, the flow of students increased to a new class every week. This also allowed a graduation every week, with the additional advantage of accommodating students who were allowed to repeat certain subjects.

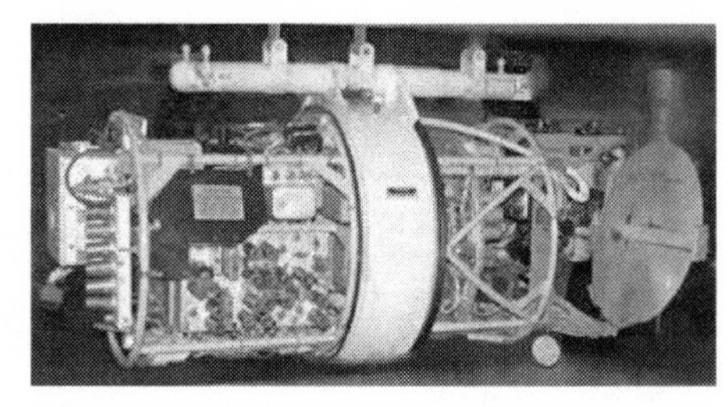

Airborne Radar Systems – Small and Large

Classified electronics, particularly microwave radar systems were covered in the classroom and laboratory courses, all given in the secure compound. Students did not have access to the training compound after hours, and could not take out their training notes, but evening study was expected. Each barrack had a study area, and this was usually filled at night. The library had an excellent collection of technical books and periodicals, but many of the students also had their own.

NATTC Training Planes

Aircraft were used for flying-classroom training in radar starting in 1943. Initially there were two JRF-4 planes, flying out of the nearby NASCC. Before the end of 1944, the Ward Island "air wing" included 9 pilots, 30 mechanics, and several aircraft of various types. The last two weeks of the instruction were taken in combined flight training and work on equipment in aircraft kept in the secure compound on the island.

In addition to the training activities, the instructional staff engaged in special projects in the laboratories – mainly experimentation and testing new types of radar and communications equipment. Certain students were sometimes involved in this, as well as in examining equipment that was not included in the regular training. Specialists from industries and other government operations such as the NRL were often brought in for special lectures to keep the staff up to date on recent developments and forthcoming equipment.

When the school first opened, the greater portion of the base remained with underbrush where personnel sometimes carried sticks as protection against rattlesnakes. Streets were unpaved, and there was a sea of mud following the often-occurring rains. Within a year, Ward Island was primarily cleared of mesquite bush, grass was planted, streets and sidewalks were paved, and the facilities had increased to 77 buildings, including 17 barracks, 6 mess halls, and 16 school buildings.

Leaving the Secure Compound

The eventual facilities in 87 main buildings were highly self-sufficient. On the northern end of the island, was the secure compound containing all classroom and laboratory buildings and an aircraft hangar (but no runway). On the southern end were barracks for students, instructors, support personnel, WAVES, and the Marine Detachment; the Marines included guards as well as students. There were also several Bachelor Officer Quarters. The central area had the headquarters building, a very complete ship's store, the dispensary with 34 beds and dentistry offices, a chapel seating 350 persons, the library holding 4,000 volumes, a large auditorium, and a visitor's reception center. Since the island was only a few feet above sea level, the buildings were very well built with many constructed on pilings; consequently, a hurricane on August 25-27, 1945, with sustained 125 mph winds, did not do major damage.

Recreational and physical training facilities included a gymnasium, two swimming pools, tennis and basketball courts, and baseball and football fields that doubled as parade grounds. A cross-country running track went around most of the island. There was regular bus service to the Naval Air Station and downtown Corpus Christi. After exams on Saturday mornings, weekend leave was customary. Other than two excellent USO clubs, Corpus Christi itself did not have much to offer, but Houston, San Antonio, and Mexico were only hours away. Hitchhiking was a common means of transportation – a sailor never had to wait for more than a few minutes for a ride from the friendly Texans.

With warm weather almost year long, weekend stays at cottages on the beach at Port Aransas were popular, followed by the sickbay being filled with sunburned patients. Big-name entertainment in the auditorium on the base, however, often kept the trainees at Ward Island. Weekend baseball and football games on the Island were also popular, particularly those pitting teams of enlisted men against those of officers.

Ward Island had an excellent band – it has sometimes been observed that good mathematicians are often musically inclined – and led the trainees, assembled by company, as they marched into the compound each morning. A weekly news bulletin, "Wardial," was written by the trainees and published under the supervision of the Chaplain's Office.

Ward Island ***Band***

The number of students under instruction at Ward Island peaked at about 3,100 during 1944. In September of that year, Stoddard returned to retirement, being relieved by Commander Proctor A. Sugg. After the end of the war, Ward Island continued the Aviation Secondary School for some time, adding a Primary School in mid-1946. This two-part training activity continued until October 1947, at which time the full operation was transferred to NATTC Memphis.

An article in the March 2000 *Corpus Christi Caller-Times* stated that more than 20,000 electronics technicians had been trained at the facility. A document from the Aviation History Branch of the Naval Historical Center, however, placed this number at about 10,000. This smaller number is possibly only those Navy personnel earning the ART / AETM rating in the regular Secondary School.

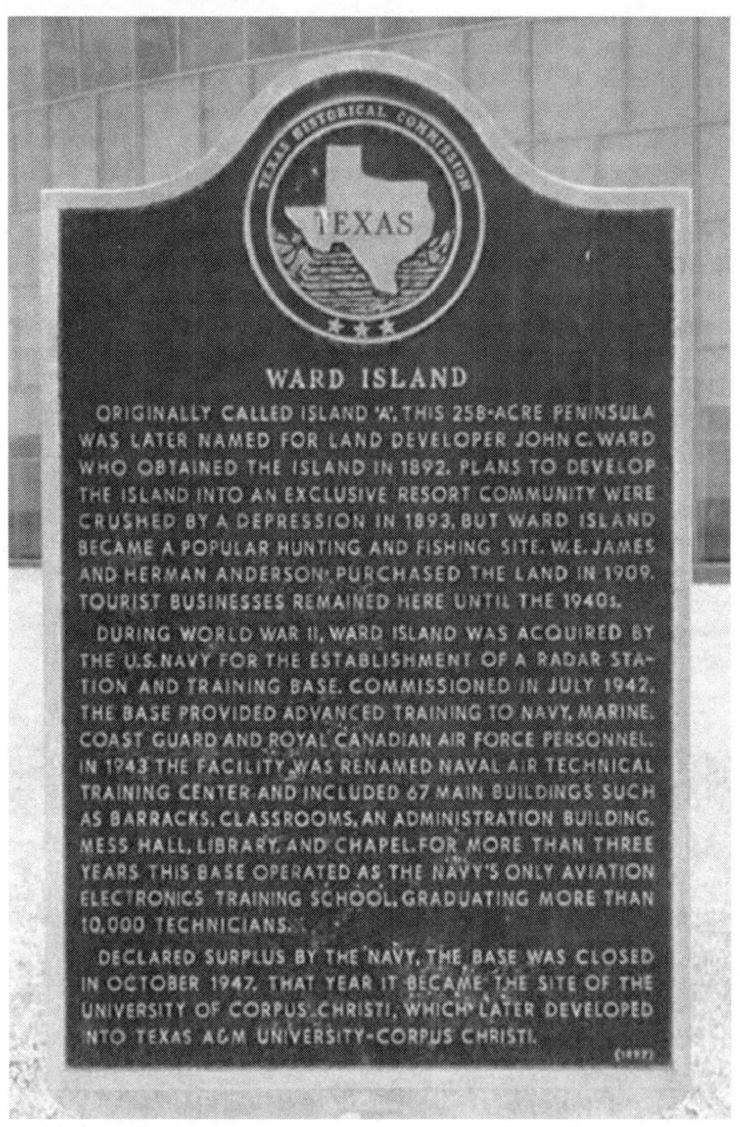

Navy Pier (Chicago, Illinois)

Navy Pier is located on Lake Michigan at the mouth of the Chicago River, just over a mile northeast of downtown Chicago. It opened in 1916 as Municipal Pier to serve as a shipping and recreational facility. Resting on a foundation of over 20,000 wooden pilings and built-up land, this tradition of Chicago's public works had twin, two-story, shed-like buildings extending 3,300 feet into the lake, with two prominent towers on large buildings at each end. The width of the extended structure was 440 feet, with the large open area between the long sheds paved with wooden blocks. On the east, the Pier expanded into the Auditorium, also known as the Hall, with a Grand Ballroom 138- by 150-feet and a 100-foot, half-domed ceiling. Entrance to the Pier was through the large three-story Terminal Building on the west end. This building included vehicle entrances to the center court area and, later, a vehicle ramp to the second floor.

Early Navy Pier

During World War I, the Pier housed Army and Navy units as well as the Red Cross, Home Defense, and a jail for draft dodgers. One of the towers on the east end served as a carrier-pigeon station. The two decades between the World Wars were called the Pier's "Golden Years." It served as a major port for steamship traffic on the Great Lakes. The Grand Avenue streetcar entered the ramp to the second level and ran the length of the facility, delivering passengers to their particular terminals. The Pier was also a center for family entertainment, theater performances, cultural events, and industrial expositions. A radio station, WCFL, operated from the north tower of the Auditorium, and, for a period in the late 1920s, Chicago's first experimental television station (mechanical scanning) operated from the WCFL studio. The facility was renamed Navy Pier in 1927, honoring naval personnel from Chicago who served in World War I.

In August 1941, Navy Pier was closed to the public and, in a five-month period, was converted to a major Naval Training School. A large hangar-type building and a Drill Hall were constructed on 20 acres just west of the Pier. The facilities were intended to accommodate up to 10,000 service personnel. Six days before Pearl Harbor, classes began for aviation machinist mates, metalsmiths, and diesel engine mechanics.

Navy Pier During WWII – Entrance View, Drill Hall on Lower Right

The USS *Sable* and the USS *Wolverine,* two coal-fired, side-paddlewheel excursion steamers converted to training aircraft carriers, operated from a dock adjacent to the Pier. These were used for practice landings by Navy and British pilots from the Naval Air Station at Grosse Ile, Michigan. Those receiving this training included the youngest pilot in naval history – the future President George H. W. Bush. It is said that many World War II planes remain at the bottom of Lake Michigan, casualties of training accidents.

In December 1943, as the need for electronics technicians continued to increase, BuPers directed that another Secondary School of the Electronics Training Program needed to be established. The physical security and availability of support functions made Navy Pier a logical choice for this activity, and during the early months of 1944, a large portion of the buildings were converted for this purpose.

Administration occupied the upper floors of the Terminal Building. In the long buildings, the first floor was refurbished as closely controlled classrooms and laboratories with Marine guards. Study rooms, a library, and chaplains' offices were also on this floor. The barracks were set up on the second floor in four large former exhibition halls, each serving 1,500 trainees when full. Bunks were three-high but each person had an adjacent locker. A small lounge area gave some separation between companies. The mess hall was also on the second level, just back of the auditorium. In addition to the tough school, the men had the challenge of dealing with birds; sparrows often roosted in the living quarters and had little respect for Navy cleanliness.

To initiate the program at Navy Pier, a number of faculty members and officers were sent there from Treasure Island. In addition, Treasure Island built many of the classroom and laboratory teaching aids for the Navy Pier school. After reviewing the ETP in late 1943, BuPers directed that strong instruction in sonar be added to the program. A sonar laboratory with a large tank of water was constructed; to supplement this, Navy vessels tied up at the Pier to provide sonar operation on Lake Michigan.

The Navy Pier Secondary School opened with its first class on June 5, 1944. The size and importance of this school was indicated by the ranks of its initial top officers. Captain Edwin A. Wolleson was the Commanding Officer, Commander C. F. Kottler was the Executive Officer, and the Educational Officer was Commander Charles C. Caveny.

Wolleson was a 1906 graduate of the U.S. Naval Academy and had an excellent career over the next 35 years. He received the Distinguished Service Metal for leading mine-sweeping activities during WWI, had three tours at the Naval Academy, and was an NROTC professor at Northwestern University. Promoted to Captain in 1931, his assignments over the next decade included Commanding Officer of the Great Lakes Naval Training School and of the battleship USS *Tennessee.* Wolleson officially retired in July 1941, but continued on active duty to serve as the Commanding Officer, Naval Training School, Navy Pier. When the Advanced RMS at Navy Pier opened in 1944, he remained to head this activity during its entire existence.

Edwin A. Wolleson

In its beginnings, Secondary School at Navy Pier had two new classes starting weekly, each with 100 to 120 students. This gave a complement of near 6,000 trainees at a time, making it by far the largest school in the ETP. Classes were divided into four groups. With time split about equally between lectures and laboratories, some 100 classrooms were needed. Students came from the college-based Primary Schools, but were mainly from the large, newly opened Navy-operated schools at the College of the Ozarks, Del Monte Hotel, Gulfport, and Great Lakes. Navy Pier also had many who attended directly from the fleet without passing through Primary School.

The activities at Navy Pier were strictly military. In initial operations, the day opened at 6 a.m. with a drum and bugle corps marching down the open area between the sheds. One trainee noted,

"You wouldn't believe the vibrations of that steel-beamed, corrugated metal structure that the blare of bugles and beating of drums could evoke." When groups of trainees moved from one location to another, they always marched. Instruction started at 8 a.m. and continued until 5 p.m., sometimes with physical exercise breaking the afternoon.

Direction Finding Laboratory

Surface-Search Radar Laboratory

Problem Session

Little space was available for homework or private study, so several nights each week students returned to the classrooms for two hours of problem sessions. Exams were on Friday, with reviews on Saturday mornings. Weekly graduation exercises were held in the Auditorium.

Leisure time and liberty for the students at Navy Pier were very good. At the facility, there was always entertainment. Off base, Chicago, with hundreds of thousands of servicemen from nearby installations, had much to offer and was an excellent host. For students with satisfactory grades, liberty was available certain evenings, and everyone had weekend liberty following exam reviews on Saturday mornings. Perhaps unlike the stereotype sailor, many ETP students were attracted to the museums, symphony concerts, and events on university campuses.

In mid-1946, the Navy closed all operations at Navy Pier and the facility was returned to the City of Chicago. In four and one-half years under the Navy, over 60,000 servicemen from the U.S. and Allied nations had trained there in the many types of schools. This included about 15,000 electronics technicians in the two years of the Secondary School of the ETP.

The portion of Navy Pier previously used by the Secondary School was immediately turned into the Chicago Center of the University of Illinois, accommodating the flood of returning veterans. This "campus" was nicknamed "Harvard on the Rocks." After being discharged from the Navy, Caveny remained as Dean of the Undergraduate Division, and Wolleson joined him ad Dean of Students. . In recent years, the Navy Pier has been used as an entertainment center and museum.

OTHER RADAR MAINTENANCE TRAINING

In addition to the Electronics Training Program, the Navy developed radar schools for upgrading capabilities of technicians already serving in the fleet. The main schools were at the Naval Air Station, San Diego, California, and the Naval Operating Base, Norfolk, Virginia. The one in San Diego was designated the Airborne Radar School, Fleet Air West Coast. Two schools were operated at Norfolk: the Radar Materiel School, located on the main base, and Fleet Service School - Radar, located at nearby Virginia Beach. To house the Virginia Beach school, the Navy took over the prestigious Cavalier Hotel from October 1942 to June 1945. There was also a Fleet Service School at Pearl Harbor, and some documents indicate that Strikers showing promise as Radio Technician Mates were sent there for a radar maintenance course a few months in length.

The Army also had great needs in electronics maintenance, actually far greater than the Navy in numbers of personnel. Earlier, as in the Navy, Army operators usually performed the maintenance, but the advent of radar changed this. The Signal Corps had long operated an advanced school for electronics maintenance at Fort Monmouth. With the start of the war, this was expanded to several other locations and specializations.

Unlike the Navy electronics schools, those for Army enlisted personnel were along the lines of traditional military training. While the Army sought very capable students, there was nothing equivalent to the Eddy Test or Pre-Radio School. Instruction in the Army schools was generally short on theory and long on standard routines. The advanced portions gave specialization in particular equipment. Nevertheless, the programs were very adequate for the Army's needs, and their radar systems gave great service throughout the war.

Camp Murphy, built in Southeast Florida, served as the Signal Corps's secret radar training facility and provided radar operating and maintenance training. Similarly, the airport at Boca Raton, Florida, was converted into the Army Air Corps's only radar training station during the war.

As the radar hardware evolved, the need for engineering officers trained in this technology was apparent. In June 1941, MIT established a Radar School for selected Army and Navy officers. The course involved two months of instruction in the basic principles of radar, followed by separate Army and Navy hardware programs of three-months length. For officers needing preparatory studies – mainly those who did not have an engineering degree – the Navy set up such a school at Harvard University. The Harvard Pre-Radar School was five months in duration and covered much of the same instruction as was given in the Pre-Radio and Primary Schools of the Electronics Training Program. Graduates of Pre-Radar went directly into the Navy's three-month hardware program at MIT. Later, a second Pre-Radar school was set up at Bowdoin College, located in Brunswick, Maine.

The instructional notes for the MIT Radar School were compiled as *Principles of Radar*, the first comprehensive book on this subject. This was first printed as a secret document in 1944, then published by McGraw-Hill in 1946.

There were significant programs for training radar technicians in other countries. A major training facility in Great Britain was at the Radio School Yatesbury in Wiltshire, where ground radars such as CH and CHL were taught. The RAF operated Radio School Prestwick in Scotland, providing maintenance instruction for airborne systems such as AI, ASV, and IFF equipment.

His Majesty's Signal School in Portsmouth, later the Admiralty Signal Establishment, gave maintenance training for naval radars. The Army's Air Defense Research and Development Establishment at Christchurch included training for radar mechanics. A number of other schools existed during wartime years, but the secrecy surrounding them led to a loss of their identification.

The critical need for radio mechanics, as they were called in the Commonwealth, was emphasized by the High Commissioner for the United Kingdom when he stated in 1941, "Radio technicians have the highest priority of all requests for manpower from the United Kingdom." In June 1940, Canada agreed to provide RDF (radar) technicians for worldwide assignments, as well as to give training for other Commonwealth nations. The Royal Canadian Coastal Services offered the first RDF course in November 1940 at Halifax, Nova Scotia.

After a group of signal officers were trained in Great Britain, an RDF school that included maintenance was established at Clinton, Ontario, in May 1941, by the Air Council of Canada. In addition to students from Canada and other Commonwealth nations, initial classes included U.S. Army, Navy, and Marine students. For Canadians and others from the Commonwealth, initial training was in 10- to 12-week preparatory courses provided by Universities and Technical Schools across Canada. Those who passed a tough exam completed their training at the highly secret #31 RAF school in Clinton. There eight weeks were devoted to ground radars and six to airborne radars.

Concerning the secrecy, U.S. Navy Radioman Eugene H. Fellers, who was sent to Clinton for maintenance training, told an interesting story:

> The girls in town were friendly and would tease us a bit, asking, 'What are you doing here?' We would reply, 'We're studying radio.' They would laugh and say, 'You mean radar, don't you?' We thought that no one outside the school was supposed to know that word! They would then say, 'Don't worry; we know the name radar and what it does.' And that was that.

During the war, about 6,500 Canadians and 2,300 Americans went through courses at Clinton. At its high point in early 1945, the #31 School had a staff of 478. The airborne radar training program set up by the U.S. Navy at Annapolis in late 1941 was patterned on this school.

In 1942, the Canadian Army set up RDF maintenance schools at Debert, Nova Scotia, and at Camp Barriefield, near Kingston, Ontario. The Barriefield school was enlarged in 1944 to become the Royal Canadian Electrical and Mechanical Engineers (RCEME) training center.

The Royal Canadian Navy's Signal School at St. Hyacinthe, Quebec, was the central location for training Canada's shipboard and shore-base communications personnel during World War II. Moved from an original location in Halifax in the summer of 1941, the St. Hyacinthe school had 73 buildings spread over a 25-acre site. At a given time, there were about 3,200 officers, enlisted men, and Wrens (Women's Royal Canadian Naval Service), involved in all phases of training in communications: visual signalmen, wireless telegraphers, communications coders, RDF (radar) operators, and radio artificers (mechanics or technicians).

The course at St. Hyacinthe for Radio Artificers was 42 weeks in length. The first 20 weeks were devoted to basic training, with studies theory of electricity, mathematics, and shop skills. During the second half of the course, they applied this basic knowledge to wireless and RDF equipment. On graduation, they were qualified as Radio Artificers Fifth Class, just below the rating of a Petty Officer. Officers taking RDF courses at St. Hyacinthe were all university graduates in electrical engineering, physics, or similar fields. Their course, lasting several months, entitled them to use the letter R after their rank on graduation

In Australia, even before their first indigenous radar was built, training programs were begun in September 1941. These provided 6-month courses for signal officers at Sidney University and for maintenance mechanics at Melbourne Technical College. As previously noted, personnel from Australia were also trained at the RAF radar school in Canada.

For comparative purposes, radar maintenance in Germany will be mentioned. Funkmess (radar) maintenance training was conducted by the three services of the Wehrmacht: Army (Heer), Navy (Kriegsmarine), and Air Force (Luftwaffe). Entry of personnel into the higher levels of training was through examinations. As in America, radio amateurs were particularly recruited.

German radar systems were highly modularized, and field repair was mainly by replacement of modules, which, in turn, were sent off for maintenance in centralized arsenals or at factories. Most of the personnel trained by the Wehrmacht worked in these facilities and, therefore, were likely specialized on particular equipment.

In the mid-1930s, the German Reichsarbeitdienst (RAD, or German National Work Service) was established. As the war began, the RAD trained young men in providing needed services prior to their entry in the Wehrmacht. Included were many, particularly radio amateurs and hobbyists, who serviced anti-aircraft (Flak) batteries under control of the Luftwaffe. These personnel were known as Luftwaffe-Flakhelfer.

Later in the war, many servicemen in non-combat roles were replaced by RAD personnel. Starting in 1943, Flakhelfer were allowed to stand the examinations for entry into extensive radar training. Subsequently, they too increasingly took part in more militarized roles. Many RAD units were encircled and forced into front-line combat, while other units were drafted directly into military service on the spot.

An interesting twist of fate is found in German radar maintenance. As the tide of war turned and technicians from the military were increasingly taken out of maintenance facilities and sent to the battlefields, the need for personnel to make critical repairs became great. Some Jewish engineers and scientists in concentration camps escaped possible death by serving this function.

Epilog

The Ryukyus Islands, long considered a part of Japan, were the last obstacle to the invasion of the Japanese mainland. Between March 26 and July 2, 1945, the campaign to capture these islands, particularly Okinawa, was the most fiercely fought of the Pacific war. America's concentration of naval, air, and ground fire was unmatched in warfare history, with over 97,800 tons of ammunition expended in this campaign. America suffered 49,151 casualties including 12,520 killed or missing, and 36,361 wounded. In the air, 763 U.S. planes were shot down. The Navy lost 36 ships sunk and another 368 damaged, with about 4,900 men killed. There were some 2,000 kamikaze attacks. On the other side, it is estimated that 110,000 Japanese died in battle, 7,800 planes were downed, 10 ships sunk, and 150,000 Okinawans – one-third of the native population – perished.

This tragic event is noted here primarily because it was the precursor to the dropping of atomic bombs a month later on Hiroshima and Nagasaki. This ended the war and stopped the planned invasion of Japan's homeland with potential American casualties projected at many times the Ryukyus losses. The Japanese death toll from the two bombings was approximately 214,000, significantly less than their total military and civilian losses in Ryukyus.

Upon cessation of hostilities, America started its return to "normal." But this was not the America prior to December 7, 1941, and was certainly not the same as when she led the world in escaping from the Great Depression a few years earlier. For many, many reasons, America would never again be the same.

America emerged from the war as militarily and economically the strongest nation in the world. As is well known, this strength was not derived just from our natural resources, which are many, or from our great monetary institutions, which were unequalled in history, but came from the work ethic, practical intellect, and achievement drive of our people. As shown in the preceding chapters, the electronic training program of the Navy carefully selected and nurtured people with these characteristics. Thus it should be expected that, although they were relatively small in numbers, graduates of this program would contribute much in Post-War America.

At the start of the war, the Navy's Electronics Training Program (ETP) rapidly arose as an unparalleled development in military traning endeavors. Having served well in its primary purpose, the program, like other similar wartime activities such as the Rad Lab, quickly phased down. Before the war, maintenance training was given by the prestigious Radio Materiel School (RMS) at the Naval Research Laboratory. Sadly, as the ETP changed to a peacetime operation, the RMS was closed.

This brief Epilog provides information on the accomplishments of the ETP and the fate of all of the leaders and schools. Of more significance, however, some of the contributions to post-war America of both the program and its graduates are described.

THE ACCOMPLISHMENTS

There are no records as to the number of people standing the Eddy Test during WWII, but this could have been 500,000 or more. In an interview, Captain Eddy made reference to about 86,000 persons receiving training; it is assumed that this is the number entering the program starting in January 1942, mainly after passing the Eddy test; many of these, however, did not make it through Pre-Radio or Primary School. Eddy's Legion of Merit citation indicates that through 1945 some 30,000 technicians completed the program. Estimates and cited numbers for the Secondary Schools indicate numbers between 40,000 and 50,000, but this would include trainees who entered Secondary School from the fleet and those involved in special and refresher courses.

Commodore Robert W. Cary, holder of the Congressional Medal of Honor and skipper of vessels involved in European invasions, gave great praise to the program's graduates. He noted in *The Naval History of Treasure Island* that the ETP involved the top three to five percent of the Navy's personnel and that more than 80 percent of the graduates received ratings of "excellent" on performance evaluations from their future commanding officers. Cary also stated that about 25 percent of the graduates during the war were promoted to become Warrant or Commissioned Officers.

Upon graduation, the technicians were posted to naval vessels and bases throughout the world. Many later reported that the electronic equipment that they found, although relatively new, was often inoperative. Thanks, however, to the RTs / ETMs and ARTs / AETMs, the Navy's electronic equipment *did* usually perform well. Radar made a decisive difference – perhaps *the* difference – in the outcome of the war.

THE SCHOOLS

Within a few weeks after the victory in Europe on May 8, 1945 (V-E Day), a number of the Primary Schools in the ETP were closed. This included the remaining college-based schools as well as the Navy-operated school at the College of the Ozarks. Following Japan's surrender on September 2, 1945 (V-J Day), the program started a transition to a peacetime, regular-Navy status. Radio Chicago, including the central operation at 190 North State Street and the remaining Pre-Radio Schools, shut down by the end of the year. These activities were added to the Great Lakes training operation.

During 1946, the Gulfport Primary School closed in March, followed by the Navy Pier Secondary School at mid-year. At the end of the year, the original RMS at the Naval Research Laboratory shut down after 22 years of operation. Great Lakes took up all of these activities, encompassing Pre-Radio, Primary, and Secondary Schools. The last stand-alone Primary School, Del Monte, ceased operations in June 1947. The Secondary School at Treasure Island continued, being fed by the Primary School at Great Lakes. The schools at both Treasure Island and Great Lakes centered on training for sea- and land-based equipment.

NATTC Ward Island added its own Primary School in mid-1946; then, in October 1947, the activities at Ward Island were transferred to NATTC Memphis at Millington, Tennessee, to become the Navy's training center for aviation electronics equipment. All of the Navy's electronics training activities were reorganized in mid-1946, but the Eddy Test continued as the admission criteria for several years.

The ETP set the pattern for a number of highly technical training activities provided for enlisted men in the Navy today – programs showing that through highly selective admission and exceptional performance requirements, outstanding results can be obtained.

THE LEADERS

William Crawford (Bill) Eddy (1902 – 1989) again retired from the Navy in December 1945. He was awarded the Legion of Merit as recognition for his contributions,. The citation includes the following:

> For exceptional meritorious conduct in the performance of outstanding services to the Government of the United States as Commanding Officer of Naval Training Schools (Radio Chicago). . . . Captain Eddy pioneered in nationwide recruiting and developing selection tests which greatly facilitated the procurement of over 30,000 men for radio technician training. . . . By his keen foresight, unwavering zeal, and meticulous attention to detail in discharging all details pertaining to the recruiting, selection, and training of radio technicians, Captain Eddy contributed materially to the successful prosecution of the war against the enemy.

He was also offered the rank of Rear Admiral in recognition of his service, but declined this, preferring to be known as "Captain Eddy."

For the next two years, Eddy led WBKB-TV, converted from W9XBK in 1945 by him while "moonlighting." This was Chicago's first commercial television station and the third oldest in the nation. His accomplishments at WBKB included initiation of the long-running puppet show, *Kukla, Fran, and Ollie.* Eddy authored *Television: The Eyes of Tomorrow* (Prentice Hall, 1945), a book that defined television broadcasting for the years to come.

In 1947, Eddy formed Television Associates, headquartered in Michigan City, Indiana. Arch Brolly, who had been the chief engineer at WBKB and led the instructors at 190 North State Street, joined Television Associates to head their engineering efforts. Among other activities, this firm performed geographical surveys using radar in low-flying aircraft, handling projects such as mapping for a 3,000-mile communications network through Turkey, Iran, Iraq, and Pakistan. An inventive genius, Eddy led his firm in many electronic developments and was awarded a large number of patents.

Eddy also continued as an accomplished cartoonist and artist. For 33 years, starting in the 1930s, he designed annual calendars for Honeywell's Brown Instruments. In later years, he won awards for his paintings, sculptures, and stained glass creations.

For his long service at the NRL-Bellevue Radio Materiel School, **Nelson Mangor Cooke** (1903 – 1965) received a Letter of Commendation from the Secretary of the Navy. When the school was moved from Bellevue to Great Lakes at the beginning of 1947, Cooke was assigned to Germany, where he received the Medal for Humane Action for his role as Technical Officer in the Berlin Airlift. In 1949, he joined the Bureau of Ships in Washington, and then retired from the Navy two years later.

In 1951, Cooke established Cooke Engineering Company in Alexandria, Virginia. Here he led a small group of researchers in diverse analytical areas such as factors affecting naval shore receiving stations and effects of radiation from buried telephone cables. His company also designed and built microtitration devices (microplates) for biological analysis.

As early as 1943, Cooke began a relationship with Allied Radio, editing their *Radio Data Handbook,* and continued this through 16 editions. He coauthored the *Electronics Dictionary* (McGraw-Hill, 1945), recognized for many years as the definitive source in this field. His outstanding, wartime book, *Mathematics for Electricians and Radiomen,* was rewritten as *Basic Mathematics for Electronics* (McGraw-Hill, 1953); this became a best-seller textbook and went through seven editions.

At the end of the war with Germany in May 1945, **Sidney Richard Stock** (1895 – 1969) was briefly assigned to the Naval Research Laboratory, and then was sent to the Army's Signal Corps Laboratory at Fort Monmouth, New Jersey, as a Navy liaison officer. In this capacity, he assisted in the standardization of electronics equipment for different U.S. military services.

Stock was discharged from the Navy in 1946, but remained at the Signal Corps Laboratory as a civil service employee. In a new career, he served as Chief of the SCL Power Branch, then later became Director of Employee Welfare and also served as an electronics consultant. Stock finally retired in 1964 and returned to Utah.

After being transferred from the NRL-Bellevue RMS in 1943, **Wallace Joseph Miller** (1903 – 1951) served the next two years commanding vessels involved in nine engagements in the Pacific. For his actions, he was awarded the Silver Star, four Bronze Stars, and the Legion of Merit. After the war, he spent 18 months at the Naval Research Laboratory and retired in 1947, at which time he was advanced to the rank of Rear Admiral on the basis of combat awards. Miller joined the staff of the Engineering Experiment Station (later the Georgia Tech Research Institute or GTRI) at the Georgia Institute of Technology, leading research in microwave transmission.

Louis Emil Denfeld (1893 – 1972), who in January 1942 authorized the initiation of the Electronics Training Program, was promoted to Rear Admiral and served as the Assistant Chief of the Bureau of Naval Personnel. Here he closely followed the program and wrote a paper, "Research Activities of the Bureau of Naval Personnel," published in the March 1944 issue of the *Journal of Applied Physics*. In early 1945, he was given command of Battleship Division Nine and served with distinction in the South Pacific.

Following the war, Denfeld was promoted to Vice Admiral and head of the Bureau of Naval Personnel, responsible for demobilization. Elevated to Admiral in 1947, he first served as the Commander of the Pacific Fleet, and then was appointed Chief of Naval Operations. Two years later, he was a key participant in what was known as the "revolt of the admirals" – an open disagreement concerning the reduced position of the Navy and the primary use of the Air Force and nuclear weapons for deterring future war. For this, he was forced to retire in January 1950. For the next 21 years, Denfeld served as an advisor to the Sun Oil Company from his home in Westborough, Massachusetts.

THE IMPACT

Much has been written about the impact of World War II electronics on the post-war world. Radio, radar, sonar, and other "electronic marvels" gave a great impetus to the post-war explosion of the electronics industry. In addition to the new products, massive amounts of surplus military equipment were converted for personal and industrial applications. The most significant impact, however, was likely in the large number of men who, having received electronics training in the military, became a new generation of technicians, engineers, and scientists.

In addition to those veterans who immediately returned to, or started in, work in electronics, millions benefited from the G.I. Bill in receiving additional training and education in this field. In 1944, anticipating the post-war recovery period, President Franklin Roosevelt signed into law the G.I. Bill of Rights, rewarding veterans for their services to the Nation. During 1947, the peak year of its use, almost 50 percent of all college students in America were attending under this funding. Eventually, 7.8 million veterans – slightly over half of those eligible – received education and training by this means, with an incalculable impact on the Nation's future.

Electronic activities during WWII had a major influence on post-war electrical engineering education. Many critics of pre-war engineering curricula said that the graduates were not adequately prepared for handling the wartime needs, noting that the significant developments, such as radar, came from physicists, not engineers. For the most part, this was an unfair criticism; the earlier programs were mainly for persons who would be designers, builders, and operators of electrical and radio equipment, not researchers in laboratories. The criticism, however, led to the undergraduate electrical engineering curricula being drastically changed, with a much greater emphasis on theory and analytical techniques.

The former electronic technicians from the military were also influential in these educational changes; having already received excellent practical training, they now demanded something more. Also, large numbers had entered the military with some level of higher education, and the G.I. Bill not only funded the completion of this but also continued into graduate study, resulting in a huge post-war increase in master's and doctoral degrees.

THE GRADUATES

Brief biographical information on representative graduates of the Electronics Training Program is given in Appendix III, showing the great diversity of persons entering the program, their activities while in the Navy, and their post-Navy careers.

There is no way to estimate the post-war contributions of ETP graduates. Some made a career in the Navy, often rising to Chief Petty Officer or Warrant Officer, or gaining entrance to Officer Candidate School. The Nationalist Chinese Air Force recruited AETM graduates from NATTC Ward Island, offering them very high pay and commissioning as Captains.

Without further training, Navy ETMs and AETMs were in great demand in the emerging television and microwave communications fields. John (Pat) Gallagher, himself a graduate of 190 North State Street and Treasure Island, noted the following in his article "Where Did the TV Engineers Come From?" on the website of The Order of The Iron Test Pattern:

> As I traveled around the country as a DuMont field engineer building TV stations, I began to notice that the 'new man who has been hired to be their TV engineer' was usually an ex-Navy Radar Technician.

Thousands used the G.I. Bill to earn degrees in engineering, physics, and related fields, receiving up to a year of academic credit for the Navy training. Typical of these is John W. Warner, U.S. Senator and former Secretary of the Navy, who was quoted in an IEEE *Spectrum* article as saying:

> Upon entering college [Washington & Lee] in the fall of 1946, I was accorded immediate credit for the work done in this program and was able to complete engineering school in three years rather than four.

In the same article, Warner gave great credit to the program for his personal accomplishments:

> What success I have had in life is in a large measure due to the training and discipline provided by the U.S. Navy during my formative ages of 17, 18, and 19.

Through special examinations, some technicians who had not completed undergraduate degrees were admitted directly to graduate school. Arthur D. Code earned his doctorate in astronomy and headed a major observatory without ever receiving a bachelor's degree. He also made major contributions to space electronics, attributing his basic capability for this to the Navy.

Although there is no actual data concerning the number of former RTs / ETMs and ARTs / AETMs who obtained post-war education, sampling in preparing this book indicates that this was possibly as high as 90 percent, including those who studied through some of the excellent correspondence and industry-based schools. While most pursued programs in engineering, many studied physics or mathematics, and others prepared for non-technical careers. In the Navy, Eugene H. Fellers became a Chief Electronic Technician, then studied liberal arts and taught Spanish and German, never returning to electronics.

As engineers and scientists, they filled positions with the Department of Defense and, later, NASA, or joined the rapidly growing electronics industry. New firms started by these professionals abounded. As electronic computers came into being, many contributed to establishing this field. Examples include Raymond J. Noorda, known as the "Father of Network Computing," and Douglas C. Engelbart, who is internationally known for his contributions in both software and hardware, including the ubiquitous computer mouse.

The ETP graduates were often found in college and university faculties and research institutes. Some, such as Robert B. Kamm, President and Professor Emeritus of Oklahoma State University (formerly Oklahoma A&M), came to love the college of their Primary School and returned to that institution for an academic career.

In a speech at Annapolis in August 1963, President John F. Kennedy said,

> Any man who may be asked in this century what he did to make his life worthwhile . . . can respond with a good deal of pride and satisfaction, 'I served in the United States Navy.'

A relatively small number of us can add to this and proudly include on our resumes, "Passed the Eddy Test and completed the Navy's World War II Electronics Training Program."

Appendix I

ELECTRONIC SYSTEMS OF WORLD WAR II

In the period from the start of serious war preparation through the end of hostilities, the electronics industry of America was essentially devoted to producing defense items, from small hand-held radios to huge ship-borne radars. In this five-year period, there was not only an enormous numbers of units produced but also the greatest advancements in electronics technology since the field opened some 50 years earlier. It was in anticipation of these quantities and technological advancements that the Navy developed the Electronics Training Program in early 1942, and it was in this unfolding environment that the trained specialists contributed to the ultimate winning of the conflict. This Appendix provides an overview of some of the most critical electronic technologies that the technicians faced upon graduation.

The primary equipment was radar, communications, and sonar. Official documents list these as follows: more than 25 separate and distinct models of radar equipment, including IFF units; 80 models of communications transmitters and 50 models of receivers; and 30 models of sonar or echo-ranging equipment. In addition, there were various models of radio direction finders, beacons, and radio navigation sets, as well as all types of auxiliary equipment such as power systems and rotating devices, including motors, synchros, and servomechanisms. Some technicians would also encounter optical and magnetic-anomaly detectors, radio-controlled aircraft, bombsights, aircraft autopilots, and even proximity fuses. Listings show more than 100 different types of standard and specialized testing equipment used in the maintenance of this equipment.

RADAR

After the introduction of the resonant-cavity magnetron in late 1940, essentially all radar development in the United States was in the microwave class. This section provides a brief discussion of the major systems (there were many more). The Radiation Laboratory (Rad Lab) at MIT had the primary responsibility for such systems, but there were also significant microwave developments, particularly for the Navy, at the Bell Telephone Laboratories (BTL) facility at Whippany, New Jersey.

The very rapid development of high-power magnetrons produced an astonishing number of these tubes. In addition to those made in Great Britain, it is estimated that by the end of the war well over a million magnetrons had been manufactured by at least six companies in the United States and one in Canada.

In Great Britain, a number of government facilities and universities were involved with microwave development, but the central point of this activity was the Telecommunications Research Establishment (TRE) at Malvern College some 100 miles north of London, where it had moved in mid-1942 for safety reasons. At that time, the TRE had a staff of about 2,000 persons. To directly coordinate developments, the Rad Lab set up a branch operation at Malvern in September 1943.

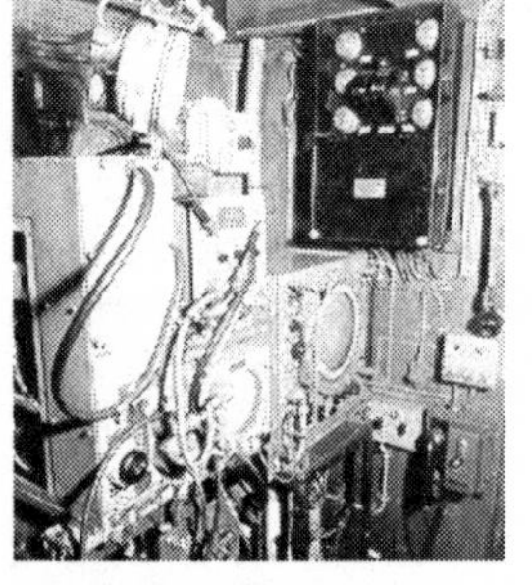

Type 291 Station

There was also some continued development in Great Britain on non-microwave radars. In 1942, the ASE introduced the Type 291, their last 200-MHz air-search system. Primarily intended for small ships, it could detect a bomber at 15 miles. By 1944, the Type 291 was installed on nearly all British destroyers and smaller escort vessels. A variant, the Type 291W, was designed for submarines.

The BTL's improvement of the original 10-cm resonant-cavity magnetron from Great Britain was previously noted, as was their development of a 750-MHz (40-cm) device and the incorporation of this in the Navy's first microwave fire-control radars: the FC (Mark 3) and the FD (Mark 4). The Mark 3 was later replace by the very similar Mark 12. The Mark 4 remained one of the the most used fire-control radars throughout the war.

3′ x 12′ FC Antenna

FD on Gun Director

Magnetron development at the Rad Lab was under Dr. I. I. Rabi. In parallel with the 10-cm developments, Rabi's team started pushing the operating wavelength still shorter, first to 6 cm, then to 3 cm, and then 1 cm. As more frequencies were used, it became common to refer to radar operations in bands. These bands and the corresponding wavelengths and frequencies were as follows:

P-Band – 30 - 100 cm (1 - 0.3 GHz)
L-Band – 15 - 30 cm (2 - 1 GHz)
S-Band – 8 - 15 cm (4 - 2 GHz)
C-Band – 4 - 8 cm (8 - 4 GHz)
X-Band – 2.5 - 4 cm (12 - 8 GHz)
K-Band – Ka: 1.7 - 2.5 cm (18 - 12 GHz) and
Kb: 0.75 - 1.2 cm (40 - 27 GHz).
The K-band was split due to strong absorption by water vapor in the atmosphere.

The designation system for American radars might be mentioned here. The U.S. Navy used two types of ship-board radars: search (Bureau of Ships) and fire-control (Bureau of Ordnance), Initially, the former were type "S" and also carried a letter showing the model number; e.g., the SC was the third model of the search radar. (Not all models reached production, so certain letters are missing.) The Bureau of Ordnance preferred the "Mark" designation, many with a modification shown; e.g., Mark 8, Mod 2 was the second modification of the type Mark 8 fire-control radar. Early fire-control radars were designated "F"; thus, the FH was also Mark 8. The "X" and "CX" designations were for experimental radars. Early aircraft radars (Bureau of Aeronautics) used the same letter-type designation as those of the Bureau of Ships, but with an "A" suffix; e.g., ASE was the fifth model of an airborne search radar.

The Army, including the Army Air Forces, used the Signal Corps Radio (SCR) series for their radars. If a radar or other type of electronic set was approved – but not necessarily used – by both the Army and the Navy, the prefix "AN" or "A/N" was used.

S-Band Systems

Another early microwave radar development at the BTL was the SJ, a 10-cm system primarily for supplementing the SD radar on submarines. The antenna of this system could be used with the vessel at either surface-level or hull-submerged, sweeping the horizon for ships and aircraft. With the antenna at a very low level, the SJ's range was only about 6 miles, but with good accuracy. This system was first tried at sea in mid-1942. Three years later, this was upgraded to the SV, increasing the range to 30 miles.

Radar Antennas on Submarines

The first project at the Rad Lab was the airborne interception (AI) radar. Although a laboratory version was first demonstrated in February 1941, it was not until 1942 that the design was completed and put into limited production by Western Electric as the SCR-520. Another version, designated SCR-517, was developed for an Army ASV radar. The BTL improved the SCR-520, developing the SCR-720 that became the major night-fighter radar for both the Army Air Forces (AAF) and the Navy. The SCR-517 was also adopted by the Royal Air Force (RAF) as the AI Mark X.

SCR-720 Radar

It was previously noted that the AI project was diverted by the Rad Lab to developing the Navy's first 10-cm radars: the SG (surface) and ASG (airborne) search systems. Put into production in mid-1941, they were first deployed about a year later. The small, highly mobile Patrol Torpedo (PT) Boat was introduced in 1942. For this, Raytheon modified the SG to a more compact version, the SO. These 3-GHz (10-cm) sets gave the PT Boats a great advantage, particularly for night operations. The future President John F. Kennedy attained hero status while commanding one of these boats.

PT Boat Electronics Suite Including CO Radar (lower left)

The second Rad Lab project was a 10-cm fire-control (gun-laying) system for the Army. Eventually designated the SCR-584, this was possibly the best-known radar produced by the Laboratory. The transmitter's magnetron generated 210-kW, 0.8-μs pulses. A parabolic dish antenna used helical scanning for air search and conical scanning for precision tracking. Although primarily intended for fire control, it also performed well as an air-search radar. Targets out to 40 miles could be detected.

SCR-584 Control Van

Initially led by Dr. Louis N. Ridenour, then by Dr. Ivan A. Getting, the SCR-584 project was an excellent example of the diversity of capabilities needed to bring a complex radar system into being. While the transmitter and receiver – the heart of the system – were developed at the Rad Lab, the other equally important units were developed elsewhere. The system included automatic tracking, a capability that required a computer as well as a highly complex mechanical mechanism for positioning the antenna.

The computer was an electronic analog device containing 160 vacuum tubes. Called the M-9 Predictor-Corrector Unit, this was developed at the BTL by David Parkinson and Dr. Clarence A. Lowell. With the M-9, the system could automatically track targets to 18 miles. A single M-9 could direct four anti-aircraft guns. Chrysler designed and supplied the highly complicated antenna mount, the parabolic reflector, and the 10-ton semi-trailer. The mount not only provided the helical and conical motion, but also accepted pointing information from the M-9. Radar work at Chrysler Engineering was led by F. W. Stock.

General Electric and Westinghouse shared (60-40) the SCR-584 prime contract. A prototype was tested in May 1942, but due to bureaucratic delays, the systems were not delivered until early 1944. Eventually, about 1,500 of these systems were used in both the European and Pacific war theaters. Together with the proximity fuse, the SCR-584 is crediting with enabling anti-aircraft guns to destroy the majority of German V-1 "buzz bombs" attacking London following the Normandy invasion.

Gun-laying 10-cm radars were also developed in Great Britain and Canada, designated as GL Mark 3B and GL Mark 3C, respectively. Their performance, however, was greatly overshadowed by the SCR-584. Eventually, the British adopted the American system, designating it GL Mark 3A. Later in the war, the Soviets were provided with the SC-584 under Lend-Lease and, after the war, produced as the SON-4.

In a development for the Navy, the Rad Lab modified the SCR-584 to have a gyro-stabilized platform and an antenna suitable for sea duty. The M-9 unit was not needed. Designated the SM and put into limited production by General Electric, the first systems were installed on heavy carriers in late 1943, actually before the SCR-584 went into service. Later, a lighter-weight model, designated SP, was built for smaller carriers.

The ASC was another 10-cm search radar. It used a large (24- to 30-inch) parabolic reflecting dish installed on large Navy patrol aircraft. Positioned by a hand crank into a clear plastic blister above or below the fuselage, the antenna could revolve 360 degrees. The range was limited to

ASC Atop a PB2Y Coronado

line of sight, 30-35 miles at normal operational altitudes or perhaps 50 miles at above 12,000 feet. It was used on PB2Y Coronado patrol planes starting in late 1943.

The most ambitions long-term effort of the Rad Labs was Project Cadillac, the first air-borne early-warning radar system. Initiated by an Inter-Service Committee in mid-1942, it was intended to provide the Navy a surveillance capability unequaled by land- or sea-based systems. With Jerome B. Weisner as the project leader, eventually about 20 percent of the Rad Labs' staff would be involved. The basic concept was to increase the range of surveillance radars by having them airborne.

Avenger with AN/APS-20

A 20-cm, 1-MW radar, the AN/APS-20, was developed to be fitted under a modified ATM Avanger aircraft. This was the only carrier-based aircraft capable of carrying the 8-foot radome and 2,300 pounds of associated equipment. The resulting system could detect bomber-sized aircraft at ranges up to 100 miles. To make the radar findings more useful, the system included a television pickup of the radar's PPI display and a VHF communications link to convey the images up to 100 miles to the Combat Information Center (CIC) on the home carrier.

The system was first flown in August 1944. Problems with signal interference delayed introduction into service until March 1945. As this activity was underway, the Navy was considering a more sophisticated method for airborne surveillance: placing the combat information operation aboard the aircraft. In mid-1944, the Rad Lab started Cadillac II with this objective. The PB-1 aircraft (the Navy's designation for the B-17) was selected. The AN/APS-20 would continue as the radar, but a CIC with three stations having 12-inch imaging tubes was built into the original bomb bay. In using the large PB-1s, the system would need to be land based. While much progress was made, the war ended before the system could be demonstrated. However, Project Cadillac was the foundation from which the later AWACS concept evolved.

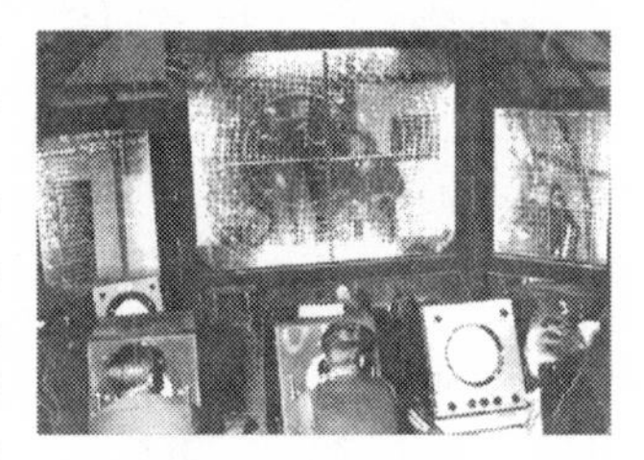

Shipboard Combat Information Center

Here it might be noted that as the war progressed, the CIC on aircraft carriers became the electronic nerve center of the vessel. With an operating staff of up to 50 persons, all of the radar information was displayed on large map-tables or hand-drawn on vertical transparent sheets.

In Great Britain, as their first 10-cm microwave radar, Type 271, was taken over by the Royal Navy, the TRE turned to developing a microwave system to guide bombers at night or on overcast days in their runs deep into German territory, beyond the 300-mile range of the existing radio navigation system. The system was designated the H2S, and Dr. A. C. Bernard Lovell led the development. Operating at 9 cm, the magnetron-based radar had its antenna directed such that the PPI display gave a map-like presentation of the forward ground, including details of the cities. The H2S was tested during mid- 1942

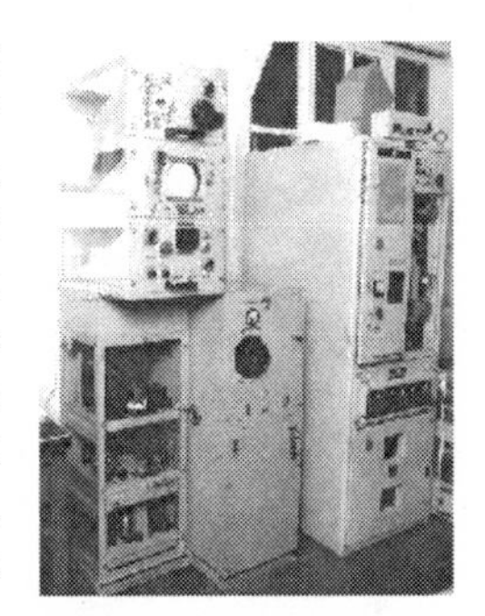

Type 271 Shipboard

There was major debate over using the new resonant-cavity magnetron in flights over Germany, fearing that it might be discovered in a downed aircraft. Prime Minister Churchill personally made the decision to proceed. The first bombing mission using the H2S took place on January 30, 1943. Three days later, a bomber went down near Rotterdam, and the Germans recovered a partial system, including the magnetron. The H2S radar was eventually reconstructed and demonstrated to Hitler. The screen display showed a detailed street map of Berlin. It was rumored that upon seeing this, Hitler considered committing suicide.

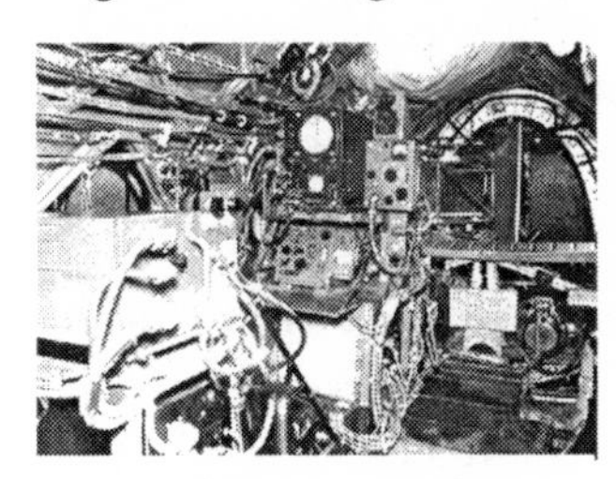

H2S Radar on Bomber

Oboe was another British navigation system to guide bombers. Developed about the same time as H2S, this used two widely spaced, ground-based radars, designated "Cat" and "Mouse," and a transponder on the aircraft. Coded signals from Cat would be used in the planes to navigate in an arc directly over the target. Mouse determined aircraft's ground speed and gave a signal when the drop point was reached. This system was conceived and developed by Alec H. Reeves at the TRE. The initial Oboe system used 100-MHz radars, but these were shortly changed to 3 GHz (10-cm). It became operational at the end of 1942. Eventual comparisons with H2S and H2X showed Oboe-equipped bombers to have significantly better accuracy.

Other C-band radars developed in Great Britain during the war included the Royal Navy's Type 173 with gyro-stabilized mounting, the

CD Mark IV or Chain Home Extra Low (CHEL), the AIMS Type 13 height-finding set, and the previously mentioned GL Mark 3B.

X-Band Systems

As the 3-cm magnetron came into being, Dr. Lois W. Alvarez, previously a physics professor at the University of California, Berkeley, used his expertise in physical optics to develop a new microwave antenna for shaping the radiation. Formed as either a "leaky" waveguide or a series of dipoles, this linear array produced a very narrow, fan-like beam that could be electronically swept.

The Alvarez antenna was incorporated in three new types of X-band systems. The first was an airborne radar in two configurations. For the Army Air Forces, the "Eagle" – later designated AN/APQ-7 – had a beam that electronically swept 30 degrees on each side of the aircraft heading. Intended for strategic bombing, the Eagle provided a map-like image of the ground about 170 miles along the forward path of the aircraft. With Dr. E. A. Luebke as the project leader, a prototype was tested in mid-1943. Put into production by Western Electric, the first deliveries were in late 1944. About 1,600 Eagle sets were built. Although too late for the bombing of Germany, it was used on B-29s against Japan.

The same technology was also used in developing the ASD, a search and homing radar for use by the Navy primarily in single-engine bombers. Eventually designated the AN/APS-3 and also known as "Dog," this system swept 150 degrees in the search mode, then switched to 30 degrees for target homing. With final design by the NRL and Sperry, it was put into production by Philco in late 1943. For carrier-based night fighters, ASD was converted to a lighter system, the ASH (AN/APS-4). This system, however, required a second crew member, and thus saw limited use. It was eventually replaced by the AIA-1 (AN/APS-6), suitable for a single-seat fighter.

AN/APS-4 on P-38M

The second application of the Alvarez antenna was the Ground Control Approach (GCA) system for military airports. This blind-landing system used a 10-cm radar with a PPI for close-in tracking and two 3-cm sets for fine azimuth and elevation positioning. The X-band radars used linear arrays of dipoles. An electro-mechanical analog computer determined the aircraft's approach path, and final guidance was by a ground controller using voice radio.

Lawrence H. Johnson, a former student of Alvarez, was the GCA project leader and personally developed the analog computer. Much of the final system design was under contract with Gilfillan Brothers. Carried in several vans, it was first field-tested in 1942. Gilfillan Brothers also built about 15 units, designated the AN/MPN-1, which were successfully used by the AAF in assisting bombers returning from Germany.

GCA *Operator*

The Microwave Early Warning (MEW) was the third X-band system using the Alvarez antenna. The system used two back-to-back rotating antennas, both 25-foot wide parabolic "billboards," one 8-feet high for low, long-range coverage and the other 5-feet high for high, near-in coverage. Each antenna was fed by a linear array of 106 dipoles. Dr. Morton H. Kanner led the project. Designated the AN/CPS-1, the radar produced 700 kW. Five 12-inch CRTs were needed to display the coverage. Requiring eight transport trucks, the system weighed almost 70 tons. Certainly the physically largest equipment of the Rad Labs, a small number of the systems were built internally. The first AN/CPS-1 went into operation by the Army Air Forces in England in January 1944. To improve accuracy in the vertical plane, the system later used a separate height-finder radar, the AN/TPS-10 ("Li'l Abner").

AN/CPS-1 MEW Radar

The Rad Lab ran tests on the British 10-cm H2S and was not satisfied with the imagery. Believing that this could be improved with a 3-cm radar, in late 1942 they initiated development of a system called H2X. Dr. George Valley led the project. The X-band Eagle project was already started, but there were problems with Alvarez's complex antennas. Valley gave up precision and incorporated a much simpler antenna for a system that would be used in area bombing – encompassing a target as much as a mile in diameter. In mid-1943, the H2X went into production, with Western Electric making the AN/APQ-13 for the Air

PBM-5A with AN/APS-15 Radar

Force and the AN/APS-15 produced by Philco for the Navy. Both versions were commonly called "Mickey."

The BTL also had successes with X-band radars. They had experimented with phased-array technology before the war, and this was used for the first time in American radars on the Mark 8 (FH), a 3-cm fire-control system for the Navy. The antenna was an end-fired array of 42 pipe-like waveguides, called "polyrods," arranged in three rows and phased such that the beam was steered. The system was produced by Western Electric starting in late-1942.

Mark 8 (FH) – First Phased-Array Radar

The original Mark 3 (FC) was deficient in determining the altitude of low-flying aircraft. In 1943, it was replaced by the Mark 12, also a 40-cm radar, but the 3-cm Mark 22 was developed as an adjunct. The Mark 22 had an "orange-slice" antenna, providing a very narrow, horizontal beam; this nodded to search the sky for targets. The land-based version of this radar used by the Army was AN/TPS-10, called "Li'l Abner" (named after Al Capp's comic strip character that liked to sit in a rocking chair).

Mark 22 (left) & Mark 12

The SS was another major X-band radar developed for the Navy by the BTL. This was a follow-on to the SJ and SV warning radars for submarines.

Closure

The last major application of wartime radar must be mentioned. On August 6, 1945, when the atomic bomb was dropped on Hiroshima, and three days later another on Nagasaki, radars strapped to the bombs were used to initiate their detonation at about 1,900 feet – and they worked exactly as planned. These trigger radars were built from older 415-MHz tail-warning systems, AN/APS-13, with the Yagi antennas directed downward. On the Nagasaki mission, there was heavy cloud cover and the B-29 was guided to the target by an AN/APQ-13 blind-bombing radar.

The Rad Lab was one of the nation's largest wartime projects, at peak involving about 4,000 people and developing more than 100 different radar designs. The operation was officially closed on December 31, 1945, but continued on a transitional basis until July 1, 1946. At that time it was incorporated into MIT's new Research Laboratory for Electronics.

Rad Lab accomplishments were documented in the MIT Radiation Laboratory Series of 28 volumes published by McGraw-Hill over the next several years. The knowledge generated by this Laboratory strongly influenced many areas of post-war science and technology. Nine former Rad Lab employees or consultants were recipients of the Nobel Prize – but for other accomplishments, not for their radar work. These included I. I. Rabi, who received the 1944 Nobel Prize in Physics, and Luis Alveraz who received the same prize in 1968.

In 1945, the BTL started developing the Nike anti-aircraft missile system for the Army. The associated radars eventually evolved to the highly accurate systems used in modern ballistic missile defense. The BTL, however, ceased doing defense work in 1970

Throughout the war, the NRL remained responsible for the Navy's non-microwave radar improvements, and cooperated with the Rad Lab and the BTL in their microwave activities. With some 900 projects ranging from radar and sonar to lubricants and uranium isotope separation processes, the NRL grew from about 400 employees in 1941 to a peak of 4,400 in 1946.

Following the war, the NRL returned to its primary mission of performing basic and applied research. Robert M. Page was named the NRL's first civilian director in 1949. When he retired from the NRL in 1966, Page had 65 patents in his name, mainly related to radar.

In Great Britain, the TRE at Malvern remained the major center for radar research and development. At the close of the war, the TBE had a staff of about 3,500 persons. Many of its wartime "alumni" applied their radar knowledge to the emerging field of radio astronomy. This included Antony Hewish and Martin Ryle, who shared the 1974 Nobel Prize in Physics for their accomplishments.

ELECTRONIC COUNTERMEASURES

Electronic countermeasures (ECM) came into existence shortly after the wireless was adopted by the British military. The basic ECM technique, jamming, involves radiating a strong radio signal that shadows the desired signal at the receiver. The term electronic counter-counter measures (ECCM) applies to antidotal techniques for ECM.

As the Chain Home system was being developed, the possibility of German ECM was considered, and ECCM techniques were incorporated in the design. The system could easily switch frequencies, the simplest ECCM for jamming. Dr. R. V. (Reginald Victor) Jones, who headed scientific intelligence in the British Air Ministry, is credited with initiating radar countermeasures. He led the efforts in identifying the characteristics of German radars and methods of countering their effectiveness.

Initially, the British used commercial short-wave transmitters as jammers. To know when ECM should be used – when they were being "illuminated" – they depended on Ham radio receivers, particularly the American-built Hallicrafter S-27 that covered 27-145 MHz.

Hallicrafter S-27

Led by Dr. Robert Cockburn at the TRE, the British developed early countermeasures equipment for use against German radars. These were both jammers and spoofers. The jammers included the broadband "Mandrel" that blinded *Freya, Wassermann*, and *Mammut* radars, and the "Shiver" and "Carpet" for countering the *Wuerzburg*, the Carpet radiating a "white noise" signal.

Serious work on ECM and ECCM did not occur until after the U.S. entered the war. Thereafter, all of the national radar laboratories in the U.S. (the NRL and the SCL) and Great Britain (the TRE and the ASE) were involved in developing ECM and ECCM techniques and equipment. The U.S. and British work was closely coordinated through a Radio Countermeasures Board.

When the Rad Lab started, a special laboratory was established for ECM/ECCM. Dr. Frederick E. Terman, previously head of the Electrical Engineering Department at Stanford, was named the director. Later, this operation became the Radio Research Laboratory (RRL), located at Harvard University just a mile from MIT. Unlike the Rad Lab, the RRL never released significant details on its accomplishments – ECM and ECCM have always been closely guarded secrets by all nations.

As an example of the RRL's work on the hardware side, a commercial field-strength meter from General Radio covering 300 MHz to 3 GHz was modified by the RRL to become the Navy's ARC-7 and the Army's SCR-587 radar intercept receiver.

One of their jammers was "Tuba," a giant system generating two continuous 80-kW signals in the range of 300-600 MHz to jam *Lichienstein* radars. Tuba used Resatron tubes, developed by Dr. David H. Shone and Dr. Lauritsen C. Marshall and manufactured by Westinghouse. It used a horn antenna built of chicken wire 150 feet long and driven through 22- by 6- inch waveguides, possibly the largest ever built.

Tuba Antenna

Tuba was placed in operation in mid-1944 on the south coast of England. The radiated energy was such that it lighted fluorescent bulbs a mile away and jammed radars throughout Europe.

Ultimately, many types of radio and radar countermeasures equipment were developed by the RRL. These usually carried the designation APT for Army units, APQ for airborne equipment, and SLQ and SPT for Navy shipboard radar jammers. Particular jammers for the *Wuerzberg* were the AN/APT-2 (airborne) and AN/SPT-2 (shipboard), and jammers for the *Freya* were AN/APT-3 and AN/SPT-3.

Using AN/SPT-3

Signal receivers for countermeasures were developed by several organizations. The XARD from the NRL covering 50 MHz to 1 GHz was one of the first, going into service aboard submarines in 1942. The RRL also developed two excellent intelligence receivers: the ARC-1 (Navy) and SCR-587 (Army) covering 100 MHz to 1 GHz, and a similar unit, the AN/SPR-2 and AN/APR-2, that allowed 8-hours of recording on a electromagnetic recorder. The ARR was an airborne search receiver, tuning from 27.8 to 143 MHz in three bands; built by Hallicrafter, it was similar in performance to their S-27 and S-36 receivers.

ARR Receiver

In 1936, R. V. Jones had suggested that metal foil falling through the air might produce false radar targets. In 1941, this concept led both the RRL and the TRE to develop chaff, lightweight foil strips of about one-half a wavelength in length. When dispersed as a cloud from an aircraft, this generated a large clutter of targets. In Great Britain, chaff cut to 0.3 meters (code named "Windows") and 1.7 meters ("Rope") was used starting in 1943 against German *Wuerburg* and *Freya* radars respectively. It might be noted that the chaff work at the TRE was led by Joan Curran, the first, and possibly only, woman researcher at this facility. About 20,000 tons of aluminum chaff was produced by the U.S. The Germans werenever able to counter this, although they offered a prize of 700,000 Reichsmarks to inventors.

Dumping Chaff

The RRL made significant contributions to the basic understanding of methods, theories, and circuits at very-high and ultra-high frequencies for radio systems, particularly in signals intelligence gear and statistical communications techniques. After the RRL closed, this fundamental work was documented in *Very High-Frequency Techniques, Vol. I and II*, McGraw-Hill, 1947.

Naturally, the Germans also had developments in ECM and ECCM. Germany's early interests, however, were primarily in offensive technologies; thus, ECM – a defensive technology – did not receive as much attention as in Great Britain or America. They did make an important ECCM breakthrough in using the Doppler effect to distinguish between falling chaff (*Spereu* or *Duppel*) and the fast-moving aircraft.

RECOGNITION SYSTEMS – IFF

Radar allowed the detection of a target, but could not identify it as friend or foe. As the CH system was being implemented, Arnold F. Wilkins at TRE developed a transponder – a combined receiver and transmitter -- that, when mounted on an aircraft, would receive the CH transmissions and reradiate a signal at the same frequency. CH operators would see this as a suddenly increased radar return. The technique was called Identification Friend or Foe (IFF), sometimes "Parrot." The initial transponder was designated Mark I and went into production by Ferranti in late 1939. One thousand of these sets were delivered in time for the Battle of Britain in mid-1940.

As other British radars became operational, the IFF units needed to respond to a variety of frequencies. A mechanical device was incorporated to sweep the receiver-transmitter across the necessary bands. This IFF transponder, designated Mark II, was put into service shortly after the Mark I. Brought to the U.S. by the Tizard Mission in September 1940, Mark II was evaluated by the NRL and the SCL. It was eventually adopted by the Army as the SCR-535 and Navy as the ABD and ABE (different frequencies for various radars); by July 1942, 18,000 sets were on order from Philco and Raytheon. This started a major and costly element of American WWII electronics; every ship and aircraft operating in a war zone needed IFF as protection against friendly fire.

The delay in adopting the Mark II was partially because the NRL and General Electric had initiated the development of an IFF set that was not dependent on the radar pulse. Designated ABA by the Navy and SCR-515 by the Army, this was scheduled for testing aboard the aircraft carrier USS Hornet on December 8, 1941. However, to ensure uniformity between American and British IFF, the Mark II was eventually selected.

When microwave radars came into being, response to these ultra-high frequencies was needed. Also, Mark II would respond, undesirably, to German radars. At the TRE, F. C. Williams solved these problems with an entirely new IFF technique. This used a separate transmitter sweeping across 157-187 MHz every 2.9 seconds and was co-located with the radar. The receiver-transmitter on the interrogated aircraft responded with a pulse of coded length, with aircraft distress indicated by a very long pulse. Designated Mark III, this system was put into British production by Ferranti in early 1941.

For flights over Europe, allied aircraft required a single type of IFF transponder. The Combined Research Group (CRG), staffed by representatives of American, British, and Canadian military services, was established in late 1941 to resolve this problem. The CRG operated mainly in a highly secure compound on the NRL campus, with testing at the nearby Camp Springs (later to become Andrews AFB), and was charged with selecting an "Allied" IFF system.

Led by Commander Frank A. Escobar, the CRG had a staff of about 200 military and civilian personnel. A major participant and assistant head from 1942-45 was Robert Hanbury Brown from Great Britain, earlier a key engineer in the AI and other radar developments. Many of the staff were graduates of the NRL Bellevue Secondary School, including Robert C. Day, who provided much of this information.

Typical Mark III Control Cabinet

After considerable analysis, testing, debate, and compromising, the Mark III was selected for all of the Allies and went into American production by Hazeltine Research Corporation as the ABJ and ABK (Navy) and SCR-595 (Army) starting in mid-1942. These were used on vessels and aircraft except fighters; these were under GCI control. Later, the Mark IIIG came into existence, suitable for all types of aircraft; this was designated ABF and SCR-695.

Typical Mark III Aircraft Unit

The more advanced airborne IFF units had Interrogator-Responsor-Transponders functions. An example is the AN/APX-2 (ABJ). The transponder scanned the frequency range in 2.5

Typical Mark III Aircraft IFF Panel

seconds looking for an interrogation signal. Its coded reply consisted of six combinations of narrow and wide pulses. It included an emergency switch for sending a very wide pulse as a distress signal. The Interrogator generated pulses at the 10- or 300-watt level in the 160-184 MHz range at 100 pulses per second. The unit contained a destructive charge that could be set off by the pilot or by an inertia switch on crash landing.

Typical Shipboard Mark III Transponder

IFF became Hazeltine's exclusive wartime project, producing thousands of the units. There were two general types of different electronic sets, and antennas were needed for essentially every type of aircraft and vessel – military and commercial – that would potentially be interrogated by radar. Antennas were very important – they had to be designed and mounted to be able to receive and reply to interrogations from any direction.

The NRL-developed system, ABA and SCR-515, was designated Mark IV was kept in "reserve" should the Mark III be compromised. Such a compromise did occur in early 1943, but rather than the great expense of converting to the Mark IV, the Mark III continued to be used with the code frequently changed on a pre-determined schedule. This led to many operating problems, particularly when the code required changing during a combat situation. There were also problems when large formations of aircraft were involved, resulting in what was called "IFF clutter."

SCR-586 with Associated IFF

Even with its problems, the Mark III remained the primary unit. The CRG continued to seek a more reliable and secure system. This was the Mark V, operating in the much higher 0.9-1.0-GHz frequency band and using pulse-coding; this set was eventually developed, but never put into production. A parallel development was the United Nationals Beacons and Identification of Friend and Foe (UNBIFF). The UNBIFF was intended to be applied at all operational wavelengths, used for land, sea, and air, and give all Allies a universal system of transponder beacons. After the war, this became the Mark X and served as the genesis of a new air-traffic control system.

IFF systems were also developed in Germany starting in 1938, the *Erstling* for the *Freya* radar, and, later, the *Zwilling* for the *Wurzburg*. IFF units were necessary for every plane and ship in America, Great Britain, and Germany; thus, this hardware became a major expense of the war.

RADIO COMMUNICATIONS

When war-preparation started, the RAA, RAB, and RAC receivers and TAV, TAZ, TBA, and TBC transmitters were the Navy's primary communications sets and little new equipment was under development. By 1940, the plans for enlarging the Navy included thousands of new ships, ranging from small, fast coastal crafts to battleships and aircraft carriers. Thus, new types or receivers and transmitters had to be quickly designed, tested, and placed into large-scale production. The earlier sets, however, very often remained in service during the conflict.

Fleet and Shore Radios

Many different types of receivers and transmitters were developed for the fleet and related short-range communications. Most followed the release of war-preparation funds in 1940. This included, but was by no means limited, to the following equipment.

Designed in 1935, the RAK (15-600 kHz) and RAL (0.3 to 23 MHz) receivers were used as a pair with the TBL transmitter. Each receiver had two TRF stages, a regenerative detector, and two audio amplification stages.

RAK and RAL Receivers

The RAO and RBH receivers were also used as a pair, the RAO covering 0.54-30 MHz in five bands and the RBH covering 0.3-1.2 and 1.7-16 MHz, also in five bands. Both were superheterodyne types and used for CW, MCW (tone-modulated CW), or voice reception. They contained a BFO for CW reception and had a crystal filter in the IF stage.

RBB Receiver

The RBA, RBB, and RBC were three receivers covering from 15 kHz to 17 MHz. All were used for CW, MCW, and voice reception in advanced base operations with the TBW transmitter. The RBA was a TRF set, covering 15-600 kHz in four bands. The RBB-RBC superheteroyne receivers were installed in pairs, the RBB covering 0.5-4 MHz and the RBC 4.0-17.0 MHz.

Some of the equipment was military versions of commercial radios. Representative of these is the Hallicrafter S-36-A receiver covering 27.8-143 MHz. Designated the RBK by the Navy, it was used mainly to monitor the VHF spectrum. It had a superheterodyne circuit with a TRF front-end and could receive both AM and FM signal. The British used both the S-36A and its sister set, the S-27, to monitor German transmissions.

RBK Receiver (S-36-A)

Another popular receiver pressed into war service was the National NC-100A. This 11-tube set used plug-in coils to cover 540 kHz to 30 MHz. Designated RAO by the Navy and used at shore stations starting in 1938, the set had a crystal filter, giving it excellent sensitivity and stability.

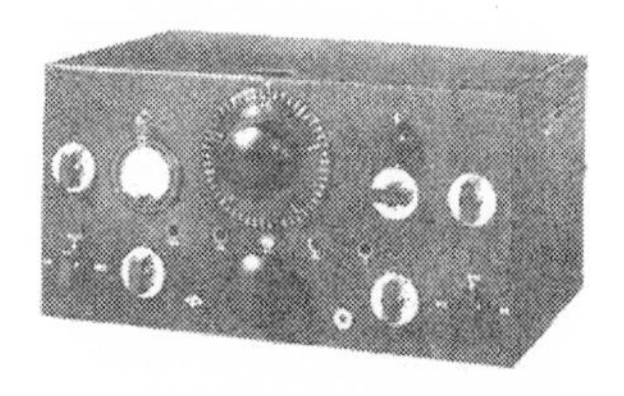

RAO Receiver (NC-100A)

Fleet and short-range communications equipment included a variety of transmitters. The TDE set was two independent transmitters in one cabinet, one covering 0.3-1.5 MHz and the other 1.5-18 MHz. The nominal output was 125 watts CW, 35 watts MCW, and 20 watts voice. The MCW mode produced an 800-Hz tone. The TDE was commonly used with the RBB and RBC receivers.

TDE Transmitter

The TBK and TBM transmitters were physically similar to each other, but the TBK was for CW only and the TBM had a modulator for MCW or voice. Both were tunable over 2.0-18.0 MHz using a Navy-designed electron-coupled oscillator followed by two doubler stages. Delivered power went from 500 to 300 watts as the frequency increased.

Possibly the most used transmitter of that time was the TBL. It had two side-by-side panels, appearing to be separate units, but one side handled 175-600 kHz and the other 2.0-18.0 MHz. The nominal power output was 200 watts CW, 100 watts MCW, and 50 watts voice. A 2,000-volt power supply was required. The low-frequency master oscillator could be "chopped" at an audio rate, making the transmitted signal audible on any receiver; this was called interrupted continuous-wave (ICW). They were built in several models and mainly used on smaller ships and submarines, but the TBL-10 and TBL-11 were for shore installations.

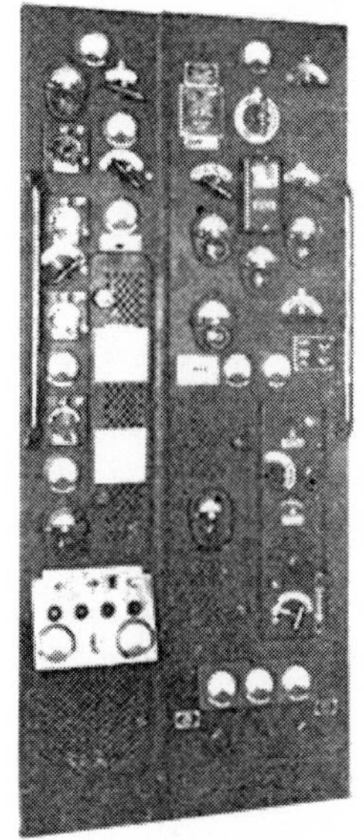

TBL Transmitter

There were rapid advancements in moving radio equipment into the VHF bands, and the Navy quickly realized the advantages of this for more secure line-of-sight communications.

The TDQ transmitter operated between 115-156 MHz for audio or MCW transmissions. Four specific frequencies were set by changeable crystals. The nominal power output was 45 watts, and the MCW used a 1,000-Hz tone. It was normally used with the RCK receiver for short-range ship-to-aircraft and ship-to-ship communications. The RCK was a superheterodyne receiver, covering 115-156 MH, limited to four predetermined channels to match the four crystals selected for the TDQ transmitter. Both the TDQ and RCK could be operated either locally or from several remote stations.

For small vessels, such and landing craft and PT boats, Collins developed the TCS-12 transmitter and receiver set. With frequency coverage between 1.5 and 12 MHz, the transmitter had AM and CW options and an output up to 40 watts. This was Collins first military receiver.

TCS-12 Transmitter & Receiver Set

The TBS radiophones (transmitter-receiver) sets provided short-range audio or MCW communications between surface vessels, particularly in convoys. With 50 watts output, they operated in the 60-80 MHz band on a specific frequency determined by an interchangeable crystal. Remote operation could be from several locations.

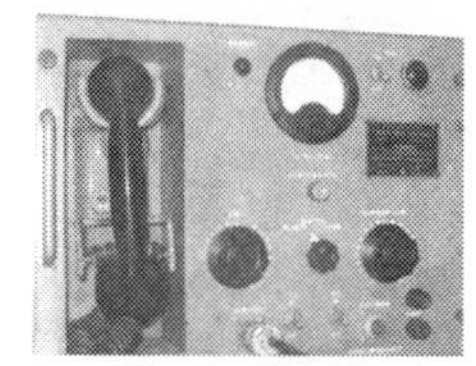

Typical TBS Radiotelephone

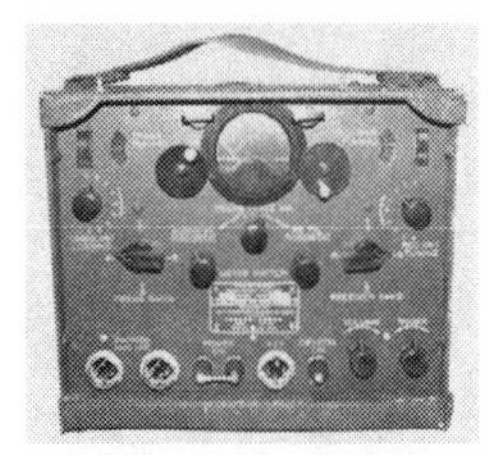
TBY-7 Transceiver

The TBY sets were portable VHF transmitter-receiver combinations (transceiver), mainly for use by the Marines. Operating between 28-80 MHz on 130 specific frequencies, they were used for either voice of MCW over short distances. The output was about one-half watt. Designed to be a knapsack load including a battery power pack, they could be removed for temporary set-up. They were built by Colonial Radio in eight versions (with minor differences) from 1939 to 1943, mainly replaced by Motorola's hand-held SCR-536 ("Handie-Talkie").

Aircraft Radios

Like fleet and shore radio equipment, new aircraft radio sets were needed for the thousands of planes required for the war. In this, however, the Navy was somewhat better off – the 1930s had been a time of considerable development in lightweight, reliable radios for civil aviation, and many of these were adapted for military use. Only a few of the many different aircraft radio sets are mentioned here.

Dr. Frederick Drake of the Aircraft Radio Corporation led a long effort in developing a high-quality receiver with removable coils for low, medium, and high frequencies. Dr. Atherton Noyes from General Radio was retained to design the companion transmitters. Designated Type K and initially intended to be commercial items, these were adopted by the Navy in 1937 as the RAT/GT covering 13.5-20 MHz and RAC/RBD covering 20-27 MHz. In 1939, these were improved and designated the ARA/ATA system. The ARA had five receivers covering 0.19 to 9.1 MHz, and the ATA covered 2.1 to 9.1 MHz with five transmitters.

ARC-5 Receiver

ARC-5 Transmitter

In 1943, the ARA/ATA evolved into the AN/ARC-5. In this, there were also five receivers covering 0.19 to 9.1 MHz, but the available transmitters increased to eight, covering 0.5 to 9.1 MHz. In configurations ranging from one receiver and one transmitter for a small fighter to three each for large bombers, the AN/ARC-5 became the main receiver-transmitter set used for air-to-air and air-to-surface communications and for aircraft reception of navigation signals. Over 100,000 of these command sets were eventually produced.

In the late-1930s, RCA developed a civilian aircraft receiver and transmitter designated ARV-50 and ATB-50. These were "militarized" and adopted by the Navy in 1942 as the ARB, a superheterodyne receiver covering 195 kHz to 9.05 MHz in four bands, and the ATB transmitter, covering 3.2 to 9 MHz using two plug-in coils. Although the receiver coverage was continuous, the frequencies in the lower two bands through 1.6 MHz were intended for direction finding (homing) while those higher than this were for communications. It had four antenna terminals, two for each type of function, as well as different IF stages for the two functional ranges. The receiver could be operated from two remote positions using motor drives for tuning. The ATB had only 20 watts power; consequently, the ARB was most often used with the more powerful ATC transmitter. Some 30,000 ARBs and 4,000 RTBs were built by RCA and several other firms.

ARB Receiver

Arthur A. Collins, founder of Collins Radio Company, invented an electromechanical device called Autotune that permitted rapid selection of up to 10 pre-tuned frequencies. Originally developed for commercial aircraft radios, in 1940 it was used in the Navy's ATC 100-W transmitter covering 1 to 18 MHz. This transmitter was also adopted by the Army Air Corps as the ART-13 made by Stewart-Warner and later designated AN/ART-13. More than 90,000 units were eventually built.

ATC (AN/ART-13)

The NRL and Westinghouse developed the GO-series of transmitters to be used on large, long-range patrol aircraft. The last in this series, the GO-9, came out in 1940 and was by far the largest airborne transmitter used in the war. A similar unit, designated TBW was for ground stations. The GO-9 had separate transmitters covering 0.3 to 0.6 MHz and 3.0 to 18.1 MHz, delivering 100 watts CW and 70 watts AM. Although the GO-9

GO-9 Transmitter

remained in service for over a decade, its much lighter successor, the GP-7, was installed on newer aircraft

The "AN" series of electronic equipment indicated that the item was qualified by both the Army and Navy, but did not mean that it was actually used by both services. "ARC" meant Airborne, Radio, Communications, and there was also "ARR" for Airborne, Radio, Receiver, and "ART" for Airborne, Radio, Transmitter. Many types of this equipment were developed over the next years. Primary manufacturers included Aircraft Radio, Bendix, Collins, Magnavox, RCA, Sylvania, and Western Electric.

A number of other types of Navy aircraft communications equipment were used during the war. These included the GF and GP transmitters: RU, RAX, and ARR receivers; ZA and ZB blind-landing systems; and the RL-series and R1-24 inter-phone equipment.

HYPERBOLIC NAVIGATION

A number of systems had been developed in the 1920s and '30s that used radio signals for navigation. These, however, were only effective over relatively short distances. In 1937, English engineer Robert J. Dippy proposed a radio navigation system using coordinated transmissions from three or more radio stations to pinpoint the location of a receiver. Central to this was the fact that all points where the time difference between any two of the stations could be graphed was a hyperbola. With a transmitter at the locus of the hyperbola, the distance between the transmitter and the receiver could be calculated. By using three transmitters – one the master and the others slaves that repeated the signal from the master – the hyperbolas overlapped and the receiver position was given from where the hyperbolas intersected. Dippy's system required the transmission of precisely synchronized pulses.

In early 1940, the British initiated the development of such a hyperbolic navigation system, with the code name "Gee" – actually a short for "Grid" – and designated AMES 7000. These operated in the short-wave frequency range 20 to 86 MHz. The initial system had a major limitation in that it required line-of-sight between the ground-based transmitter and airborne receivers, and was thus limited in range to about 400 miles. Since there was an urgent need for navigational aides for bombers, a number of Gee systems were implemented. The Gee transmitters in England provided navigation signals that covered the Netherlands and the Ruhr Valley industrial district in Germany.

Long-Range Navigation – LORAN

As chairman of the Microwave Committee of the National Defense Research Committee (NDRC), Alfred Lee Loomis was a major host of the Tizard Mission from Great Britain. At a second meeting with the Tizard members in October 1940, the British technical representative, Dr. Edward Bowen, mentioned Gee as a hyperbolic navigation system for blind-bombing but limited in range by the curvature of the Earth and also lacking accuracy due to timing problems. Loomis quickly conceived an improved system using long waves that could be reflected by the ionosphere, and with stations – perhaps a thousand miles apart – synchronized by the recently developed quartz clocks. The concept was turned over to the Rad Lab for investigation. Scientists from the Rad Lab and the Bell Telephone Laboratories visited England and gathered some, but not all, additional information on Gee and returned to establish Project 3 for developing a radio navigation system.

Gee Receiver

Preliminary plans were developed for a transmitting system that, like Gee, had a master and two slave stations with pulse emissions. In the receiver, they turned to radar for a cathode ray tube to display the received pulses. The system was initially called LRN (Loomis Radio Navigation), but Loomis objected to using this name. The first laboratory model, built with the assistance of the Bell Telephone Laboratories, was tested in the summer of 1941 using a receiver in a station wagon and transmitters at several universities. Comparative trials were made at different frequencies to evaluate ground-wave and sky-wave performance; this eventually led to the choice of 1.950 MHz as optimum. Initially, there was little success in synchronizing the different transmissions, but this changed when additional information was obtained on the Gee system, including details of an improved delayed and strobed timing technique

Dippy came to America to assist in the project, and the Rad Lab team soon realized how far the British work had advanced. Project 3 was abandoned, and the better points of Gee were incorporated in a new system. Coast Guard Captain Lawrence M. Harding, who represented the Navy at the Rad Lab and played an important role in the system testing and implementation of land stations, suggested LORAN, an acronym for Long-Range Navigation. (Thereafter, the acronym was used for this technology and, unlike radar and sonar, was not converted to a name. The LORAN Division, led by Melville Eastham and later Donald G. Fink, was formed. Other leading participants included Dr. John A. Pierce, a Harvard professor with extensive experience in low-frequency radio propagation, and Dr. Julius A. Stratton, later President of MIT.

It was decided that LORAN would initially be for maritime navigation, highly important at that time for providing Atlantic routes to avoid known German submarines. Also, the initial receiving equipment was too heavy for extensive installations in aircraft. Temporary transmitter stations were set up along the Northeast coast and tested using a receiver aboard Navy blimp *K-2*. Initial tests at sea on the Coast Guard weather ship USS *Manasquan* were underway at the time Pearl Harbor was bombed. By the spring of 1942, the system was ready for environmental testing. These tests were satisfactorily performed, and the construction of a permanent northeast Atlantic chain was started.

Typical LORAN Station

The first LORAN pair (Montauk Point, New York, and Fenwick Island, Delaware), went on the air in June 1942. The following November, a test flight was made to Bermuda in a PBY, demonstrating the performance to the Army Air Forces. Additional stations were added along the Canadian east coast, and the system became operational in early 1943. Later that year, the North Atlantic chain was completed with stations in Iceland, Greenland, the Hebrides, the Faeroes, and the Shetlands, this last station was requested by the RAF to cover bomber runs as far as Warsaw. The Rad Lab originally intended to construct, operate, and maintain the chain, but this proved to be impractical, starting with construction. The U.S. Coast Guard mainly had this responsibility, with the Royal Canadian Navy operating and maintaining stations on Canadian shores.

With the extension of chains, the Rad Lab had let equipment contracts without provisions for adequate inspection. Contractors included General Electric, Western Electric, Fada, RCA., Philco, Emerson, Sperry Gyroscope, and various others. Some of the transmitters, receivers, or timers were so poorly manufactured that they had to be practically rebuilt in the Laboratory before being released to the units or vessels for

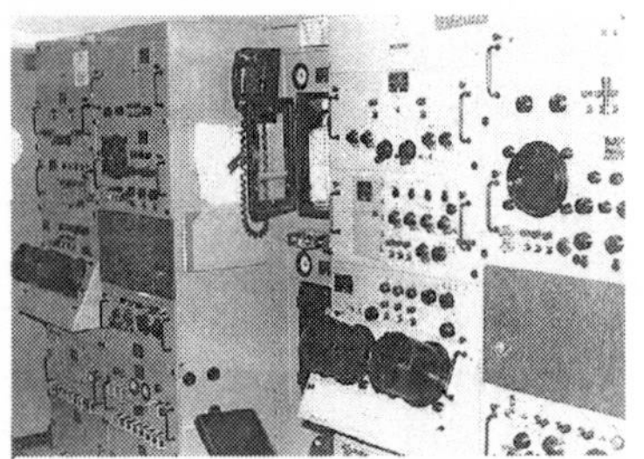

LORAN Station Equipment

which they were intended. The Navy became responsible for procuring future LORAN shore equipment and receivers for shipboard use, while the Army took on the procurement of airborne receivers. Construction and operation of new LORAN stations was assigned to the Coast Guard. Eventually, Fada was the main contractor building LORAN receivers for ships, and aircraft receivers were primarily built by Philco.

The DAS series of LORAN receivers was built for shipboard use. During the war, there were four version, DAS-1 through DAS-4. Except for DAS-2, these were essentially identical in appearance; the differences being mainly in improved components. They were able to measure pulse positions to within a microsecond, using a CRT screen for data

LORAN DAS-4

LORAN receivers for aircraft were in the APN series. Their weight excluded their application to large bombers and patrol aircraft, including blimps. APN-4, released in mid-1943, was the first version to have major use. It consisted of two units with a total weight of some 80 pounds. In 1945, a 40-pound airborne unit, the APN-9, was placed into service, greatly extending the application. Dippy had insisted that the LORAN and Gee receivers be made physically interchangeable so that any RAF or American aircraft fitted for one could use the other by simply swapping units.

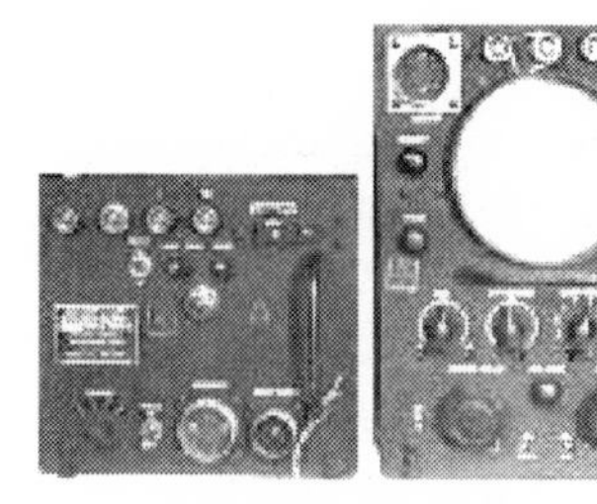

LORAN APN-4

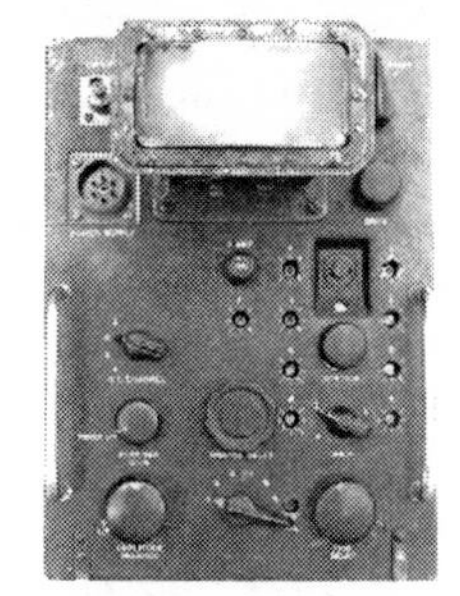

LORAN APN-9

Although LORAN could operate with one master and two slave stations (like Gee), in practice up to six slaves were used in a chain. The master and slave stations could be separated by as much as 1000 to 1200 miles. They transmitted on four frequencies: 1.75, 1.85, 1.90, and 1.95 MHz. Receivers could select the frequency providing the best signal. Each transmission pulse lasted about 40 microseconds with an accurately controlled pulse repetition rate between 20 and 34 pulses per second – different for each station and chain and used for identification. A variable delay was used between the master and slaves, changed on a schedule that was provided to the receiver users.

Since signals were pulsed, a relatively small 200-watt transmitter could deliver peak power in excess of 200 kW. The reliable operating range was about 800 miles in daytime and 1,400 or more miles at night (the height of the ionosphere varies between day and night), with an accuracy of about 1% of the range. This accuracy was about the same as that of celestial navigation, but while celestial navigation required a skilled navigator, LORAN was very simple to operate.

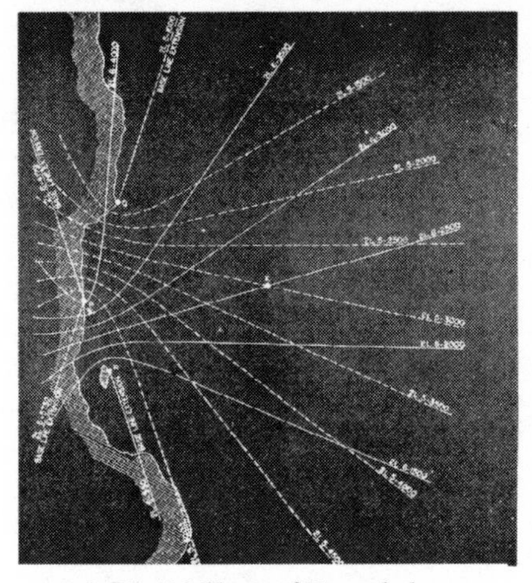

LORAN Chart of Hyperbolas

The arrival times of pulses from a pair of stations were displayed on a cathode ray tube and the difference in time read directly. Since only time difference was needed, the receiver time base did not need to be synchronized with that of the transmitter. Each fix required two observations and normally took three to five minutes. The readings were then transposed to a LORAN lattice chart and position could be plotted. In some, cases readings were referenced to special LORAN tables. With two successive fixes, ground speed, drift, and ETA could be determined. LORAN was available day and night and in essentially any weather (although severe electrical storms could disrupt the transmissions).

While LORAN was important in the Atlantic and Eastern Europe, it was in the Pacific that the system made its greatest direct contribution to winning the war. As American forces moved westward, airfields were built on many of the small islands and atolls that dot the ocean beyond Hawaii. Most aircraft at that time had a range that required frequent refueling, and LORAN was ideal for finding these small airfields in the broad expanses of the Pacific Ocean. As bases were established close enough to Japan to allow a round trip, LORAN was invaluable for precision bombing of the mainland. The Japanese either failed or never tried to jam any of the LORAN transmissions.

At the end of hostilities, navigational coverage of about 30% of the Earth's surface was provided by about 70 LORAN transmitter stations, and some 75,000 LORAN shipboard and aircraft receivers had been installed.

Other Hyperbolic Navigation Systems

Late in the war, the U.S. developed a hyperbolic navigation system based on a transponder. Called SHORAN (an acronym for Short-Range Air Navigation), it operated at around 300 MHz; while it had excellent accuracy, it was not completed in time for service in the conflict. A number of years after the war, LORAN C was developed, eventually replacing the original version then called LORAN A (there was an unsuccessful LORAN B version).

The British introduced another hyperbolic navigation system to follow Gee. It was actually invented by an American engineer, William J. O'Brien, in the same period that Dippy was developing his pulsed system in Great Britain. O'Brien's design used continuous-wave low-frequency transmissions with phase-comparison of signals from a master and slave stations.

Unable to gain the interest of the Army or Navy, O'Brian turned to the Decca Record Company in England. In 1942, the Admiralty Signal Establishment, seeking a highly accurate system to assist in the eventual invasion of Europe, invited a demonstration through Decca. In initial trials in the English Channel, the system worked well and was classified as secret. Officially designated as QM, the system was more commonly called Decca Navigation or simply Decca. The master and slave transmitters operated in the band between 72 and 197 kHz, each a precise multiple of a 14.2-kHz base frequency. At the receivers, the phase differences between the signals were displayed on clock-style dials known as "decometers" that were used to determine location.

Decca Display Unit

After more comprehensive trials, in early 1944 the Admiralty ordered 27 Decca receivers, and a chain of Decca transmitters was set up for the Normandy invasion. The chain included a third slave to decoy the Germans into believing that Calias would be the invasion point. On June 6, 1944, Decca was successfully used on minesweepers and other vessels in D-day landings. After this, the system was kept in reserve for the remander of the war. Later, Decca Navigator saw widespread service as a commercial system for off-shore oil drilling and similar activities, particularly in areas of British influence.

Another British system for navigating bombers was called Oboe. This radar-based system, developed in 1942 at the TRE, was ultimately shown to be the most accurate navigation system in the war for blind bombing. Oboe was previously described in the section on Radar.

During WWII, the Germans never built a hyperbolic navigation system. The Japanese did develop an elementary hyperbolic system but never placed it into service – it required the navigator to perform stopwatch timing as part of the procedure.

SOUND NAVIGATION AND RANGING – SONAR

Shortly after the acronym RADAR came into being, the acronym SONAR was also coined by the Navy. This stood for Sound Navigation And Ranging, and soon came to be the name sonar. This covers all types of underwater sound devices used for listening, depth indication, echo ranging, ship-to-ship underwater communication, and other functions.

The equipment described earlier in this book was developed for opposing submarine threats of the 1930s. The German U-boats of WWII, however, were faster, quieter, and deeper running than their predecessors. Despite earlier improvements made by the NRL Sound Division, the Admiralty Research Laboratory in Great Britain, and their supporting industries, sonar technology and general antisubmarine warfare techniques were in great need of improvement at the start of WWII.

The Battle of the Atlantic throughout 1942 was disastrous to the Allies. In his report for that year, Dr. Harvey Hayes, superintendent of the Sound Division, noted:

> The conclusion is reached that we are losing the present antisubmarine war by such a large margin that measures should be immediately taken to outline, codify, and put into action a more effective antisubmarine program.

In addition to equipment improvements, Hayes recommended specific tactics, including dropping a series of rapid-sinking munitions such as contact and proximity depth charges instead of the slow-sinking pressure-fuzed munitions.

Some technical improvements included the Bearing Deviation Indicator (BDI) to assist sonar operators in aiming the equipment more effectively; Maintenance of True Bearing (MTB), which incorporated a gyro to keep sonar beams directed on submarines as the surface ship underwent complicated defensive maneuvers or attacking turns; and Generated Target Tracking (GTT) that allowed the operator to track a specific target designated from the fire-control station.

These and other responses to the submarine problem began having positive effects in 1943. By then, radio direction-finders could pick up submarine transmissions and then send radar-equipped planes to pinpoint the U-boats. If the U-boats submerged, the planes could drop

Radio-Sono buoys that would send out radio alerts to ships for zeroing in for the kill.

During the 1941-1945 period, work done by the NRL Sound Division and various industries on transducer materials, the associated electronic components, and the processing and displaying of raw signals into interpretable information resulted in many improvements. The basic technology, however, remained essentially the same.

Sonar equipments that provided listening, ranging, and sounding were extensively used on surface vessels and submarines throughout the war. Echo-ranging might disclose the presence of the sending; thus, listening devices were normally used for detection. Echo-ranging was used on submarines mainly for navigation, but single-ping ranging was used for final target setting. There were many different models of sonar equipment, and only a few will be briefly described.

Submarine Sonar

The JP was an early system for sonic listening aboard submarines, the J indicating listening only and the P meaning sonic. The hydrophone was a 3-foot-long magnetostrictive device that could be rotated by the operator using a handwheel. Although the hydrophone had a flat response from 100 Hz to 40 kHz, the receiver-amplifier included filters that could be switched to 5 audio bands through 15 kHz. With a Magic Eye display, the operator could train on a source to an accuracy of 2 degrees.

Magnetostrictive Hydrophone

JP Receiver-Amplifier

The JT was an improved listen-only system that could receive both sonic and supersonic signals. It had a converter that could be switched in to change the ultrasonic signals to sonic frequencies; the JP amplifier was then used for processing both types of signals. The JT had a 5-foot-long magnetostrictive hydrophone, and its bearing was relayed by synchros to the control unit. The hydrophone could also be switched to operate as a Right-Left Indicator (RLI), comparing signals from two halves of the hydrophone to show whether it was trained to the right or left of the on-target position; this allowed bearing accuracies within one degree. Targets at up to 20,000 yards could be detected using

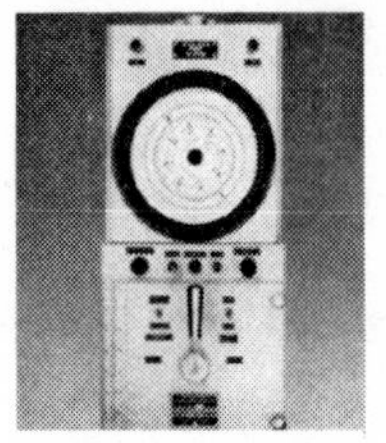

RLI Control Unit

the JT; the bearing accuracy at long ranges, however, was degraded. This could be improved by placing a hydrophone at each end of the submarine, then using triangulation for improved bearing determination.

The supersonic listening-only gear was designated JK. It used a Rochelle-salt hydrophone that was more sensitive than the magnetostrictive transducer on the JP. The supersonic signal was converted to a sonic frequency, then processed by a JP-type receiver.

QB, QC, & JK Projector Spheres

QB and QC were echo-ranging systems, the QB using a Rochelle-salt transducer and the QC a magnetostrictive transducer. These transducers, called projectors, were mounted in spherical containers that were lowered beneath the hull for operation. QC and JK transducers were usually mounted in the same housing, with the combined sound head used for both listening and echo ranging.

NM Transducer

The NM was a third type of ranging system that was used for depth finding. The NM used a magnetostrictive transducer in a housing mounted within the hull and directed downward. A single-ping driver produced a current pulse, causing the NM transducer to function as a transmitter and emit a single supersonic pulse. This was reflected by the ocean floor and detected by the NM transducer acting as a detector. The control unit contained an electronic timer that measured the time between a ping and the echo return; its indicator was marked in distance (yards). The overall system was called a fathometer.

Fathometer Indicator

NM Driver

The standard Warnings/Cautions/Advisories (WCA) suite of supersonic sonar gear on submarines included QB, QC/JK, and NM systems. Much of this equipment was grouped in the conning tower, where it was called the WCA Stack. Although consisting of three separate systems, each used one or more units in common with the other systems. The QB and QC/JK used the same PPI range indicator, and the

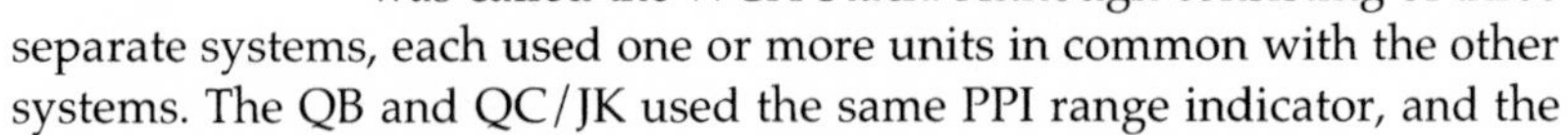

NM and the QC/JK used a common ping-driver unit. A Submarine Bathythermograph (SBT) unit was used with to determine variation of water temperature with depth. Overall, the WCA included a large amount of electronics that was state-of-the-art at the start of the war.

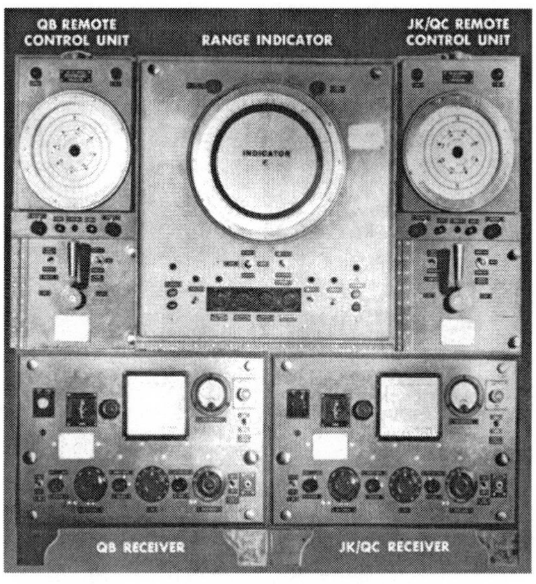

WAC Sonar Stack

The Torpedo Data Computer (TDC) was of major importance in the submarine's armament. It was located in the conning tower usually close to the WCA sonar stack. On its front panel, hand-operated dials show six critical settings: the course and speed of the submarine and the course, speed, range, and bearing of a given target. Although it was nowhere close to a modern digital computer, the TDC made analog and mechanical calculations and, upon being so directed and using the six input settings, calculated a firing solution in terms of relative bearing. Since the inputs were from various sources – particularly from the sonar – and would likely undergo changes, the TDC officer constantly monitored the dial settings. When there was agreement, the TDC made a new calculation. Thus, the success of the torpedo attack was entirely dependent on the accuracy of the TDC; maintenance of this unit was vital to the operation.

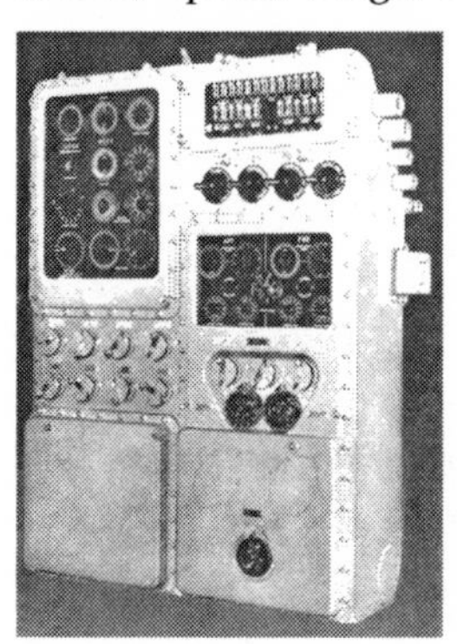

Torpedo Data Computer

Near the end of the war, the WFA combination sonar gear came into being. A highly complex electronic system, the WFA could be used for listening, echo-ranging, torpedo-detection, mine-detection, and communications. Listening was from 200 Hz to 100 kHz. The frequency range for echo-ranging was 17.2 to 46 kHz, somewhat higher than that of the earlier equipment. Higher frequencies used in enemy waters increased secrecy and bearing accuracy but shortened maximum range.

The WFA had two sound-head assemblies, one on the deck and the other near the keel. The topside head contained three units: a sonic hydrophone for listening only and two ultrasonic transducers for either listening or echo-sounding, one operating between 12.5 and 35 kHz and the other covering 31 to 100 kHz. The lower sound head included a single crystal transducer operating between 22 and 32 kHz. Either the top-side or bottom transducer could be used for mine detection, operating in a short-pulse mode that provided high resolution for small objects. Mines and other small objects could be detected up to about 600 yards. For torpedo detection, the lower transducer was rotated constantly at 12 rotations per minute. Telegraphic communications with other vessels having echo-sounding equipment was accomplished by keying the transmitter.

Other submarine sonar systems coming into being near the end of the war were the QHB-1 and the QLA. The QHB-1 was a submarine version of the shipboard QHB capacitive-scanning system described in the next section. Using a new hydrophone design, it provided video presentations of all acoustic reception from any direction. The QLA was an FM scanning system that could be used either on submarines or shipboard. The transmitted signal time-varied in frequency, and the range to a target was determined by the frequency received.

Shipboard Sonar

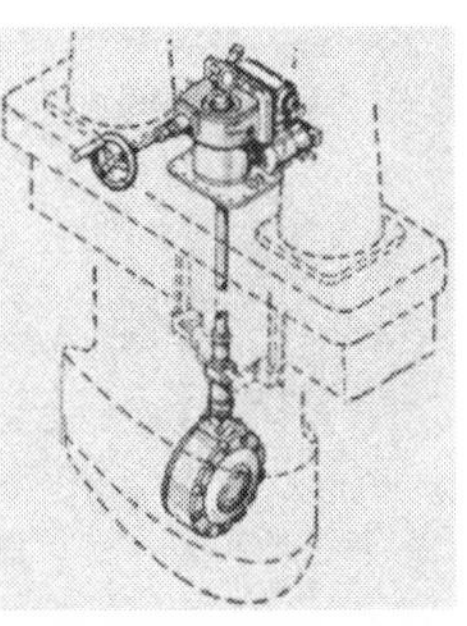
QGB Projector & Hoist/Trainer

Several types of sonar equipment were used on surface ships. The Model QGB was typical of these for destroyers and destroyer escorts. Called "searchlight" equipment, this consisted of a driver (transmitter), a receiver, an indicating range recorder, a bearing deviation indicator, and a magnetostrictive projector with a hoisting mechanism. It had two consoles: one for the operator and one on the bridge. The operating frequency was between 17 and 26 kHz, determined by the resonant frequency of the transducer. The transmitter, producing 400 watts, was

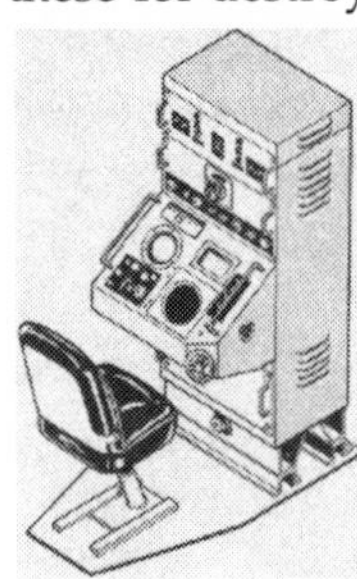
QGB Console

keyed, with the pulse rate depending on the range setting. As the name implies, the emitted beam gave a narrow, searchlight-type coverage.

The transducer, in its sound dome and hoisting mechanism, were contained within a sea chest in the bottom of the hull. When the dome was raised, it sealed off the sea chest, allowing the maintenance of the transducer while the ship was at sea. The transducer was trained through a handwheel at the operator's console. Training was also possible from the bridge console through a servo.

The QJB was similar to the QGB but with a single console. The QGA was another unit intended for destroyers. This had two complete sonar systems, one operating at 14 kHz and the other at 30 kHz. Although normally operating independently, they could be slaved. The two magnetostrictive transducers were mounted on concentric shafts that were hoisted and lowered together. The 30-kHz transducer could be tilted downward to 45 degrees, a feature of value for a deep target. The two independent consoles were similar to the one for the QGB.

To provide rapid sweeps around the ships, the most complex sonar to evolve during the war was developed. Called the QHB scanning system, this echo-ranging and listening unit provided video presentations of all acoustic reception from any direction, as well as audio of reception along a selected bearing. A single transducer, used for both transmission and reception, had 48 independent magnetostrictive hydrophones arranged along a mechanically untrainable cylinder that was lowered below the hull. For echo-ranging, a 25.5-kHz, 35-ms pulse was emitted simultaneously by all of the transducers. The transducers were then rapidly scanned for returned signals, with signal only on the hydrophones facing the return path.

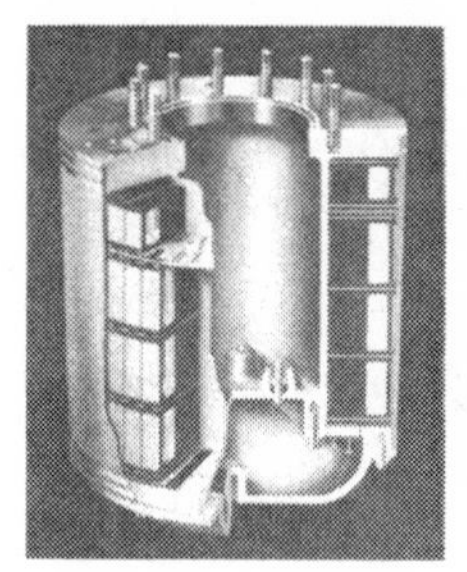
QHB Hydrophones

A scanning switch, capacitively coupling the output from hydrophone preamplifiers, was driven at 1,750 rotations per minute. (Capacitive coupling of the scanning switch was needed to allow interpolation of the signals, otherwise only a 7.5-degree bearing accuracy would be possible.) The rotation of the beam of the cathode-ray tubes in the PPIs was synchronized with the switch rotation, thus allowing the signals to be displayed in map form, giving range and azimuth.

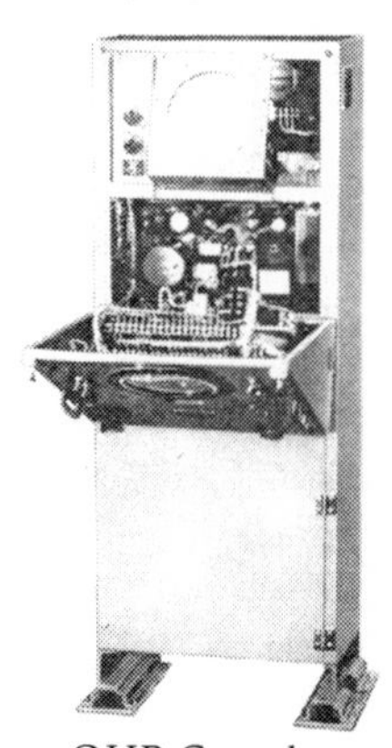
QHB Console

The QDA was a depth-determining system, used primarily as an attack instrument in conjunction with either a QGB or QHB. The transducer emitted a fan-shaped, vertical beam in the 50- to 60-kHz band. To produce the narrow beam, the transducer was made with an array of crystals. Range was determined by the same techniques as ordinary echo-rangers. For determining the depression angle, the beam was tilted, with the tilt angle indicating the depression.

Other Sonar Systems

The AN/UQC-1 system was used in both submarines and surface ships for voice or Morse-code communications through the water. The transmitter provided 400 watts of single-sideband carrier power at 8.0875 kHz. This was modulated at voice frequencies, or used with a 712-Hz keyed signal for code. The magnetostrictive transducer was mounted in a dome and had an omnidirectional, horizontal pattern. The maximum communication range was about 2,000 yards.

Harbors were protected from intrusion by enemy submarines through the use of several types of systems. One was a network of cable-connected hydrophones, monitored at a central location. Spaced up to 1,000 feet apart and at depths up to 400 feet, the hydrophones were on a tripod set on the ocean floor. Such systems were normally able to detect the sounds produced by a slowly moving submarine even when noisy surface vessels were present.

Patrick M. S. Blackett, head of a special committee in Great Britain for anti-submarine measures, first proposed an expendable buoy for detecting underwater sounds in early 1941. The idea was that these buoys would be dropped from convoy ships to detect trailing submarines, transmitting the signals back to the ships by radio. The Underwater Sound Laboratory at New London revived the concept for supplementing MAD (described in the next section), and the first production of AN/CRT-1 sonobuoys began in October 1942.

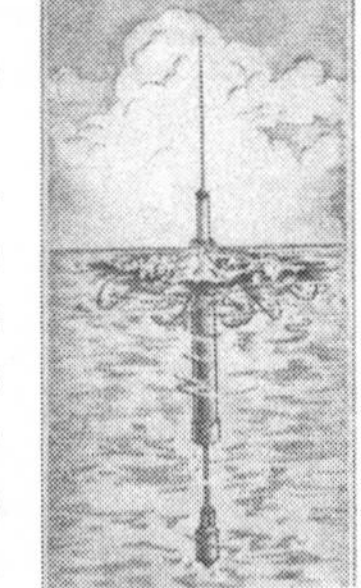

Sonobuoy

The AN/CRT-1 sonobuoy was composed of a non-directional hydrophone suspended on a 20-foot line below a floating buoy containing an FM radio transmitter and raised antenna. Underwater sounds received by the hydrophone were sent to a receiver (AN/ARR-3) in a tending plane or blimp, or a shore station. The radio range was approximately 10 miles when the receiver was airborne at 300 feet or 19 miles for a shore station. The underwater range varied from

200 to 3,000 yards, depending on the water conditions. To separate different sonobuoys in the same area, the transmitters operated on four frequencies. The batteries had a lifetime of about 3 hours, then the sonobuoy was scuttled.

Sofar (Sound Fixing and Ranging) was a long-range position-fixing system that used explosive sounds in the permanent sound channel of the ocean. (The sound channel is formed by layers of ocean water having different velocity gradients, resulting in any acoustic signal being reflected up and down (channeled) between the layers.) The sound was generated by exploding 4 pounds of TNT that was dropped to the desired depth and fused by the corresponding pressure. A fix was determined from the differences in arrival times, at known geographic positions, of a signal that was sent from any given point. The useful range from the signal source to the monitor stations could exceed 3,000 miles.

OTHER ELECTRONIC SYSTEMS

In addition to the systems previously described, there were many other electronic systems used by the Navy during World War II, and all of these involved electronic technicians for development, maintenance, or specialized operations. This section will address four of these types: magnetic anomaly detection systems, beacon radio systems, direction-finding radios, and radio-controlled aircraft (drones). There is also information on proximity fuses – mainly because these were another application of radar – and bombsights. This latter was not really in an electronic category – although it interfaced with the electronic automatic pilot – but is included because Navy electronic technicians were sometimes involved with this highly classified apparatus.

Magnetic Anomaly Detection Systems

When operated from airborne platforms, early radar, sonar, and optical systems were ineffective for detecting submerged submarines. The radio, sound, and light waves starting in the atmosphere were distorted when passing into and returning from the water. Lines of force from the Earth's magnetic field, however, are unaffected in traversing the two media. If there is an intervening ferromagnetic object – such as a submarine – the field will be disturbed, and detection of this anomaly can indicate the presence of the object. In addition, the steel of the submarine has a permanent magnetic field. The total of these fields gives the magnetic "signature" of the submarine. The operating principle of

Magnetic Anomaly (or Airborne) Detection (MAD) systems is to detect this signature.

Since their invention in 1832 by the German scientist and mathematician Dr. Carl Friedrich Gauss, simple magnetometers had been used for measuring the intensity and direction of magnetic fields. During the First World War, attempts were made to develop ship-towed MAD systems, but the available sensors and associated electronics were insufficient for detection at suitable ranges and, in addition, the metal of the towing ship greatly affected the sensitivity.

In the 1930s, significant improvements were made in magnetometers, particularly in the oil-drilling and mining industries where they were used for mapping of underground features. The flux-gate magnetometer, conceived in Germany in 1936 then developed by Victor Vacquier at the Gulf Research Laboratories, was the most significant improvement in these devices in a century. In this magnetometer, the ambient magnetic flux is converted (gated) into a time-varying field that is then sensed to provide the magnitude and direction of the ambient flux. There was also major research in this technology at the Department of Terrestrial Magnetism of the Carnegie Institution of Washington, as well as several universities.

When the National Defense Research Committee (NDRC) was formed in mid-1940, Dr. John T. Tate, chairman of the Physics Department at the University of Minnesota was asked to head a committee for antisubmarine warfare research. The Underwater Sound Laboratory at New London, Connecticut, and the Sound Division of the NRL were charged with revisiting the magnetic anomaly detection concept. They quickly concluded that technology improvements made the concept feasible and that airborne sensors – unused in the trials during World War I – would eliminate the interference in the earlier ship-towed system.

The NDRC evolved into the Office of Scientific Research and Development in June 1941 and contracted with the Division of War Research at Columbia University to develop a MAD system. For this, they set up a top-secret operation at the Quonset Point Naval Base, Rhode Island, where several PBY Catalina amphibious aircraft were based. As the U.S. entered the war, Tate asked Dr. Otto H. Schmitt, one of his most capable electronics researchers at the University of Minnesota, to lead the hardware development.

Using a flux-gate magnetometer suspended on a boom extended from the PBY, the prototype MAD equipment was tested in February 1942, and detected a submerged submarine during the first trials. Additional testing was aboard a ZNP-K blimp recently delivered by

Goodyear to the Navy at NAS Lakehurst, New Jersey. The changes in the magnetic field due to a submarine were small; thus, the indication of detection required recording of the output on a chart recorder, with interpretation of the results made visually by the operator.

In mid-1941, Columbia University established the Airborne Instruments Laboratory (AIL), with Dr. Hector R. Skifter as the director. Early the next year, the operation was moved out on Long Island to Mineola, New York. Upon demonstration of the prototype, the development of production-level MAD systems was assigned to the AIL Special Devices Division. Vacquier was hired to further refine the flux-gate magnetometer. Schmitt and other researchers were also transferred there. By mid-1942, the AIL began production of the Mark IV-B2 MAD System. Somewhat later, this was designated AN/ASQ-1. To support the operation of this equipment, a special MAD maintenance course was set up at NATTC Ward Island, Texas. ART-1/c Ralph M. Caldwell was the lead instructor for this highly secret instructional activity.

MAD Control & Recorder

MAD Sensor Pod

The range and depth of detection by MAD is a function of the size of the submarine and the height of the sensor over the water. In general, it was believed that a 750-ton submarine could be detected reliably at distances up to 400 feet. Initial operation in patrols by blimps and PBYs was quite limited by the sensitivity of the magnetometer, as well as by the "noise" generated from wrecked ships and other sources of magnetic variations. Improving the magnetometer proved to be extremely difficult; the magnetic field decreases as the third power of distance, so doubling the range required an eight-fold increase in sensitivity. Consequently, it was believed that the primary usefulness of MAD would be in rapidly surveying an area and using the results for further detection by other means. The previously described sonobuoy came to be paired with MAD systems for this purpose.

To further develop the MAD System, the California Institute of Technology developed a retrorocket to fly backward from the releasing aircraft, compensating for the forward motion and allowing a depth bomb and marker to drop directly downward into the water above the submarine. This system was designated the AN/ASQ-2. As the MAD-sonobuoy combination went into operation, it was found that the sonobuoy was more useful for the initial detection of targets, with MAD used for final target location, the reverse of the initial plan of use.

By 1943, most anti-submarine patrol blimps and airplanes were equipped with MAD systems. In January 1944, MAD-equipped Catalinas began patrolling the Straits of Gibraltar, effectively closing that important channel to enemy submarines. By the end of the war, a total of 13 submarines had been sunk by aircraft equipped with MAD systems.

Beacon and Direction-Finding Radio Systems

Like a light beacon, a radio beacon is used to show its location. After the loop antenna was developed, civil and military pilots used radio direction-finding and automatic direction-finding radios to locate the line to a known radio station or beacon transmitter. This type of system could tell the pilot what "quadrant" the aircraft was in relation to the beacon, but it could not provide a true compass bearing, or radial, to the beacon.

In the late 1930s, the Germans develop a radio beacon system called *Sonne* (Sun) that allowed determination of the radial. The beacon operated at 300 kHz (1-km wavelength) and used three fixed antennas in a row, spaced one wavelength apart. If an aircraft was on a path roughly perpendicular to the antennas, the navigator could take two fixes at different positions and use triangulation to find the aircraft's location. The system was known as "Consol" to the Allies, and was so useful that it remained in service well after the war.

Beacons can also transmit coded information for navigational purposes. The two most used in the 1930s were the "four-course radio range" that was the standard throughout America and the similar *Ultrakurzwellen-Landefunkfeuer* (LFF) sold worldwide by Lorenz. Both of these systems allowed pilots to navigate along presecribed airlanes. By 1941, the system in America had over 200 beacon stations and covered about 31,000 miles of airways. All military aircraft had receivers for the beacon frequencies (190 to 534 kHz), and the system was extensively used for flights within the United States.

As Germany started night-time bombing of Great Britain in 1940, the Lorenz System was modified to have two 30-MHz beams crossing each other, the second providing bomb-release positioning. Called *X-Leitstrahlbake* (Direction Beacon) and nicknamed *Knickebein* (Bent Leg), this was analyzed by Dr. R. V. Jones in British intelligence, and Dr. Robert Cockburn at the TRE led a team in developing a countering beam system. The British had code-named the German beams "Headaches," so the countermeasures transmitters were called "Aspirins." The Battle of the Beams followed and the *Knickebein* system was soon abandoned. Later in 1940, however, a more sophisticated modification of Lorenz System – the *X-Geraet* (Device) using several beams – was used in the incendiary bombing of Coventry. Again, countermeasures transmitters (called "Bromides") soon neutralized the system.

As early as 1937, work began on converting the American airways navigation system from low-frequency to the VHF band. In 1940, the Civil Aeronautics Administration (CAA – predecessor to the current FAA) was formed and it began modernization of the radio-navigation system as part of the war-preparation effort. Called the Visual-Aural Range (VAR), this included a rotating radio pattern as well as a light beacon. Operating in the 112-118-MHz band, this had the potential of a large number of navigational courses. Because of shortages of VHF equipment, major installations did not start until 1944.

In early 1940 at the Telecommunications Research Establishment (TRE) in England, Dr. John W. S. Pringle (a biologist), assisted by Robert Hanbury Brown and E. K. Williams, developed a responding-beacon system called Rebecca-Eureka to assist in dropping, with pin-point accuracy, supplies to French partisans and for other covert operations in Europe. The airborne Rebecca had an interrogating transmitter, receivers with antennas on opposite sides of the aircraft, and a display unit. Upon receiving the interrogating signal, Eureka, the ground unit, would turn on an associated transmitter and emit a series of coded pulses. The outputs of the Rebecca receivers, together with the interrogating pulse, were displayed on a CRT, by which both the approximate distance and direction to the beacon could be determined.

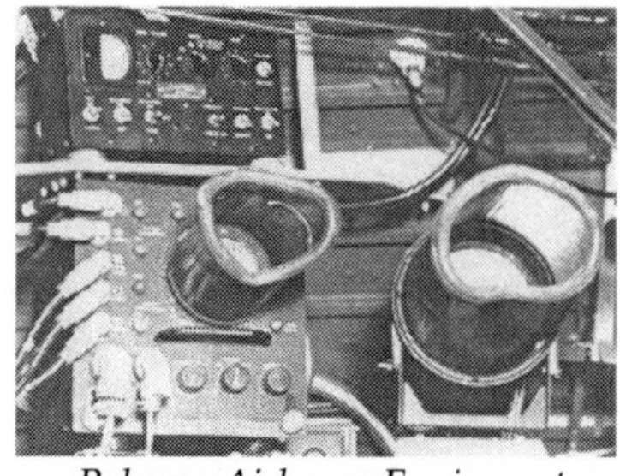

Rebecca Airborne Equipment

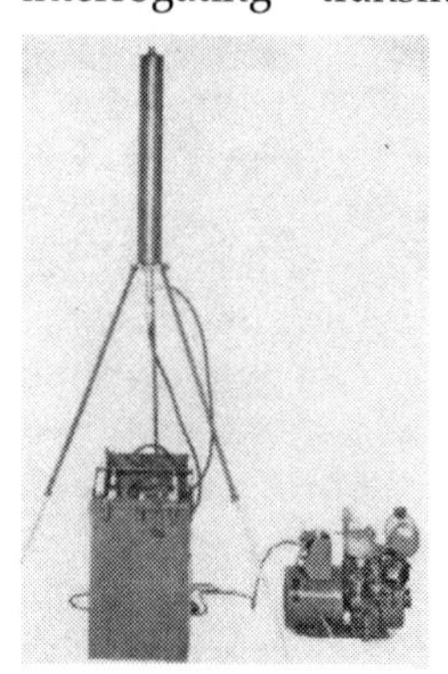

Eureka Ground Set

The system operated on frequencies in the 135-210 MHz band and had a useful range of about 80 miles. Hanbury Brown came to the NRL in 1942 to assist in developing the American version of the system, designated AN/APN-2 and AN/PPN-2. Rebecca-Eureka was used throughout the war, and became particularly beneficial during and after the Normandy invasion.

When aircraft carriers were first developed, there was no means of guiding the flyers back home. The NRL developed such a system, called Model YE, and in 1938 it was required to be installed on all carriers, landing fields, and the associated aircraft. The system used coded information to identify the carrier or field and had a directional antenna that was aimed at the incoming aircraft to reduce enemy interception. In the aircraft, the pilot used a beacon receiver with a directional antenna (described below) to determine his return bearing. The system was later upgraded to the Model YG and used two different frequencies. The British also adopted this system for their aircraft.

In addition to the YE and YG, several other beacon systems were developed in America for aircraft carriers and landing fields to guide flyers back home. In one system, a signal would be emitted from a rotating antenna, keying a different Morse character for each 30-degree segment. A pilot could use a beacon receiver to determine his general bearing to home.

Near the end of the war, a 30-cm (1 GHz) transponder-beacon system was developed to determine the distance between an aircraft and a base station. Known as the Distance Measurement Equipment (DME), an inquiry pulse was sent from the aircraft. In less than a microsecond after the base station received this pulse, it was retransmitted. Thus, the airborne set could determine the distance using the time delay between transmission and return reception, less the microsecond delay. The AN/APN-34 was the airborne DME set and the AN/GPN-4 the base set. After the war, the system was used with a VOR directional beacon to serve as a VOR/DME station.

Direction-finding (DF) radios are used to determine the direction of a line between the receiver and an independent transmitter. In this, there is no requirement for a cooperative effort between the receiving and transmitting stations. These were used in intelligence to find the direction of an enemy radio emission, but they were also used for navigation purposes. In the first application, the triangulation of the directional lines from two different DF radios could pinpoint the transmitter location. For navigation, a DF radio could monitor two different transmitting stations and use triangulation to find the position of the receiver.

During and after WWI, the Navy had low- and medium-frequency DF equipment using loop antennas installed on many ships and around important ports. By the mid-1930s, initiation of high-frequency (HF) communications drove the need for DF equipment at these frequencies. In 1936, the NRL developed a HF DF system using two opposed vertical dipole antennas with the receiver in between. It was first tested at sea in

1937 and located transmissions up to 200 miles distance, but it was not implemented by the Navy. Great Britain, however, had developed a similar HF DF system, and by 1939 had set up a widespread network of these stations. Commonly called Huff-Duff, they were located across the British Isles, Northern Africa, Iceland, Greenland, and Bermuda, as well as along the Canadian Atlantic coast and a station in Maine. Shipboard versions were the FH-3 and, later, the improved FH-4.

In 1941, shipboard HF DF systems based on the NRL design began to be implemented by the Navy. From that time onward, a wide variety of new DF systems covering from medium to very-high frequencies were developed, mainly by the NRL. These carried a DA designation and eventually used all of the alphabet from A through Z: DAA (2-18 MHz) to the DAZ (100-160 MHz). Major manufacturers included the Washington Navy Yard, Collins Radio, RadioMarine, Sargent Radio, Airplane & Marine Instruments, International Telephone & Radio, Federal Telephone & Radio, Garod Radio Corporation, and General Electric.

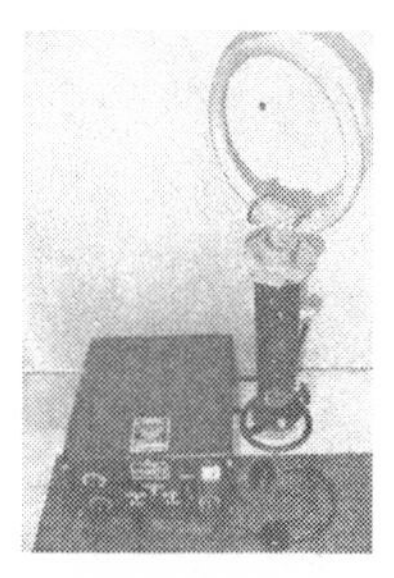

DAE DF Set

Direction-finding sets were installed on most American vessels and, together with the British FH-3 and FH-4 sets on British and Canadian vessels, played a very important role in protecting the Atlantic shipping. U-boats that located a convoy would often first shadow it and radio homing signals for other U-boats, forming them into "wolfpacks." It was through these signals that the escorts could track down and then attack the enemy submarines. Two sets finding the direction of a transmission could be used in triangulation for determining the position of the emission.

Several types of FD receivers were used on Navy aircraft. Some had the loop antenna was mounted in an aerodynamic housing on the exterior of the fuselage; a remote drive control provided a readout of the loop's relative position. Typical of these was the DZ-2, first supplied by RCA in 1939. This set covered 15-70 kHz and 0.1-1.75 MHz. A line antenna between the cockpit and tail was used in finding "true" direction. Larger planes and blips commonly used an ARB receiver with an auxiliary DU-1 loop antenna unit. The ARB was described in the previous section on Aircraft Radios. The DU-1 had a 12-inch rotatable loop for determining the line to the transmitter and an outside antenna for finding the specific

DZ-2 DF Receiver

DU-1 Loop Antenna Unit

direction. The DU-1 units were mainly built by Bendix. (A 1942 DU-1, looking like it just came out of the Bendix factory, presently sits on the author's credenza.)

The SCL also developed a number of DF systems, primarily for use on land. These ranged in frequency coverage from 100 kHz to 156 MHz and in physical size from the SCR-504, fitted into a suitcase to be carried inconspicuously by an agent, to the large SCR-502, used for intelligence and ground-aided air navigation. A number of these were Army versions of the Navy units.

Radio Controlled Aircraft – Drones

The Navy had long had an interest in remotely controlled aircraft, with the NRL demonstrating a pilotless N-9 aircraft in 1924. After a decade lapse, this activity was revived in 1935 with the initial development of an anti-aircraft target designated the NT. The aircraft would be developed at the Naval Aircraft Factory in Philadelphia, and the necessary radio equipment furnished by the NRL Radio Division. Lieutenant Commander Delmar S. Fahrney, sometimes called the "Father of Guided Missiles," was assigned to lead the project. Following the British in naming a similar radio-controlled target the "Queen Bee," Fahrney called his aircraft a "drone." Converted from the Curtis N2C-2, the drone was first flown in March 1937, and the next year it was successfully used as an aerial target for the USS *Ranger*.

Remotely controlled target aircraft also came from the civilian sector. One that became very popular was developed by movie actor Reginald Denny, who had been a British pilot in WWI. In the early-1930s, Denny (actually named Reginald Leigh Dugmore), became fascinated with radio-controlled model planes. In association with Walter H. Righter, they developed the RP-1 with a wingspan of 12 feet and speed of 60 mph. In 1935, this was demonstrated to the U.S. Army as a possible artillery target, with the speed-to-size ratio corresponding to a full-size aircraft. This went through several versions, and in 1939 an initial order was placed by the Army; an order from the Navy quickly followed.

The RP-5 version had a flight duration of a little over an hour and a speed of 85 mph. It was catapult-launched and recovered using a 24-foot-diameter parachute. The Army designated this the OQ-2 and the Navy the TDD-1 (Target Drone, Denny). The Denny-Righter firm, Radioplane Company, set up a factory at the Van Nuys Airport in June

Target Drone by Denny

1941. The radio control system was supplied by Bendix. A total of 14,891 of these drones were built for the Navy and Army Air Force through 1945.

As a side note, Denny gave permission to *Yank* magazine for a war industry photo-shoot at the Radioplane plant. During this, a pretty young assembly-line worker, Norma Jean Mortensen, became a magnet for the photographers. She is better known by her later screen name, Marilyn Monroe.

Early success with the radios on the NT drone led the Bureau of Aeronautics to investigate using radio for testing new aircraft. A contract was let to RCA for developing a television system for real-time monitoring of onboard instruments. The television was first field-tested in early 1941, providing usable picture information from a plane up to 30 miles from the ground receiving station. This was augmented by other sensors, leading to the now commonly used telemetry monitoring systems.

In August 1941, at the request of the Bureau of Aeronautics, the NRL investigated using a manned aircraft with radar to control drones as guided missiles. Terminal guidance would be by a miniature television system, called "Block," being developed by Dr. Vladimir Zworykin at RCA. Again led by Fahrney, a top-secret activity known as "Project Option" emerged in February 1942. In April, the system was successfully demonstrated, launching a dummy torpedo against the destroyer USS *Aaron Ward* while under guidance from a pilot aboard a control aircraft eight miles away. The "fish" ran dead on target and passed directly under the hull of the *Ward,* despite the fact that the destroyer made evasive maneuvers. Fahrney, in his report, wrote exuberantly,

Airborne TV Camera

> Weaponeers down through the ages had sought a missile that could be launched from a safe distance and guided unerringly to a target. There off Narragansett Bay such a missile was successfully demonstrated.

With the highly successful demonstration, orders were placed for 162 control planes and 1,000 drones. Hundreds of control-plane pilots would be needed and over 1,000 radio technicians assigned when the system went into combat. The first assault drones were delivered in December 1943, and the order increased to 2,000. This drone, however, proved to be difficult to produce and was replaced by the TDR, a converted twin-engine Beechcraft JRB-3. These drones, built by Interstate Aircraft, were adapted to carry a 2,000-pound general-purpose bomb, or several 500-pound bombs.

Beechcraft JRB-3

In early 1944, the head of the Bureau of Aeronautics was replaced, and the new leader questioned the wisdom of committing to an "unproven weapon." A part of the opposition came from the fact that two similar activities – Army Air Corps' Project Aphrodite and the Navy's Project Anvil – were underway in Europe and were plagued by failures but few successes.

In Project Anvil, an unmanned PB4Y-1 (Navy version of the B-24) loaded with 12 tons of explosives was to be remotely guided to a high-priority German target in occupied France using images from an on-board Block camera. Pilots would get this "flying bomb" into the air, arm the explosives, then parachute to safety as control was taken over from an accompanying aircraft. On August 12, 1944, the first Project Anvil mission was attempted. One of the two PB4Y-1 pilots was Navy Lieutenant Joseph P. Kennedy Jr., son of the U.S. Ambassador to the Court of St. James and brother of future President John F. Kennedy. Shortly after take-off, something malfunctioned and the explosives detonated, destroying the plane and instantly killing the pilots. A second mission was moderately successful, but the project was cancelled.

PB4Y1 Aircraft

Project Option was drastically scaled back, but a Special Task Air Group (STAG-1) of TBF Avenger control planes and their TDR drones finally saw combat in the South Pacific in late 1944. Over a one-month period, 46 TDRs were launched, with 37 reaching target areas and at least 21 with successful executions. The greatest problem was inadequate television contrast between the targets and their backgrounds, particularly over land. Assignment of STAG-1 in the South Pacific was discontinued at the end of 1944, when the initial inventory of TDRs was expended, drawing to a conclusion America's first wartime application of the now-indispensable assault Unmanned Aerial Vehicle (UAV).

It might be noted that during the period between 1917 through 1945, the Navy ran tests on 20 different aircraft configured for radio control, most of them during WWII. After the war, Fahrney drafted plans for 18 types of Navy guided missiles.

VT Proximity Fuse

Although the variable-time proximity fuse never required electronics maintenance, this was a highly specialized, WWII radar-like device and set the stage for post-war miniaturized electronics. Also, its development and production was one of the most costly activities of the war. Thus, it has been included in this Appendix.

In Great Britain, while he was testing the coastal defense radar (the CD) in 1938, William Butement conceived the idea of a radar fuse on the projectile to improve the kill probability. He made an excellent design, but was unable to interest defense officials in the development.

When the National Defense Research Committee (NDRC) was formed, Dr. Merle A. Tuve of the Carnegie Institution was asked to head a section devoted to developing a proximity fuse. In August 1940, a laboratory devoted to this project was set up in the Department of Terrestrial Magnetism. The Tizard Mission brought Butement's design to the U.S., and his basic circuit was adapted.

Designated the VT (Variable Time) Fuse, this was a substitute for the conventional fixed-time (set by the expected time of flight) or contact (detonation upon impact) fuses. The designation indicated that the time to detonation was determined by the variable flight-time to the target. This miniature, continuous-wave radar operated in the frequency range of 180-220 MHz. Using the body of the fuse as both a transmitting and receiving antenna, the Doppler-shifted RF return from the target beat with the original RF to form an audio signal, the amplitude of which was used to trigger the detonation. Screwed into the head of a projectile, it needed an operational range of only a few feet. By early 1941, the concept was proved through firings of unminiaturized devices.

Conventional (left) and VT Fuses

Dr. Richard Roberts led the efforts on electronic miniaturization and shock resistance. Central to this was the development of tiny vacuum tubes that could withstand the initial acceleration and the high-rotation rate while on the trajectory. Starting with existing tubes for hearing aids, a suitable triode, pentode, and thyatron were designed by Raytheon and Hytron. The circuit used a triode in the oscillator/detector circuit, a pentode and a second triode as Doppler-signal (200-800 Hz) amplifier, and a thyatron output trigger.

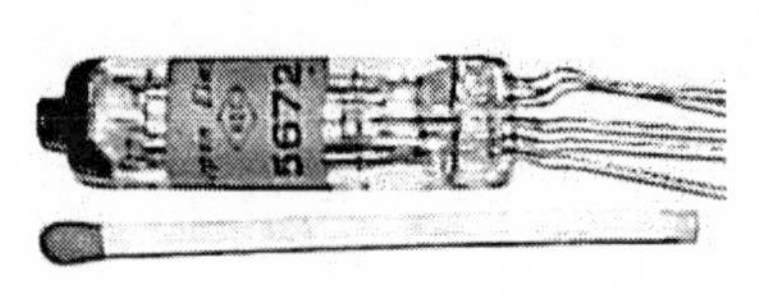

VT Fuse Tube
Compared With Matchstick

A tiny battery with a long shelf life was another major problem. National Carbon came up with what was called a "reserve" battery -- a small wet cell containing the needed electrolyte stored in a glass ampoule that broke upon launching. Premature detonation was a serious safety problem; this was solved by using a mechanical device that turned on the electronics in about a half-second after launch

Many industrial and academic research organizations became involved in various parts of the project and, because of very tight security, most were not aware of each other or their accomplishments. Consequently, a number of organizations lay claim to developing the proximity fuse. The overall project was later transferred to the newly formed Applied Physics Laboratory of Johns Hopkins University.

The first device, the Mk 32 for the Navy's 5-inch antiaircraft weapon, was test fired in April 1942, and production by the Crosby Corporation started the next September. The first Japanese plane shot down using the Mk 32 was in January 1943. As the initial problems on the Mk 32 were solved, work began on the Mk 33, a smaller device for the British 4-inch antiaircraft gun. The Army pressed for a suitable VT fuse for their field artillery, and the Mk 41 for a 3-inch projectile evolved. This was followed by a number of other types for a variety of Allied weapons. In most of these, as well as the improved Mk 32, the reserve battery was replaced by a small wind-driven electric generator.

Throughout the war, over 100 companies and research institutes were involved; 110 plants manufactured parts and assemblies for in excess of 20 million devices. In 1942, the Mk 32 cost $732 each; by the end of the war, the average device had an incremental cost of $18. The total program is estimated to have cost about $1 billion, one of the largest outlays of the war.

Germany had experimental work in proximity fuses at least as early as 1939. Several types of devices were designed and tested for anti-aircraft missiles being developed at Peenemunde, but none were put into production before the war ended. The American and British military units were initially prohibited from using VT fuses where there was a possibility of the enemy recovering an unexploded warhead.

Bombsights

Bombsights for aircraft came into being during World War I with a device made by the Edison Phono Works. This unit was mounted outside the fuselage, and the bomb aimer had to lean over the side to use it. In 1911, Dutch engineer Carl Lucas Norden joined Sperry Gyroscope and is credited, along with Elmer A. Sperry, with developing the first gyrostabilizing equipment for ships. While with Sperry, Norden conceived of using the gyroscope for use in an accurate bombsight, but resented Sperry's requirement that all patents related to gyrostabilizers be assigned to his company. In 1913, Norden left to work on his own, and this was the beginning of a bitter rivalry that lasted for decades.

In 1923, Norden teamed with Theodore H. Barth for developing these devices. They received their first bombsight contract in 1928 and formed Carl L. Norden, Inc. In this period, Norden also served as a consultant to the NRL on the Navy's radio-controlled airplane project. By 1931, Norden had developed the Mark XV, a much-improved bombsight. In a demonstration for the Bureau of Ordnance, it was used in bombing the hulk of the USS *Pittsburgh*, a retired heavy cruiser. The accuracy so impressed Navy officials that they ordered forty sights, initiating a Navy-Norden relationship that lasted for decades. By 1935, minor improvements to the Norden unit allowed accuracies to within 100 feet from drops at altitudes over 12,000 feet.

To obtain good bombsight accuracy, the aircraft must correct for drift while maintaining a constant air speed and altitude; even minor fluctuations could cause a miss. To solve this flight problem, Norden devised a gyrostabilized automatic pilot that could be actuated to take over to reduce the effects of turbulence and pilot over-controlling. The bombsight computed the release point based on ground speed, altitude, bomb weight, and drift information programmed into it by the bombardier. Once the target was visually identified, the bombsight was "locked" onto the target and the bombardier could take over aircraft controls and keep the cross-hairs of the sight centered on the target. If the target drifted from the crosshairs, the bombardier could correct using controls on the bombsight. The sight would then calculate the proper "time" of release. Upon bomb release, a light would signal the human pilot to resume control of the aircraft.

The Norden bombsight was not an electronic device; it was a mechanical analog computer composed of gyros, motors, gears, mirrors, levers and a telescope. It had two principal parts: the Stabilizer and the Sight Head mounted atop. The critical part of the Stabilizer was a large gyroscope that sensed deviation about the yaw (vertical axis) of the airplane. The Stabilizer was an essential part of the aircraft autopilot and was usually on a vibration-isolating mount supplied by the aircraft manufacturer.

The Sight Head, mechanically coupled with the Stabilizer, would pivot about the yaw axis and remain directed toward the selected target should the aircraft deviate from its course. By this, the bombardier controlled the aircraft through the autopilot during the final bomb run. The Stabilizer included an Automatic Bombing Computer, another analog device. After the bombardier made his calculations, they were entered into this computer. In turn, the computer compensated for any evasive lateral maneuvering of the aircraft without any need for the bombardier to make recalculations.

Initially, the bombsights were all produced at the Norden factory in Manhattan. Skilled craftsmen hand-made essentially every part of what might be compared with a large Swiss watch. Hundreds of ball bearings had to be ground and polished for each unit. Between 1932 and 1938, the factory produced an average of only 121 bombsights each year, with Norden personally inspecting essentially every piece. As preparations for war were made, production was greatly increased and other manufacturers were brought in, all working in some of the most carefully guarded facilities of their time. Norden himself was always accompanied by bodyguards.

During use of the Norden bombsight, great precautions were taken to guard its secrecy. Under armed guard, the unit was loaded onto its aircraft just before takeoff. After the aircraft landed, the unit was immediately removed and secured, again under armed guard. Over 45,000 Army Air Forces and Navy bombardiers were trained in its operation, each swearing under oath to protect its secrecy. If an aircraft had to be ditched, the crew was required to jettison the bombsight to keep it out of enemy hands.

Despite all of this security, the Germans had obtained the plans for the Norden unit as early as 1938. Hermann W. Lang, a not-yet naturalized German immigrant, was allowed to work at the Norden Manhattan plant as a machinist, draftsman, and inspector; he was also a traitor and spy. Code named "Paul," he sent full drawings of the bombsight to Germany and even went there to assist in assembling a unit. A double agent betrayed him, and, after being tried in one of the grandest espionage cases in U.S. history, was sentenced to 18 years in prison. The Germans, however, never built the Norden unit; this bombsight was for high-altitude bombing, and the Luftwaff essentially used only low-altitude and dive-bombing.

The basic configuration of the Mark XV (called M-1 by the Army) remained the same over the years, but various attachments were made to improve the accuracy. The ultimate model during the war years was a complex assemblage of more than 2,000 components. With it, a fixed speed and altitude had to be maintained for only 15 to 20 seconds while setting the bombsight. The manual cited the bombsight's accuracy as allowing a bomb to be placed inside a 100-foot circle from four miles up; bombardiers, however, said it could "put a bomb in a pickle barrel from 20,000 feet." When asked if that was true, Carl Norden often responded, "Which pickle would you like to hit?" (Here it should be noted that in practice, with bombardiers working in the freezing cold of high altitudes and under stressful assault from enemy aircraft and ground fire, the theoretical accuracy could never even be approached.)

Norden bombsights were used by the Navy on the PBY Catalina, the TBM Avenger, and several other large aircraft, and the Army Air Forces had them on most types of their bombers. The radar altimeter and other radar systems on the aircraft were coupled with the bombsight to provide automatic inputs. For select students, the ETP Secondary School at NATTC Ward Island included the Norden bombsight in association with these electronic systems.

Norden Bombsight

Carl Norden did not personally profit from his bombsight; he sold the rights to the U.S. Government for one dollar. By the end of hostilities, about 43,300 units had been built by Norden and other firms, of which the Navy used some 6,500. A closely held test and calibration bench was available for certain facilities (NATTC Ward Island had one), but the bombsights were mainly sent to a government depot for any needed refurbishment.

In the early 1930s, the Army Air Corps had contracted with Sperry Gyroscope to develop a bombsight called the S-1. Following Norden's 1931 demonstration to the Navy, the Army recognized the superiority of this unit and decided to also adopt the Norden unit for their bombers.

When an Army colonel approached Norden about this, his response was, "No man can serve the Lord and the Devil at the same time – and I work for the Navy."

Sperry Gyroscope continued on its own with developing the S-1 Tecometric bombsight. The Army, seeking to circumvent purchasing the Norden units through the Navy, accepted the S-1 and in March 1942 designated it as "standard" equipment. It was used on some Army Air Forces bombers early in the war, but in late 1943 all S-1 contracts were cancelled.

Bombsight development had also taken place in Great Britain. In 1940, Patrick M. S. Blackett, scientific advisor to the Air Ministry and future Nobel Prize winner, volunteered to lead this development. After an initial concept was proven, Dr. H. J. J. Braddick took Blackett's place and was later named as the co-inventor of the MkXIV bombsight. For production, the British turned to America and Sperry Gyroscope. Sperry was not only interested, but proposed that the Army Air Corps would welcome an alternative to the Norden unit. The unit, Blackett-Braddick MkXIV, was turned over to Sperry for further improvements and production, to be called the T-1 when sold to the Army. Actual production of the T-1 was subcontracted by A.C. Spark Plug Company, a division of General Motors, at its plant in Flint, Michigan.

The primary units of the MkXCIV/T-1 were the Computing Cabinet and the Sighting Head. The units did not have to be matched and were actually interchangeable with units produced in Great Britain. A total of about 23,500 of these bombsights were produced at the Flint plant between November 1942 and July 1945. They were divided between Great Britain and the Army Air Force; it does not appear that any were used by the U.S. Navy.

Appendix II

RADAR DEVELOPMENT IN OTHER COUNTRIES

Pre-war radar development in the Axis countries was primarily in Germany, although there was some independent work in Italy and Japan. Activities in The Soviet Union, France, and The Netherlands should also be mentioned. Here it should be noted that most of these developments were for detection only and did not include range determination; thus, they were not full radar systems.

In 1939, Australia, Canada, and South Africa, being major members of the British Commonwealth, were invited to send representatives to England where they were briefed on the RDF developments. Domestic projects in this technology quickly followed.

GERMANY

As early as 1904, patents were registered to Christian Hülsmeyer for a radar-like apparatus, the *Telemobilskop*, intended to prevent collision of ships. Nothing came from this invention, and little further work was done in Germany for many years. In 1923, five years after their defeat in World War I, the Kriegsmarine (German Navy) established the Torpedoversuchsanstalt (Torpedo Research Establishment) to investigate electronic detection systems. In 1933, while developing sonar at this laboratory, Dr. Rudolf Kuenhold conceived a similar apparatus using radio beams. Using a 600-MHz, 50-watt split-anode magnetron purchased from Philips, in March 1934 he demonstrated a continuous-wave system that detected a battleship at a short distance in the Kiel harbor. This was much the same as the earlier observations in the U.S. at the NRL. Nevertheless, Kuenhold is often credited in Germany as the inventor of radar.

It was then that work in Germany on *Funkmessgerat* (radio measuring device) for *Erknnung* (reconnaissance) began in earnest. A firm, Gesellschaft fur Electroakustische und Mechanische Apparate (GEMA), was established in 1934. Kuenhold joined GEMA and, funded by the Kriegsmarine, worked to improve his CW system by making it pulsed. In May 1935, he demonstrated a 375-MHz, pulse-modulated radar, a few months behind the Americans and the British. Called *Seetakt*, this was the first system built by GEMA.

Accomplishments of Dr. Hans Erich Hollmann were central to the development of radar in Germany. In 1936, he completed the first comprehensive treatise on microwaves, *Physik und Technik der ultrakurzen Wellen [Physics and Technique of Ultrashort Waves]*; included was the application of microwaves in measuring devices. The previous year, he had filed German and U.S. patent applications (U.S. No. 2,123,728 granted in July 1938) on a multi-cavity magnetron.

Hollmann

Dr. Wilhelm T. Runge, laboratory director at Telefunken, also conducted early research in radar. Actually, Kuenhold had discussed radar with Telefunken before going with GEMA, but felt snubbed by Runge. Subsequently, the two radar developments were totally independent. Runge, with Hollmann as a consultant, built a 150-MHz, pulse-modulated experimental radar. The system included high-frequency "acorn" tubes purchased from RCA and the split-anode magnetron from Philips. It also had a cathode-ray tube display developed in the late 1920s by Dr. Manfred von Ardenne. In May 1935, the system observed a Doppler reflection from an aircraft. An article in the September 1935 issue of the U.S. magazine *Electronics* was headlined as follows: "Telefunken firm in Berlin reveals details of a 'mystery ray' system capable of locating position of aircraft through fog, smoke and clouds."

In late 1935, GEMA successfully demonstrated a 575-MHz pulsed radar (code named *Dezimeter Telegraphie* or *De Te*) using a "mattress" antenna (similar to the CXAM's "bedspring"). Trials were held aboard the vessel *Welle*, believed to have been the first warship ever equipped with radar. In late 1937, the radar developed by Kuenhold became the *Freya*, a ground-based radar operating around 125 MHz with a range of 80 to 150 miles. GEMA also developed *Stichling*, an IFF system for this radar. *Freya* was used effectively by the Heer (German Army) and Luftwaffe (German Air Force) against British bomber raids. GEMA ultimately led German radar work, with central operations growing to over 6,000 employees.

Freya Radar

Many types of radars were developed in Germany throughout the war; a few will be described. The major firms producing radars were GEMA, Telefunken, Lorentz, Siemens, and A.E.G. (Allgemeine Elekrizitats Gesellschaft, a German operation of General Electric).

Wurzburg Radar

The *Wurzburg*, a 560-MHz system, was built by Telefunken. With 10-foot diameter reflector antennas, this was used to direct anti-aircraft guns. Lorentz supplied the *Kurfurst*, a gun-laying system with a range of about 20 miles. This operated at 470 MHz using 7-foot parabolic antennas. Telefunken also developed *Klein-Heidelberg*, a very ingenious passive radar system. Used along the English Channel and North Sea opposite the Chain Home installations, this system's stations picked up signals directly and as reflected from aircraft, with the timing of two signals used for a simple early-warning system.

In 1940, the Germans initiated a well-organized air-defense network. The Allies called this the "Kammhuber Line," after the developer Colonel Josef Kammhuber. This used *Wurzburg* radars with the dish antennas enlarged to over 24 feet. Over the years, the Kammhuber Line was enlarged and included *Freya* radars for early warning. The *Freya* was continuously improved by GEMA, with over 1,000 systems built throughout the war. One version, introduced in 1941, was the *Mammut* (Mammoth) using 16 *Freyas* linked into a giant (100 by 33 feet) antenna using phased-array beam-directing. Another version, the *Wasserman*, used eight *Freyas* with phased-array antennas stacked on a steerable, 190-foot tower and giving a range of about 150 miles. About 150 of these were built starting in 1942.

Mammut Radar

Telefunken added phased-array antennas to their *Wurzburg* in 1943. The improved system was called *Mannheim*, and about 400 sets were built for gun-laying in anti-aircraft batteries *Jadgschloss* was a system designed to provide long-range, wide-area tracking of Allies' bombers as they swept across Germany. Built by Siemens, this radar operated at 150 MHz and was placed atop a tall tower. Put into service late in 1943, the system had Germany's first PPI display, with the output communicated to a command center via a cable or RF link

In the Kriegsmarine, most capital vessels had a gun-ranging radar, the *Flakleit*, evolved from the original *Seetakt*. This system, built by GEMA, operated at 375 MHz and had a 50-mile range against large vessels. In 1939, after an epic battle with a British cruiser pack, the battleship *Graf Spee* was scuttled off the coast of Uruguay. Salvage

Flakleit Radar

operations later gave the British valuable information concerning German radars.

Airborne radars were available for nightfighter aircraft in the Luftwaffe. In early 1942, Telefunken started supplying the *Lichtenstein B/C,* based on the *Wurzburg* and operating at 560-MHz. The range was limited to about 3 miles. To increase the range, the *Lichtenstein SN2,* operating at 91 MHz and with much greater power, was introduced later in 1942. The Luftwaffe's interest, however, was more on radio navigational systems for bombers. Several types of radio-based guidance systems were developed and supplied by Lorentz, using its technology that had dominated the aircraft guidance market worldwide for many years.

In early 1943, a British bomber was shot down near Rotterdam, and the Germans found an H2S radar containing a 10-cm resonant-cavity magnetron. Telefunken was given the mission to reverse-engineer the equipment. (It might be noted that six years earlier the Japanese had developed a similar magnetron, but never shared this with the Germans.) The magnetron and the T-R switch were fabricated by Sanitas GmbH, and a 10-cm radar, called *Rotterdam,* was demonstrated by Telefunken in mid-1943. Sometimes called *Marbach,* this system was produced in limited quantities and used in anti-aircraft batteries around a number of large industrial cities.

Attempts were made to develop several other 10-cm systems, but none made it into mass production. One was called *Berlin,* and an airborne version for nightfighters saw very limited service in early 1945. The Germans also salvaged American blind-bombing radars containing 3-cm magnetrons. These systems were duplicated as the *Meddo* and *Rotterdam-X* (the latter indicating that the Germans might have been aware of the "X-band" designation used by the Allies), but they were never produced. An important application of both the 3-GHz and 10-GHz equipment was in developing countermeasures, particularly warning receivers.

JAPAN

Prewar work on radar technology in Japan was primarily conducted in their universities. Professor Hidetsugu Yagi of the Osaka Imperial University was a leader in this activity, having learned much from his studies in Germany and visits to research laboratories in Great Britain and the United States.

Yagi and his Antenna

One of Yagi's students, Kinjiro Okabe, developed a split-anode magnetron in 1929. Yagi and another of his students, Shintaro Uda, developed a "wave projector," patented in 1932 as the Yagi-Uda antenna. Often called "Yaggi," this was licensed to RCA and used in early U.S. radars, as well as in various systems throughout the world. In 1936, Yagi demonstrated a 10-cm Doppler radar using Okabe's magnetron tube. The Japanese military, however, showed little interest at that time.

Tsuneo Ito and Kanjiro Takahashi, working in the Electronics Laboratory of the Imperial Navy (ELIN), developed a series of resonant-cavity magnetrons starting in 1937. The power of these devices, however, was not sufficient for a radar. The magnetron concept was eventually revealed by ELIN to Japan Radio for improvement. In late 1938, Shigeru Nakajima of this firm built a magnetron delivering 500 watts at 10 cm. Experimental work had been underway on CW microwave radar, but, even with the higher power, the system was not adequate for applications.

A major hindrance to Japan's radar development was the intense rivalry between the Army and Navy; essentially nothing was shared. For example, in early 1941, separate delegations from the two services went to Germany. Unlike in the Tizard Mission, little secret information was exchanged between Germany and either of the Japanese groups. The Army group learned almost nothing about German radar. The Navy group did learn about the benefits of pulsed modulation, as well as the advantages of the British in using radar in naval battles with the Germans. They did not, however, tell the Germans anything about their resonant-cavity magnetron development.

Mark I Model 1

Upon returning to Japan, researchers from ELIN initiated major efforts in radar development for shore and shipboard applications. With development assistance from Japanese industries NHK and NEC, a 100-MHz warning system for shore positions was put into production as the Mark I Model 1 just before the attack on Pearl Harbor. Mark I Models 2 and 3, transportable versions of Model 1, were put into service in 1942 and 1943, respectively.

In late 1941, experimental 200-MHz search radars were installed on two vessels, the *Oi* and the *Kitagami*. These were the first Japanese naval ships with radar, and the only ones as the war began. Continued development led to pulse modulation, more sensitive receivers, and microwave radar. A 10-cm shipboard surface-search/fire-control radar, designated Mark II Model 2, was put into service in mid-1942. Two large, side-by-side, round-horn antennas characterized this system. About 400 of these microwave radars were built and used on several types of surface warships as well as submarines.

Mark II Model 2

In early 1942, the Japanese captured a British GL radar and American SCR-268 and SCR-270 sets. The ELIN reverse-engineered the SCR-268, putting it into production as the Mark IV Model 1 and, later Model 2. Operating at 200 MHz, some 2000 of these were built during the war and used for base protection. The ELIN also developed an airborne search radar, the 150-MHz Mark VI Model 2, which went into service in early 1944.

While the ELIN was the initial and primary developer of Japanese radar, the Imperial Army soon followed, but depended largely on industry for the work. Most of this activity was centered in Tama, an advanced industrial region near Tokyo, where the Tama Army Technology Research Institute was eventually set up. Radars for the Imperial Army were mainly designated "*Tachi*," with the "*Ta*" indicating Tama, and the "*chi*" indicating earth or ground-based. Similarly "*Taki*" would be for Army air-borne radar, but none of these were deployed.

For the Imperial Army, NEC and Toshiba developed *Tachi* 1 and *Tachi* 2, 200-MHz gun-laying systems based on the SCR-268, but these were generally found to be unsuitable for field use. Both NEC and Toshiba produced improved versions of the British GL. Designated *Tachi* 3 and *Tachi* 4 and operating at 80 MHz and 200 MHz, respectively; these were widely used as searchlight-directors for anti-aircraft guns.

Tachi 3

At the same time, the Tama Institute developed for the Imperial Army a ground-based, 75-MHz air-warning radar similar to the British CH system. This transmitted a broad beam, with the reflected signals picked up by receivers at several locations using directional antennas. Designated the *Tachi* 6, the system went into service in late 1942. *Tachi* 6 was made into the transportable 100-MHz *Tachi* 7 that was deployed in 1943, followed by *Tachi* 18, a portable version, in 1944. *Tachi* 20 and *Tachi* 35 were height-finding radars, but only a few were built.

As the war continued, the Imperial Navy greatly accelerated radar development, but also changed the designation system, going to a Type number for shipboard sets. To add to the confusion, they also redesignated existing systems in this manner. Although a number of new systems reached the experimental stage, only a few became operational before the end of the war. These included the H-6, a 150-MHz ASV system, operational in late 1942 on large patrol aircraft, and the Type 22, a 10-cm search system released in mid-1944 and widely used on surface vessels and submarines. Some Navy radars that did not make it into operational status included Type 23, a 60-cm (500-MHz) system that was powered by a resonant-cavity magnetron, and Types 31, 32, and 33, additional 10-cm systems. There was also the TH, a 200-MHz set intended for *Shinyo* suicide boats to defend the homeland against landing forces.

In 1943, the Germans finally made available information on the *Wurzburg*. Based on this system, the Imperial Army initiated the *Tachi* 24, and the Imperial Navy started development of the Mark II Model 3. Prototypes were completed in 1945, just before the war ended. Altogether, about 30 different types of radars were developed in Japan during the war, with approximately 7,000 sets of all types produced.

With the development of high-power tubes during the war, interest arose concerning the use of microwave beams for anti-aircraft purposes. Designated Project Z, this was conducted in great secrecy and involved some of Japan's best physicists; these included Drs. Hideki Yukawa and Sin-Itiro Tomonaga, both future Nobel laureates. Little progress was made before the end of hostilities.

THE SOVIET UNION

The Soviet Union has a great heritage in radio development. In fact, Russian engineer Alexander A. Popov is often credited with the first demonstration of wireless telegraphy in March 1896. During the 1920s and early 1930s, D. A. Rozhanski and A. A. Chernyshev at the Leningrad Physical-Technical Institute (LPTI), directed major work on radiophysics. Rozhanski is credited with the initial proposal for velocity modulation of electron flow to increase the operating frequency of vacuum tubes. An associated laboratory, the Ukranian Physical-Technical Institute (UPTI), was involved in microwave research. The Leningrad Electro-physical Institute for Communications (LEIC) was split off from LPTI in 1930.

Popov

Based on Rozhanski's laboratory studies, in early 1934, Yu. K. Kobzarev at the Red Army's Central Radio Laboratory used a 150-MHz transmitter to demonstrate Doppler signals reflected from an aircraft at close range. P. K. Oshchevkov, working in a radio research team led by B. K. Shembel at the LEIC, continued with Kobzarev's work. By mid-1934, Oshchevkov developed a 60-MHz bi-static system with antennas separated about 7 miles. Called *Rapid*, it was demonstrated in detecting Doppler reflections from aircraft at a 50-mile range.

Control of the LEIC programs was taken over by the Red Army's Scientific Research Institute 9 (SRI-9) in 1935. Shemble's team continued there, and in 1936 initiated development of a 1.7-GHz (18-cm) system called *Storm* for anti-aircraft gun-laying. Stalin's Great Purges led to Shemble's dismissal in 1937, and Oshchevkov being sent to the Gulag. *Storm* used a resonant-cavity magnetron developed at the UPTI by N. F. Alekseev and D. D. Malairov. Although the continuous power was possibly as much as 50 watts, the microwave radar project was eventually abandoned. Their magnetron, however, was described in the open literature in 1940.

RUS-1 Transmitter

In early 1939, the bi-static apparatus, *Rapid*, was developed into RUS-1, *Rhubarb*, the Soviet's first deployed system. This had a truck-mounted transmitter and two truck-mounted receivers spaced about 25 miles. Some 45 units were produced and placed into service in the Caucasus and the Far East. This was actually a radio-location system, not a full radar, and had little value.

The SRI-9 also started development of a pulsed air-warning system in 1936. This was also a bi-static system using a 75-MHz, 50-kW transmitter with a receiver located about a half-mile away. This eventually became RUS-2, *Redut*, the Soviet's first true radar system with pulsed transmission. It was placed into limited production in 1939. In 1940, the RUS-2 was improved as the RUS-2S with a single antenna. This system had a power of around 100 kW and a range of about 90 miles. In the Soviet Navy, the *Redut-K*, a shipboard version of RUS-2S, was installed in 1940 on the cruiser *Molotov*, the first Soviet ship with a radar system.

RUS-2S System

When the Soviets began the war with Germany in June 1941, the RUS-2S was successfully used in the defense of Moscow and Leningrad, and radar finally gained the interest of the military. With Leningrad in danger of invasion, SRI-9 set up an operation in Moscow in late 1941. Several experimental versions of a 2-GHz (15-cm) radar using another UPTI resonant-cavity magnetron were built by a team led by M. L. Sliozberg, but none were satisfactory.

During 1944-45, a number of shipboard radars were developed and tested, but did not go into production until after the war; these included an air-warning series called *Gyuis*, and a series called *Redan* and *Vympel*, both for fire-control.

Most of the radars used by the Soviets during the war were from Great Britain and America, primarily provided under the Lend-Lease Plan. These started to arrive in 1942, and by the end of the war most of the large warships were equipped with radar. Naval radars from Great Britain included the Types 281, 282, 284, 285, 286, and 291, the last becoming the most widely used radar in the Soviet Navy. Also included was the Type 271, holding the very secret, high-power, resonant-cavity magnetron. Radars from the U.S. included the SG, SK, and FC.

For the Red Army, the British provided units of their GL radars, as well as American SCR-268s and 545s. The Sliozberg team copied the GL and it was put into production as the SON-2, but only about 100 sets were built. Late in the war, a few SCR-584 systems were provided; this system was eventually produced as the SON-4.

THE NETHERLANDS

In 1933, Klass Posthumus at the Physics Research Laboratory (NatLab) of Philips Electronics in Eindhoven developed a magnetron with an anode split into four elements that generated 50 watts at 2 GHz (15 cm). This tube was manufactured by Philips and sold to researchers in several other countries. The Dutch Royal Navy sponsored work by C. H. J. A. Staal at NatLab on a detection system using this magnetron and other components developed at Philips. By 1939, the system could detect a ship at a distance of two miles, but further work was terminated upon the invasion by Germany.

Following World War I, the Dutch Parliament had set up the Committee for the Applications of Physics in Weaponry. In 1934, this Committee established a laboratory to develop a radio listening device for anti-aircraft guns. At this facility, a team led by J. L. W. C. von Weiller and assisted by S. Gratama successfully demonstrated a pulsed system operating at 425 MHz in 1938 that detected an airplane at 12 miles. The transmitter was built at the University of Delft and the receiver at the University of Leiden. The system used a common antenna array of 16 dipoles for transmitting and receiving, switching at the 10-kHz pulse rate.

Impressed by the demonstration equipment, the Dutch Royal Navy added Staal to the laboratory and ordered 10 sets to be assembled in great secrecy under J. J. A. Schagen van Leeuwen at the firm Signaalapparaten. The first set was put into service at The Hague just before The Netherlands fell to the Germans in May 1940. The system worked very well, spotting the incoming aircraft during the few days of fighting, but there were no associated anti-aircraft guns so the operation was frustrating. Von Weiller and Staal fled to England, carrying two of the initial sets with them. The remaining sets, as well as all plans, were destroyed. Later, both Gratama and van Leeuwen also escaped to England.

Model of Dutch 425 MHz Radar

ITALY

The "father of radio," Guglielmo Marconi, initiated radar research in his native Italy. He had presented a paper to the Institute of Radio Engineers in 1922 that included reference to using reflected radio waves for navigation purposes. In 1931, while participating with his Italian firm in experiments with a 600-MHz communications link across Rome, Marconi noted transmission disturbances caused by moving objects in its path. This led to a 1935 demonstration for the military of a 330-MHz Doppler detection system, but the output power was insufficient for practical use.

Tiberio

Ugo Tiberio, a naval professor and radio researcher at the Regio Instituto Electrotechico e delle Comunicazioni (E.C.), continued with Marconi's work, demonstrating an FM system (E.C.1) in 1936, followed by a pulsed 200-MHz system (E.C.2) in 1937. Fellow researcher Nello Cararra, supported by the vacuum tube firm FIVRE, developed a magnetron producing 10 kW at 425 MHz (70 cm). By 1940, a prototype Radio Detector Telemeter (RDT) designated E.C.3 had been designed for coastal defense and shipboard applications. Italy entered the war, however, without deployed radar systems. The Italian Navy sent a commission to Germany in June 1940, but they were given no technical information on radars.

In March 1941, the Italian Navy suffered greatly in an engagement with radar-equipped British warships. The E.C.3 prototype RDT was quickly reevaluated and sets placed into service as the *Folage*, a 200-MHz system for coastal surveillance, and the *Gufo*, a shipboard system operating between 400 and 750 MHz. The vessel *Littorio* was the first Italian warship with radar. The Italian firm Marelli received an order for 150 *Folaga* RDTs, while SAFAR was to build 50 of the *Gufo* sets. The lack of skilled personnel and damages due to Allied bombing, however, resulted in only a few sets being delivered to the Navy by the time Italy surrendered in September 1943.

Gufo Radar

The Italian Army and Air Force had also ordered RDTs to be designed and built for land and airborne applications; several Italian industries participated. Except for a small number of the *Lince*, a ground-based air-search system developed by SAFAR for the Air Force, no other systems went beyond prototypes.

FRANCE

In the late 1920, Pierre David at the Laboratori National de Radioelectricite had experimented with reflected radio signals at about a meter wavelength. During the same time, Professor Camille Gutton of the Faculte des Sciences de Nancy studied reflections at 16 cm. Both activities gained the interests of Dr. Maurice Ponte, research director at the firm Societe Francaise Radio-Electrique (SFR).

Ponte

In 1934, David demonstrated Doppler detection of an aircraft using a 300-MHz apparatus with widely spaced antennas. He called this a "barrage electromagnetique" (electromagnetic curtain). This led the French Navy to fund SFR for a basic system that was tested in 1936, including techniques developed by David for estimating aircraft speed and direction.

David's demonstration eventually led to the Army's funding SFR to develop a network of Detection Electromagnetique (DEM) stations in the region between Paris and Germany. Similarly, the Navy contracted with SFR for several stations at seaports. A number of these DEM stations were installed between 1938 and 1940. The Army's stations to the East of Paris were completed in May 1940, but were almost immediately destroyed so as to not fall into German hands. The Navy's DEM station at Richelleu detected a formation of invading Italian planes on June 15.

Microwave work in France took a very different direction. Dr. Henri Gutton, son of Professor Camille Gutton and assistant to Ponte at SFR, initiated work on a collision-avoidance system based on his father's studies. On July 20, 1934, SFR filed a patent for "a new system of location of obstacles and its applications." The description is for a 16-cm, CW system with parabolic transmitting and receiving antennas separated by about 20 feet. A magnetron (likely from Philips) was indicated as the transmitter and a Barkhausen tube as the receiver.

In late 1934, tests of Gutton's system were made on the cargo liner *Oregon*. Two systems were tried: one microwave set operating at 1.9 GHz (16 cm) using Barkhausen tubes and the other at 375 MHz (80 cm) using a split-anode magnetron developed by Dr. M. R. Warneck at SFR. The coast and other vessels were detected at distances up to 12 miles by the 375-MHz set, but the narrower beam of the microwave system concentrated the power and also gave better directional accuracy. This shorter wavelength was selected for further tests.

"Obstacle Locater" on the SS Normandy

The first demonstration at sea of the microwave set was on the new trans-Atlantic passenger liner, SS *Normandy*. The system was used for iceberg warning during a crossing in 1935, but it was not well received by the ship's officers. It also became inoperative due to water damage. No further trials were made before the war.

Warneck's Magnetron

Earlier, during tests at the SFR, a Doppler signal was observed from an aircraft. Based on this, Ponte prepared a proposal in March 1935 for "the location of mobile objects by microwaves and its immediate applications to national defense." The Army and Navy showed some interest, but believed that the needed power of several kilowatts at the proposed microwave (4- to 6-cm) operation could not be achieved. Toward this end, Warneck greatly improved his split-anode magnetron. To obtain range, Henri Gutton developed a system using a pulse-modulated, 1.9-GHz (16-cm) magnetron. Demonstrated in early 1939, it could determine the range to large vessels at 4-miles distance, but was unable to range aircraft.

As war in Europe drew closer, France and England set up a technical exchange for defense information. Under this exchange, Robert Watson Watt and Arnold Wilkins visited France in April 1939. Following this, Pierre David developed a pulse-modulation system with a 50-MHz, 12-kW transmitter. Produced by the firm SADIR in late 1939, the system could range out to 35 miles.

France declared war on Germany on September 1, 1939. In May of the next year, Ponte personally carried one of Warneck's latest magnetrons to England, ensuring that the researchers there knew this device. After months of waiting, in June 1940 the Germans circumvented the Maginot Line (France's "impenetrable" defenses along the Eastern border) by invading from the North through Belgium, and the Italians came in from the South. An Armistice was signed on June 22. Further French development of radar terminated for the duration of the war.

AUSTRALIA and NEW ZEALAND

In February 1939, the British Government requested that representatives from Australia and New Zealand be sent to England to be briefed on development in radio direction finding (radar). Dr. David F. Martyn, professor of electrical engineering at the University of Sidney, represented Australia, and Dr. Samuel E. Marsden, secretary of the Department of Scientific and Industrial Research (DSIR), represented New Zealand. Upon Martyn's return in August, he convinced Dr. John P. V. Madsen, chairman of the Radio Research Board, to establish the Radiophysics Laboratory (RPL) as a unit of the Council for Scientific and Industrial Research, devoted to developing RDF systems for Australia.

Marsden remained in Great Britain for several months, gaining knowledge of the RDF developments. When he returned to New Zealand in October, he established an RDF research facility within the DSIR. Dr. Frederick W. G. White, a senior scientist with the DSIR, was assigned to develop a gunnery RDF for the New Zealand Navy and also to train RDF scientists. After over a year in this effort, the New Zealand Government decided to depend on Great Britain for defense RDF systems. Research continued, however, with the developed equipment devoted to exploring characteristics of the ionosphere. Dr. Elizabeth Alexander, touted as the world's first female radio astronomer, was a leader in these efforts.

RDF Research Station, Whangaroa, New Zealand

When the RPL was formed in Australia, Martyn was named director and immediately set up highly secure facilities on the campus of the University of Sidney. Dr. John H. (Jack) Piddington, a Sidney University Fellow, was made responsible for system development and Dr. Joseph L. Pawsey, who had previously been with EMI's Television Group in England, was hired for technology research. Their first efforts were in assessing the British RDF for application in Australia. They concluded that the large coast of Australia needed a different type of system. On September 3, 1939, war with Germany was declared.

Piddngton

In 1938, Piddington had developed a pulse-transmission system intended for ionoshperic studies, but his appointment to the RPL changed the direction of his work. His first RPL project was developing a 200-MHz shore-defense RDF system for the Australian Army. Designated ShD, the system was tested at Dover Heights in Sidney in September 1941. Sets were eventually installed at 17 ports around Australia.

ShD Radar at Dover Heights

The Air Council of Canada agreed to open a school for training RDF personnel from throughout the Commonwealth. This school (RAF #31) opened in Clinton, Ontario, in May 1941, with a number of officers and enlisted personnel from Australia in attendance. Realizing the need for indigenous training, Australia established its own programs in September 1941. These included six-month courses for signal officers at Sidney University and for maintenance mechanics at Melbourne Technical College.

With the Japanese attack on Pearl Harbor, the need for an air-warning system became urgent. For the Royal Australian Air Force (RAAF), Piddington's team designed and built a 200-MHz system in five days. Designated AW Mark I, the set was briefly tested using the existing antenna at Dover Heights, then sent to Darwin for installation. Unfortunately, it was not yet operational when the Japanese devastated that city in their initial attack on Australia on February 19, 1942. Before the end of March, the set was in operation and detected a large incoming raid. While still 20 miles offshore, the raiders were successfully intercepted by U.S. fighter aircraft. Eventually, the RAAF established an integrated network of AW sets and successfully repelled the Japanese attacks.

After a few AW sets were made by the RPL, production was turned over to HMV Gramophone (an EMI company). The NSW Railways Engineering Group developed a lightweight antenna, making the set air transportable. Called simply the LW/AW Mark II, 56 of these sets were used by the Australian forces, 60 by the U.S. Army in the early island landings in the South Pacific, and 12 by the British in Burma. The antenna was hand-cranked, and the operators sat in a tent under the structure.

LW/AW Radar

LW/AW Uncovered

In early 1942, U.S. Army troops (including the brother-in-law of the author) started arriving in Australia. Several SCR-268s brought by these troops were turned over to the Australians, who rebuilt them to become Modified Air Warning Devices (MAWDs). These 200-MHz systems were deployed near major cities. In time, many other radar systems were supplied from Great Britain and America. The Royal Australian Navy had essentially all types of British-built radars on their vessel.

After New Zealand had abandoned its defense RDF development, White went to Australia to participate in its projects. In June 1942, Martyn became the director of an Operational Research group for the armed services and White became director of the RPL. During 1943-44, the RPL involved a staff of 300 persons working on 48 radar projects, many associated with improvements on the LW/AW. Height-finding was added (LW/AWH), and complex displays converted it into a ground-control intercept system (LW/GCI). There was also a unit for low-flying aircraft (LW/LFC). Near the end of the war in 1945, the RPL was working on a microwave height-finding system (LW/AWH Mark II).

Dr. Edward G. Bowen, one of the three original developers of RDF in Great Britain, came to Australia in January 1944 and joined the RPL as chief of the Radiophysics Countermeasures. Projects to detect, locate, and jam Japanese radars ensued.

CANADA

Dr. John T. Henderson, head of the Radio Section of the National Research Council of Canada (NRCC), visited Great Britain in early 1939 for briefings on RDF developments. Following this, with almost no assistance from Great Britain, he led a group in developing a 200-MHz surface-warning system using commercial components for the Royal Canadian Navy.

In mid-1940, the system, designated Surface Warning 1st Canadian (SW1C) and often called Night Watchman, was successfully tested, and a prototype was demonstrated in Halifax early the next year. A rather primitive apparatus, it used a Yagi antenna that initially was rotated by a mechanism driven by an automobile steering wheel.

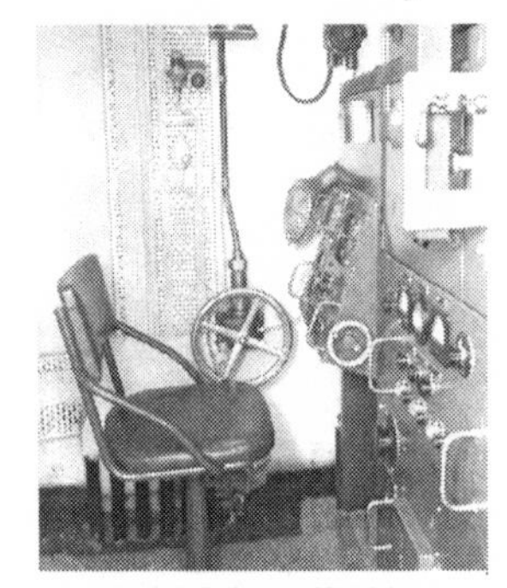

SW1C Installation

There were considerable difficulties in placing the SW1C into production, and by the end of 1941 only a few corvettes and merchant ships had been outfitted. In early 1942, the frequency was changed to 214 MHz to make it compatible with the recently adopted IFF sets. With other minor changes, it was known as the SW2C and produced for corvettes and mine sweepers. A lighter version, designated SW3C, followed for small vessels such as motor torpedo boats.

The Tizard Mission also visited Canada in September 1940, and disclosed the resonant-cavity magnetron. The NRCC then developed an S-band (10-cm) gun-laying system, initially with little technical information from Great Britain. In production, this system was designated the GL Mark 3C, joining the British GL Mark B, and the American GL Mark A (SCR-584).

In 1944, an X-band (3-cm) surface-search radar was developed in Canada for the Royal British Navy. Designated Type 268 (not to be confused with the American SCR-268), it was designed to detect German submarine schnorkles. Only a small number were produced before the end of the war.

Type 268 Radar

British and some American radars were also supplied to Canada. This included equipment for the Atlantic Region Air Defense System with 25 stations along the east coast, and, starting after Pearl Harbor, the

Pacific Coast Air Defense System on the west coast with 11 stations. These were mainly British GCI and CHL radars, but American MEWs were included on Vancouver Island and Newfoundland. Most radars of the Royal British Navy were also used on Canadian vessels.

SOUTH AFRICA

The lead in RDF development in South Africa was taken by the Bernard Price Institute (BPI) for Geophysical Research, a unit of the University of the Witwatersrand (Wits) in Johannesburg. Dr. Basil Ferdinand Jamieson Schonland, a world-recognized authority on lightning and instrumentation for its detection, headed the effort. As a member of the Commonwealth, South Africa was invited to send a representative to Great Britain in early 1939 for briefings on their RDF developments. No representative was sent from South Africa, but Dr. Ernest Marsden, who represented New Zealand at the meeting in Great Britain, went by South Africa on his return trip in August. Schonland boarded Marsden's ship in Cape Town and was briefed by Marsden on the RDF activities as they traveled around the coast to Durban.

Schonland

With nothing more than copies of some "rather vague documents" and notes received from Marsden during his brief sea trip, Schonland assembled a team and started developing an RDF system. Those from the BPI included Dr. Christopher P. Gane, Dr. Guerino R. Bozzoli, and graduate student (later Dr.) Frank J. Hewitt. Other participants included Professor W. Eric Phillips from the University of Natal and Noel Roberts from the University of Cape Town.

Wits Campus Late 1930s

Before the end of 1939, the team used locally available components to build an 85-MHz, 5-kW, pulse-modulated system. It had a steerable array that initially had separate transmitting and receiving antennas with 30 degree beam widths. A commercial oscilloscope was used for an "A-scope" display of range. Designated JB (for Johannesburg), the RDF system (using the British name) was taken to Durban on the coast and tested over a three-week period, detecting ships on the Indian Ocean as well as aircraft to a range of up to 50 miles.

There was a pressing need to protect supply shipping coming down the east coast of Africa. In early March 1940, the first JB was deployed to Mombrui on the coast of Kenya, joining an anti-aircraft Brigade in intercepting attacking Italian bombers. Frank Hewitt joined the South Africa Defense Force and was responsible for JB radar operations throughout the war. Improved JB systems were built, and in mid-1941 several were deployed to Egypt and used in protecting RAF installations near the Suez Canal.

JB systems were also placed at the four main South African ports, guarding the highly important shipping lanes around the Cape of Good Hope. Eventually, British and American radars were also installed, but the JBs remained in operation throughout the war.

Paintings by Geoffrey Long of Improved JB Radar System in Operation

Shortly after the JB went into service, the University of Natal initiated special training courses in radar operation and maintenance. This effort was led by Professor Hugh Clark, head of the Electrical Engineering Department.

In late 1941, Schonland, as a Colonel in the South Africa Defense Force, went to Great Britain to serve as Superintendent of the Army Operational Research Group and scientific advisor to Field Marshall Bernard Montgomery. Returning to South Africa after the war, he founded the Council for Scientific and Industrial Research and served as its president. (Dr. Hewitt was later the CSIR deputy president.) In opposition to the apartheid policies of the South African government, Schonland went back to England in 1954. There he headed the Atomic Energy Research Establishment at Harwell and was knighted by the Queen in 1960 for his services to British science. It might be noted that in 1999, 27 years after his death, Sir Schonland was elected South Africa's "Scientist of the Century."

Appendix III

REPRESENTATIVE STUDENTS

In any educational endeavor, the students are the critical element. The most knowledgeable and highly eloquent instructors are but "sounding brass" unless understanding students are there to absorb their lectures. Similarly, the finest of laboratories are totally worthless unless the students are capable of using them. This was fully recognized in early 1942 when the new Electronics Training Program (ETP – the author's designation) was initiated at a television facility in downtown Chicago. In a few days, an excellent instructional staff had been assembled, and within a short time the station's existing test equipment was quickly expanded to serve in laboratories. The "prototype" Primary School (officially the Elementary Electricity and Radio Materiel School) was ready to open on January 12. Who were the appropriate students and how would they be found?

The Radio Material School (RMS) at the Naval Research Laboratory, prior to 1942 the only full school for training radio maintenance personnel, had long administered a very comprehensive examination for admission. As the war loomed and maintenance needs greatly increased, the RMS was opened to new enlistees who were amateur radio operators (Hams), with those holding a Class A license entering with Radioman 2/c ratings. This carried over when the ETP began in January 1942. For persons who were not Hams, Bill Eddy and his staff devised a classification examination – popularly called the Eddy Test – that soon became the standard for admission.

Nelson Cooke, a long-time instructor at the original RMS and major contributor to the evolving ETP, described the essential standards for selection of trainees:

> The candidate for radio technician training must first indicate an interest in this type of work. Secondly, he should be a highly intelligent individual with mechanical aptitude and a good background in mathematics. In addition, he must be able to get along with others and be physically capable and willing to work a 15-hour day during his training – and like it.

The Eddy Test was designed to identify candidates with the requisite intellectual capabilities, and, although only a small fraction of the test-takers passed, many lacked the ability to stand the pace of instruction and were eliminated within a short time. The Pre-Radio Schools were set up in early 1943 to serve as an additional filter, passing on to the Primary Schools only those persons with the will and ability to complete the work. Reviews of the successful graduates of Primary Schools indicate that they ranked in the top two percentile of native intelligence. These same intellectual and drive characteristics are also recognized in persons who might become high achievers in their careers.

This Appendix provides brief biographical information on representative graduates of the ETP. Shown herein is the great diversity of their pre-Navy backgrounds, their training and assignments while in the Navy, and their post-Navy activities. The collection of information and preparation of these write-ups has been the most enjoyable part of the author's work on this book. The relatively small number of ETP graduates is a special subset of the generation – outstandingly heralded by Tom Brokaw in his book, *The Greatest Generation* – that matured during World War II. They made significant contributions to winning the war and in the ensuing growth of our Nation; however, until now, they have not been recognized as a group.

GENE M. AMDAHL

Many graduates of the Electronics Training Program had a highly creative, entrepreneurial spirit. Gene Amdahl, a major contributor to the computer revolution, is a good example. Although educated in theoretical physics, his Navy experience provided a very practical background and assisted him in becoming – in *Fortune* magazine's words – "the wizard of high tech."

Born in 1922, Amdahl was raised on a farm near Flandreau, South Dakota. His education started in a one-room school that covered the first eight grades. He attended high school in Egan, then graduated in 1940 from Augustana Academy, a Lutheran school in Canton. The next year was spent working on the family farm. He then enrolled at South Dakota State College (SDSC), studying a potpourri of math, physics, and electrical engineering courses, and paying his way by working as a hospital janitor.

In May 1944, the draft caught up with Amdahl, and, having earlier passed the Eddy Test, he entered the Navy and was sent to Great Lakes for Boot Training. Pre-Radio at Wright Junior College was followed by Primary School at 190 North State Street. He then went across the country to Treasure Island for Secondary School. Upon graduating, he returned to Chicago for Teacher Training, followed by an instructional assignment at Herzl Junior College. Eventually he was sent to Gulfport where he taught in the Primary School until discharged in June 1946.

Amdahl returned to SDSC and resumed his studies as an engineering physics major. Immediately upon receiving his B.S. degree in June 1948, he entered the University of Wisconsin for graduate study in theoretical physics. During a research project, he was frustrated in attempts to use a mechanical desk calculator to determine forces between nuclear particles and began to think about how computing could be done better. This led to a summer job in 1950 at the Aberdeen Proving Grounds with exposure to the Electronic Discrete Variable Automatic Computer. Then still under final development, this was the first binary, stored-program, electronic computer.

Upon returning to Wisconsin, Amdahl started his own computer design using a magnetic-drum memory with re-circulating registers, a paper-tape input, and a teletypewriter output. Although it was an unusual topic for a theoretical physics student, he was allowed to use this design for his doctoral research. His dissertation was titled "The Logical Design of an Intermediate Speed Digital Computer." Amdahl's was the first electronic computer to have floating-point arithmetic, pipelining, and concurrent input and output independent of computing, and launched his career as a leading computer architect.

Amdahl received his Ph.D. degree in 1952, and, based on his dissertation, started work at IBM's research center in Poughkeepsie, New York. There he was placed in charge of designing the IBM 704, built around their newly developed magnetic core memory. His next project was designing the IBM 709. This was followed by an assignment to design Stretch; using transistors, this would be IBM's first supercomputer. After conflicts over the design approach, he resigned in 1955, but after a year with Ramo-Wooldridge and Aeronutronics, he rejoined IBM in late 1960. There he was asked to design an entirely new line of computers: the IBM 360 family. This was highly successful, and remained IBM's largest revenue-producer for decades.

In 1961, Amdahl was named an IBM Fellow, entitled to work on any project of his choosing. He elected to work at IBM's new Advanced Computer Systems (ACS) laboratory in Menlo Park where he became the laboratory director. His primary work was in designing the ACS 360, a very-high-speed machine. During this period, he left a mark on the industry with Amdahl's Law, used to predict possible speed advantages in using multiple parallel processors. When the 360 design was rejected by IBM as being too great a technology upgrade, he decided to form his own company and compete head-on with IBM for advanced computers.

Amdahl Corporation, headquartered in Sunnyvale, got underway in December 1970. Many of the ACS staff soon joined Amdahl, being careful not to use intellectual property belonging to IBM. Their first computer, the Amdahl 470 V/6, had a 100-gate, large-scale-integration, air-cooled chip – revolutionary in that day. Under a license, it used the IBM operating system. The first sale was in 1975 to NASA for its computing center in New York. Other early sales were to American universities, then quickly extended into a number of European countries.

By 1979, Amdahl Corporation had captured 22 percent of the market for large systems. Sales were over $1 billion, and the corporation had some 6,000 employees worldwide. Recognizing the competition from Amdahl, IBM announced the Model 3030, equal in speed to the 470 and 30 percent lower in price. Amdahl responded with improvements and price reductions, maintaining the competition. The personal stress on Amdahl, however, resulted in health problems that led him to resign.

Amdahl was relatively inactive for almost a year, then bounced back with a new approach for a wafer-sized chip using very-large-scale integration. Trilogy Corporation was formed for its development, and major investors were readily found. A number of problems, however, caused serious delays in its becoming operational. With most of the investment capital expended, Trilogy merged with Exlst Corporation, an existing maker of mini-supercomputers. Amdahl resigned in 1989 to form Andor International, pursuing mid-sized mainframe machines. Because of the Trilogy misfortune, adequate investment could not be found, and Andor was bankrupted.

Although then in his mid-70s, ever-determined Amdahl co-founded Commercial Data Servers in 1996 to develop super-cooled processors for next-generation mainframe servers. Their Model 2000 Enterprise Server focused on mainframe users that faced Y2K-related shutdown.

Amdahl retired from developmental work in 2000, and now serves on the Advisory Board of Massively Parallel Technologies. He has received honorary doctorates from four universities, has been a member of the National Academy of Engineering since 1967, and was recognized as the Centennial Alumnus of South Dakota State University in 1986. *The Times of London* named Amdahl one of the "1,000 Makers of the 20th Century," and *Computerworld* called him one of the 25 people that "changed the world."

BENNETT L. BASORE.

Some of the graduates of the radio technician program had extended relationships with the institution providing their college-based Primary School. Bennett Basore is an excellent example, having attended Oklahoma A&M before joining the Navy, training there in the program, returning there to complete his undergraduate education, and eventually becoming a leading electrical engineering faculty member at the school.

Basore was born in Oklahoma City, Oklahoma, in 1922. After high school, he began electrical engineering studies at Oklahoma A&M, but, with the start of the war, enlisted in the Navy in January 1942. Sent to New Orleans, he joined men in a special Boot Camp at the former Algiers Coast Guard Station, awaiting the opening of college-based radio technician training. One hundred men from this group arrived at Oklahoma A&M on March 2, forming the first class in their Primary School. Basore graduated in May at the top of his class with a grade of 97.8. He then went to Treasure Island for Secondary School and in October completed the overall program.

Basore was sent to New London, Connecticut, for a special course in submarine radar. In mid-December he was assigned to the submarine USS *Hoe* (SS-258), serving during her shakedown along the Atlantic coast. In April 1943, she sailed to Pearl Harbor via the Panama Canal, then departed in May on her first combat patrol. Between then and March 1945, the *Hoe* engaged in eight wartime missions, sank or severely damaged enemy vessels in five of these patrols, and received seven battle stars. Her patrols were primarily in the South China Sea, operating out of either Pearl Harbor or Fremantle, Australia. On several patrols, the *Hoe* received depth-charge attacks, in one case sustaining damage sufficient to require refitting.

In February 1945, on her eighth and last patrol, the *Hoe* collided with another submarine, the USS *Flounder*, at a depth of 66 feet. The damage was light, and the *Hoe* continued her patrol, then in March returned to the United States for an overhaul.

During most of his sea duty, Basore was the only rated radio technician aboard the *Hoe*. For his activities, he was twice awarded the Bronze Star Medal, and was promoted to Chief Electronic Technician. With the award of his second Bronze Star, Basore was recommended for a commission. He received this in July 1945 and was assigned as an engineering duty officer on the USS *Fogg* (DER-57), then in Philadelphia being converted to a radar picket destroyer. Sent to MIT for the officer's three-month Radar Course, he was there when the war ended. He briefly served on the *Fogg* in a training cruse along the East Coast, then left active duty as 1946 began. Basore remained in the reserves until 1973, retiring with the rank of Lieutenant Commander.

Upon being discharged, Basore immediately returned to Oklahoma A&M to continue his studies under the G.I. Bill. He was President of the Student Senate and graduated in 1948 with highest honors, receiving the B.S.E.E. degree, as well as a master's degree in mathematics. Following this, he began doctoral studies at MIT and was awarded the D.Sc. degree in electrical engineering in 1952.

After receiving his doctorate, Basore was employed by the Sandia National Laboratories in Albuquerque, New Mexico, as a weapons delivery systems engineer. Following the Cuban Missile Crisis, Basore decided to work in disarmament and in 1963 took a position with the Arms Control and Disarmament Agency in Washington, D.C. There he was instrumental in developing the "Hot Line" – the direct communications link between Washington and Moscow. He was also involved in the ABM Treaty, participating in verification negotiations in Geneva.

In 1967, Basore returned to Stillwater, joining the Electrical Engineering faculty at the Oklahoma State University (OSU – changed from Oklahoma A&M College in 1957). For the next 34 years, he was one of the most productive and respected professors on the campus. In addition to teaching in electrical engineering, he was also the administrative head of the General Engineering Department. An active Ham (W5ZNT), he was the faculty advisor to the OSU Radio Club. Basore was eventually awarded the title of Emeritus Professor of Electrical Engineering.

DAVIS L. BREWER

Technical Representatives (Tech Reps), often called Field Engineers, have played a vital role in the military's electronic operations. Functioning between the factory and the user, their activities cover everything from preparing requirements to making major repairs. Navy electronic technicians were well suited for this function, and many went into this field after leaving the service. Davis Brewer is highly representative of persons in this activity.

Brewer was born in 1926 at McComb, Mississippi, and spent much time from age 12 onward "hanging around Ham shacks and reading radio magazines." Upon seeing an advertisement in a *QST* magazine, he took and passed the Eddy Test, then enlisted in the Navy in June 1944. After Boot Camp at Great Lakes, he attended Pre-Radio at Wright Junior College, then Primary School at Gulfport. This was followed by Secondary School at Treasure Island, where he graduated in October 1945, just after the end of the war.

Brewer boarded the *Dawson* (APA-79), an attack transport, and departed San Francisco for Guam, expecting an assignment in the Pacific Fleet. They arrived at Guam only to find that there were no open billets for electronic technicians on available ships. Eventually, Brewer was assigned to work at the Joint Air Communications Service, Pacific, operated by the Air Force, and remained there until his discharge points came up. He returned to the States on the USS *Mitchell* (AP-114) and was discharged in mid-June 1946.

Before entering the Navy, Brewer had briefly apprenticed as a printer, and returned to this upon being discharged. In a short while, however, he learned that American Overseas Airlines was hiring former Navy RTs to work at Air Force bases. He joined this firm, and in September was sent to Bluie West One AFB in Greenland to maintain radioteletype equipment. This work ended in June 1947, and, after six-months study at Capital Radio Engineering Institute, he joined Philco, again as a Tech Rep. With Philco, he had assignments with the Davis AFB, Adak, Alaska; then at Barksdale AFB, Shreveport, Louisiana; and then with the Navy in Norfolk, Virginia, supporting auxiliary ships.

Having stayed in the Naval Reserves, Brewer was recalled to active duty in 1951 and assigned as an instructor at Fleet Training Groups at Chesapeake Bay and Guantanamo. Released in 1952, he returned to Philco and continued working at Norfolk where, in 1957, he was promoted to the elite Naval Aviation Engineering Service Unit (NAESU), specializing in MAD and other ASW systems. Three years later, still with NAESU, he was assigned to the Naval Air Forces, Atlantic (NAVAIRLANT), as advisor to the Staff Avionics Officer. In 1964, Philco NAESU personnel at NAVAIRLANT were converted to Civil Service positions. Brewer was eventually in charge of all "common avionics," and continued in this activity until he retired in December 1988.

Following retirement, Brewer remained at Virginia Beach, where he had lived since 1953. For the next 15 years, he and his wife spent much time on the road in an RV, mainly touring the Western States, Canada, and Alaska. In 2003, they moved to Okeechobee, Florida. He had operated as a Ham overseas while a Tech Rep, but his first U.S. Ham license (W4WBT) was in 1952; in 2000 this changed to KN4US.

MORTON H. BURKE

Many of the program trainees were first-generation Americans. Morton Burke is typical of these. He was born in New Jersey in 1924 and raised in Howell Township by parents who had emigrated from Russia to escape the violent attacks against Jews. His father served in the U.S. Army during the First World War and received the Purple Heart.

Burke started at Rutgers College when he was 16. As draft age approached, he took and passed the Eddy Test, then was allowed a deferment to complete his degree in electrical engineering. Entering the

Navy in September 1944, he took Boot Camp at Great Lakes, then attended Pre-Radio at Wright Junior College. Primary School at Gulfport was followed by Secondary School at Navy Pier, and he graduated just as the war ended.

For the next year, Burke served as an Electronic Technician Mate aboard the USS *Card* (CV-11), the USS *Botetourt* (APA-136), and the USS *Latimer* (APA-152); all of these vessels were undergoing decommissioning. Holding an engineering degree, he was asked to take a commission and remain in the Navy, but he elected to return to civilian life and was discharged in July 1946.

Burke returned to Rutgers and earned a master's degree in electronics. In his industrial career, he held positions with seven electronics firms and became a licensed professional engineer. His activities included Army thermoelectric generators, Navy transmitters, analog computers, and television, but he eventually specialized in power-handling equipment. A major project was when he was with Teledyne INET, installing and maintaining uninterruptible power systems for large computer data centers, including the New York Stock Exchange.

Did his Navy experience help in his professional career? "Absolutely! Working on my own from manuals, I had to understand and service equipment that I had never studied in college or in the Navy electronics courses. Later, in my engineering work, I had to put diverse information together to make decisions. My training and experience in the Navy was invaluable for this."

RALPH M. CALDWELL

While most of the program trainees were in the 17-21 year group, some were older and entered with technician ratings based on their experience in radio servicing. Ralph Caldwell is an excellent representative of these people.

Born in 1916 at New Albany, a very small town in northeastern Mississippi, Caldwell's ambition following high school was to be a radio serviceman. After preparing by way of a correspondence course from the National Radio Institute, he opened a service shop. Without success in rural Mississippi during the depression, he moved to Sardis, Mississippi, in 1938, and operated a radio repair shop at an appliance store.

When the war started, Caldwell learned that the Navy was in great need of radio technicians, and he enlisted at the beginning of March 1942. Entering as a Radio Technician 2/c, he went to New Orleans where he and about 30 other volunteers were given a special one-month Boot

Camp at Algiers, a former Coast Guard station. In early April, he was sent to Oklahoma A&M, entering the second class in their newly opened 12-week Primary School. On July 1, he started in the first class at the Aviation Radio Materiel School at Ward Island, Texas (later renamed NATTC Ward Island, one of the Secondary Schools in the Program). At that time, the 20-week course was limited to airborne radar (ASB and ASE) and radar recognition (IFF) equipment. His performance in school was such that, upon graduating in November, he was assigned to the Ward Island instructional staff.

In 1942, the Navy completed development of a new system for detecting submerged submarines, the Magnetic Anomaly Detector (MAD). A highly classified instructional program on maintaining the Mark IV-B2 MAD System was established at Ward Island in early 1943. Caldwell was assigned as the initial laboratory instructor, then later became the Petty-Officer-in-Charge of the program. This special training activity was so secure that even Ward Island's commanding officer was not initially briefed on the content.

The MAD training program closed in late 1944, and until May of the next year, Caldwell remained at Ward Island teaching refresher courses in communications and LORAN. He was then transferred to the Naval Auxiliary Air Facility at Charlestown, Rhode Island, maintaining radars on F6F night fighters. In July, he was sent to Boca Chica Field, Key West NAS, to train for boarding the carrier USS *Antietam*, but the war ended in September. He was promoted to Aviation Chief Electronic Technician Mate and remained at Boca Chica until discharged in December 1945.

Caldwell returned to Sardis and his radio service shop. In a short while, he was able to buy the appliance store where his shop was located. While he was operating this store, he used the G.I. Bill to study part-time at the nearby University of Mississippi, eventually earning the B.Ed. degree. In 1960, he was invited to start a program in electronics at the Northwest Mississippi Junior College. After a very successful four years at this school, Caldwell joined the Mississippi Department of Education in Jackson as State Supervisor of Technical Education. In his work, Caldwell was responsible for developing Junior and Community College technical programs throughout the State.

In 1972, Caldwell received the M.Ed. degree from the University of Mississippi, and in 1982, he was elected president of the American Technical Education Association. He retired from the State in 1983, and, for a number of years, assisted the Southern Association of Colleges and Schools in their accreditation evaluations.

ARTHUR D. CODE

In addition to gaining a good knowledge of electronics, some participants in the Navy program used this period to continue their studies on an individual basis, thus accelerating their academic progress. Art Code is an excellent example of such a person.

Born in 1923 in Brooklyn, New York, Code viewed a solar eclipse in 1930, and in a short while his future was set – he would be an astronomer. Following high school, he entered the University of Chicago. At the end of 1942, with the U.S. at war, he took the Eddy Test and enlisted in the Navy. From Boot Camp at Sampson, New York, he was sent to the Michigan City (Indiana) Armory for Pre-Radio, then to Chicago for Primary School at 190 North State Street, being quartered at Navy Pier. In October 1943, he started Secondary School at the Naval Research Laboratory.

Upon completing the program with high grades in April 1944, Code remained at the NRL-Bellevue School as an instructor. Over the next two years, he taught in the sonar and special circuits segment of the course and wrote materials to supplement the textbooks and manuals. Even while in the training program, he continued studying mathematics and physics, intending to use the University of Chicago practice of allowing students to receive credit by testing. During his period as an instructor, this private study intensified, and he took examinations from the University of Chicago and night courses at George Washington University.

Discharged in April 1946, Code petitioned the University of Chicago for direct admission to graduate school. Based on his examinations, this was granted and he began studies at the University's Yerkes Observatory, Williams Bay, Wisconsin. There he completed an M.S. degree in 1947 and the Ph.D. degree in 1950. His mentor was Subramanyan Chandrasekhar, who later received the Nobel Prize in Physics. After teaching for a year at the University of Virginia, Code joined the faculty of the University of Wisconsin and performed research in measuring the brightness of stars and galaxies. In 1956, he accepted a position with the California Institute of Technology and conducted research at the Mount Wilson and Palomar Observatories.

Code returned to the University of Wisconsin in 1958 as Director of the Washburn Observatory. Based on a proposal that Code had made for a satellite-based ultraviolet telescope, NASA initiated the Orbiting Astronomical Observatory (OAO) program. Code then started the University's Space Astronomy Laboratory to develop the Wisconsin Experiment Package (WEP) – a narrow-field, ultraviolet photometric system for the OAO. The first successful OAO was launched in 1968. During its 50 months of operation, WEP was the first true stellar space observatory and the source of many new discoveries.

The instrumentation and operating systems of the OAO represented in their day a greater technological leap forward than that of the Hubble Space Telescope of two decades later. Because of his experience with OAO, Code played a major role in the development of the Hubble. The Space Astronomy Laboratory also developed many other astronomical instruments, including the High-Speed Photometer for the Hubble, the Wisconsin Ultraviolet Photo-Polarimeter Experiment flown in two Space Shuttle payloads, and instrumentation for the WIYN Observatory in Arizona.

Code was elected to the National Academy of Sciences in 1971, and served as the 1982-84 President of the American Astronomical Society. He has published widely in astronomy and astrophysics journals, and has received many honors, including the NASA Public Service Award and Distinguished Service Medal. Code retired from the University of Wisconsin in 1994 but remained active as the WIYN Telescope Scientist, as an Adjunct Professor of Astronomy at the University of Arizona, and as Hilldale Professor Emeritus at the University of Wisconsin.

Throughout his career, Code's activities centered on developing highly advanced electronic systems for space instrumentation, and he credited the Navy electronics training for providing the primary foundation for his work. A Ham operator for many years, he holds AD7C (Extra) and was previously W9AD.

PERRY F. CRABILL, JR.

As co-ax cable and microwave-relay systems spanned the Nation following World War II, many former Navy radio technicians found that they were well trained for the technologies involved and established life-long careers in this field. Perry Crabill is an excellent example of these persons.

Crabill was born in 1920 near Waterlick, a crossroad in Northern Virginia, and raised in Washington, D.C. He had an early interest in building radios, and easily obtained his Ham license (W3HQX). Upon finishing high school, he received a scholarship to Bliss Electrical School in nearby Tacoma, Maryland. Graduating from Bliss in 1939, he was employed by the Chesapeake & Potomac (C&P) Telephone Company in its Washington facility. The C&P provided telephone services to the White House and other Government agencies, including the giant Pentagon when it was constructed in 1942. His work was such that draft deferments were given, but he took the Eddy Test and was finally inducted into the Navy in September 1944.

Following Boot Camp in Great Lakes, Crabill attended Pre-Radio at Wright Junior College, and then returned to Bliss for Primary School, where he graduated first in his class. Secondary School at the Naval Research Laboratory followed; here he finished in January 1946 near the top of his class. Having married prior to entering the Navy, he was fortunate in being near his family in Washington while attending at Bliss in nearby Tacoma Park and the NRL-Bellevue. In fact, his apartment was within walking distance of the NRL.

After a short assignment at the Algiers Naval Station in New Orleans, Crabill was discharged in April 1946 and immediately returned to Washington to work for C&P Telephone. Two years later he transferred to the C&P Television Center; then in 1953, he was promoted to a management position in the C&P Engineering Department. His activities initially centered on microwave and television transmission systems. Concerning this, he noted, "My radar training in the Navy provided an excellent foundation for this work." As the technology evolved through the years, his work primarily involved data transmission and high-speed digital systems for Government applications.

Crabill retired in 1981, and moved to Winchester, in the Virginia area of his childhood. There he continued interests in Ham radio and the collection of Navy electronic equipment, as well as photography, amateur astronomy using an 8-inch telescope, and hiking on the Appalachian Trail and in other nature areas in the region.

ROBERT C. DAY

Most graduates of the electronics maintenance training saw duty aboard Navy ships or at Naval bases throughout the word, but a few were assigned to special centers and laboratories, where they were involved with the ongoing research and development of new equipment. These were usually men who had shown a special aptitude for creative work during their training. Bob Day well represents such people. In addition, he is one of the many people who later made maximum use of the Navy training and experience in their civilian careers.

Day was born in San Francisco in 1923, and raised in Reno, Nevada. He began radio experiments at an early age, winding tuning coils on Quaker Oats cartons and converting a doorbell to a spark-gap transmitter. After starting at the University of California in Los Angeles in 1942, he found that the Navy urgently needed persons for radio maintenance training. He must have done exceptionally well on the Eddy Test; although he did not have a Ham license or extensive work experience, he was offered a Radio Technician 2/c rating.

Entering the Navy late in the year, Day was sent to San Diego for Boot Camp, then to Treasure Island for Pre-Radio. Houston University for Primary School was next. As he neared completion, he was asked to remain at Houston for an extra month to assist in the repair shop. Finally graduating in early September, he was one of the first three from Houston to be sent to Secondary School at the Naval Research Laboratory. He finished the program in March 1944.

While in the NRL Secondary School, Day built a receiver with parts from the scrap heap. The Officer of the Day saw the radio and was impressed. It is likely that this show of creativity led to his being selected to remain at the NRL after graduation, assigned to the Combined Research Group (CRG). This was a top-secret activity, operated jointly by the U.S. Navy, U.S. Marines, U.S. Army, and Canadian and British military services. The CRG was charged with developing advanced IFF systems, but, because of the security, the participants were not allowed to discuss their work until many years later.

In 1945, the CRG developed the Mark 5, a system involving one of the first applications of digital circuits. Day was responsible for building and testing the prototype shipboard transponder, including the delay lines used in the coding/decoding circuits. He took his initial breadboard unit to Hazeltine Research Corporation in New York for use in developing a production version. (The CRG is described under Recognition Systems in Appendix 1.)

After being discharged in April 1946, Day returned to California and joined the United Geophysical Corporation in Pasadena. Working under the renowned geophysicist Dr. Raymond A. Peterson, he was engaged in developing equipment such as an ultra sensitive altimeter used with gravity measurements in oil exploration. His activities took him to northern Alberta, Canada, and to Amchitka Island in the Alaskan Aleutian Chain.

In 1951, Day joined the Missile Systems Division of General Dynamics in Pomona, California, and spent the next 30 years with this firm. He was initially involved in the design of automatic text equipment for missile systems. Through the next several years, he progressed to Section Head of a Test Equipment design group, followed by activities in the program office specifying requirements for support equipment. He provided liaison with the Navy, and developed documentation for shipyards in installing weapon systems aboard vessels.

On an Army project at General Dynamics, an IFF system was involved, and in coordinating with Hazeltine, Day encountered some of the same persons from his CRG days. In 1970, he was part of a team providing consultation with Telefunken on a weapon system for a new class of German naval vessels. In his final work, he was involved in providing installation information for the Phalanx Close-In Weapon System. Day retired from General Dynamics in 1981.

DOUGLAS C. ENGELBART

Sometimes people can identify a single occurrence that led them to their careers. This is the case for Douglas Engelbart. Shortly after the war with Japan ended, Engelbart was relaxing in a Red Cross facility at Laiti in the Philippines and picked up a copy of the July 1945 *Atlantic Monthly* magazine. In this was an article, "As We May Think," by Dr. Vannevar Bush, the primary scientific advisor to the President. In the article, Bush discussed where electronics was headed, describing future computers that could handle words and pictures, not just numbers. He predicted screens for displaying computer information, massive electronic storage files, and small machines for individual use. Engelbart immediately determined to be a part of this future. He would later write to Bush acknowledging the influence of the article on his career.

Born in 1925, Engelbart grew up in Portland, Oregon, where his father ran a radio repair shop. After finishing high school, he went to Oregon State University in 1942 to study electrical engineering. In June 1944, he was drafted into the Navy, where he took the Eddy Test. Following Boot Camp at Great Lakes, he attended Pre-Radio at Hugh Manley, went on to Primary School at Oklahoma A&M, then Secondary School at Treasure Island. Upon completing the program in July 1945, he awaited transportation to the Philippines and, likely, the invasion of Japan.

On September 2, as Engelbart's transport ship was leaving San Francisco, it was announced over the speaker system that Japan had surrendered. The ship, however, sailed on to the Philippines, where he was temporarily assigned to maintain equipment at a communications base on Laiti Island. It was while stationed there that, by chance, he read the article that started him on the path to becoming one of the most creative computer science engineers of the 20th century. He was later assigned to Manila at the Navy headquarters for the Philippines.

Upon being discharged from the Navy in June 1946, Engelbart returned to Oregon State and completed the B.S.E.E. degree two years later. He joined the staff at Ames Laboratory of the NACA (predecessor of NASA), but soon began to think about a long-term goal for himself. Based on his experiences as a radar technician, he envisioned using computers as a technology to interact with information displayed on screens. He saw the connection between a cathode-ray screen and an information processor, as a medium for representing symbols to a person.

Leaving Ames after three years, he went to the University of California, Berkeley, to pursue graduate study. There he participated in the development of a general-purpose digital computer, the CalDiC, the first such machine on the West Coast. He was awarded the M.Eng. degree in 1952 and the Ph.D. degree in 1955, and also received several patents on bi-stable gaseous plasma digital devices, the subject of his doctoral research.

After a period on the faculty at Berkeley, in 1957 Engelbart joined the Stanford Research Institute. Over the next two years, he received a dozen patents on computer components, digital-device phenomena, and miniaturization theory. In 1962, he published a seminal report for the Air Force Office of Scientific Research, "Augmenting Human Intellect: A Conceptual Framework." Contracts resulting from this report led to the formation of the Augmentation Research Center (ARC) at SRI, out of which for 15 years flowed many of the concepts, tools, and processes of computer science

In early 1967, the Advanced Research Projects Agency (ARPA) funded 13 research centers to be networked for rapidly exchanging information. Engelbart's ARC at Stanford became the second host on the new ARPANet, forerunner of the Internet. Also at the ARC, Engelbart personally led in developing the computer mouse, hypermedia, windows, and groupware. In the fall of 1968 at the Joint Computer Conference, Engelbart used a 20-foot screen in the first public demonstration of the computer mouse, hypermedia, and on-screen video teleconferencing.

Tymeshare bought commercial rights to most of the ARC developments, and Engelbart joined them in 1978. Six years later, Tymeshare was acquired by McDonnell Douglas. In 1989, Engelbart and his daughter formed the Bootstrap Institute, devoted to research and consulting. He has received many honors and awards, including two honorary doctorates and the 1997 Lemelson-MIT Prize, carrying with it an award of $500,000. In 2000, President Clinton awarded him the National Medal of Technology, the highest recognition of this type in the Nation. Engelbart credited the Navy electronics program as a major stimulant in initiating his career.

EUGENE H. FELLERS

For some persons, the electronics training and subsequent experience while in the Navy was, throughout their lives, essentially their only involvement in this activity. While they might look on the Navy period as important in shaping their future, their earlier backgrounds and post-Navy careers had no relationship to electronics. Gene Fellows, former Chief Electronic Technician Mate, is representative of such people.

Fellers was born in 1919 at New Philadelphia, Ohio. As the Great Depression started, his father abandoned the family, and his mother moved them to a farm to survive. Having played football in high school, Fellers obtain an athletic scholarship at Baldwin-Williams College and took a liberal arts curriculum for two years. He also obtained his private pilot license and made unsuccessful attempts to become a military pilot. In January 1941, he signed up for a six-year enlistment in the Navy and soon became an Aviation Radioman stationed at NAS Jacksonville, Florida.

To give Fellers some capability in radio maintenance, he was sent to an RAF school that had just opened in Clinton, Ontario. There, in a six-week program, he received elementary instruction in both communications and radar maintenance. After returning to Jacksonville, he served at a communications station for over a year. Realizing a need for more thorough training, he applied for and was accepted in the Navy's regular electronics maintenance program. Fellers attended Pre-Radio at the Chicago Naval Reserve Armory in early 1943, then, while still being housed at the Armory, went to 190 North State Street for Primary School. In April, he crossed the country to Treasure Island for Secondary School.

Upon graduating from Treasure Island in October, Fellers was sent to Pittsburgh where a destroyer escort (DE-669) was beginning to be built. In December, the ship was launched and ferried to Orange, Texas, to be outfitted as the USS *Pavlic* (APD-70), a high-speed transport with minimal defensive weapons. Finally commissioned in December 1944, the *Pavlic* first went to Portsmouth, Virginia, for amphibious training, then steamed to Hawaii by way of the Panama Canal, arriving at Pearl Harbor in March 1945. After a month of training exercises, she joined Task Force 51 and the capture of Ryukyus Islands.

During the first part of May, the *Pavlic* was stationed on the picket line off Okinawa, performing rescue work and fighting off bombers and long-range fighters. Later in May, she was designated a special rescue vessel and continued operations while undergoing frequent air raids, including a repelled kamikaze attack. For the remainder of the war, the *Pavlic* served in the Pacific war zone. At the end of August, she steamed into Tokyo Bay to demilitarize two guarding forts and raise the colors; on September 2, 1945, she was present at the surrender ceremony.

From the time of her launching, Fellers had been the only rated electronic technician aboard the *Pavlic,* and was promoted to Chief Electronic Technician Mate near the end of the war. He remained with the *Pavlic* in Japan for several months, then was assigned to the USS *Missoula* (APA-211) as she returned to the States in early 1946 for mothballing. (It might be noted that the *Missoula* was the ship that carried the Marines and the well-known flag that was raised over Iwo Jima.) Fellers was then assigned as an instructor in the Primary School at Great Lakes, remaining there until his six-year enlistment was up in January 1947.

Fellers returned to Baldwin-Williams and completed his B.A. degree in 1948, majoring in languages. For the next several years, he remained at Baldwin-Williams teaching German. In addition to studying summers in Germany and Mexico, he attended the University of Virginia, receiving the M.Ed. degree in 1954. After serving as a Principal at several schools, Fellers became a teacher of German and Spanish at the high school in Glens Falls, New York, and remained there until he retired in 1974.

In the ensuing years, living in Greenwich, New York, Fellers farmed, built houses, and developed his artistic abilities. He proudly displayed in his living room a painting done by him of the USS *Pavlic.* None of his activities and interests after leaving the Navy, however, involved electronics.

WILLIS ARNOLD FINCHUM

Graduates of college-based Primary Schools often returned to those schools to continue their post-Navy education. Arnold Finchum not only did this, but returned a second time for graduate school, then a third time for a final academic position.

Finchum was born in 1921 at Indianapolis, Indiana, and grew up in that city. While in high school, he worked under the National Youth Authority program, and, after graduation, became an apprentice electrician. He attended a "Code School" – paying the $1.50 fee by serving as the janitor – and obtained his first Ham license (W9FVO) in 1940; this was upgraded the next year to Class A.

On the afternoon of December 7, 1941, Finchum turned on his receiver to hear the message: "Pearl Harbor has been attacked by Japan. All radio amateurs are ordered off the air immediately!" The following

February, he enlisted in the Navy, and, like many other Class A Hams, received a rating of Radioman 2/c. He was sent to Great Lakes for Boot Camp, but the training was cut short in March when he and 100 other men went to Logan, Utah, to form the first Primary School class at Utah State Agricultural College. Three months later, he was sent to Treasure Island for Secondary School, finishing in December 1942.

Finchum was then sent to the Teacher Training School, operated at 190 North State Street in Chicago. Completing this in February 1943, he was assigned as an instructor in the Primary School at Bliss Electrical School, Takoma Park, Maryland. The following year, he attended a Radar Refresher Course at the NRL Secondary School, then became a part of the skeleton crew for the USS *Benner* (DD-807), a destroyer being built in Bath, Maine. The *Benner* was launched in late 1944, commissioned the next February, and reported (via the Panama Canal) to the Pacific Fleet. In July, she joined the 3rd Fleet off Japan, screening carriers during the last strikes against the Japanese homeland. After the end of the war, Finchum was sent back to the States and received his discharge in October 1945.

Finchum returned to Utah State under the G.I. Bill, and also worked as an engineer at radio station KVNU. He completed the B.S. degree in Radio and Electronics in 1949, and for the next nine years was employed in industry as an electronics engineer. Until 1952, he worked for Sandia Corporation in Albuquerque, New Mexico, developing telemetry and radar equipment used in atomic bomb testing. Then for two years he was with Bell Aircraft at Point Mugu, California, testing the Meteor missile guidance system. Another two years were spent with Raytheon in Oxnard, California, developing the FMCW radar for the Sparrow III missile. This was followed by a short period in Van Nuys, California, with the Radioplane Company on the Cross-Bow missile, then another short period with Coleman Engineering in Los Angeles as the instrumentation supervisor on a supersonic rocket sled for testing pilot escape systems.

Finchum went back to Utah State in 1958 to attend the school's first graduate program in electrical engineering, receiving the M.S.E.E. degree in 1959. He then joined the faculty and taught at Utah State over the next decade. During the 1960s, he had a year of further graduate studies at Purdue University as a NSF Science Faculty Fellow.

In 1969, Finchum was appointed Head of the Electrical Engineering Department at the University of the Pacific in Stockton, California. In 1974, he became Head of the Electrical Engineering Technology Department at Purdue University, then two years later was selected for a similar position at California Polytechnic University in San Luis Obispo.

Retiring from this university in 1984, he again returned to Utah State University, serving as a Professor as well as assisting in the administration of the Electrical Engineering Department. He was one of the authors of "The Navy Training Station at Utah State Agricultural College During World War II," an archive document for the Utah State University Library.

Finchum finally retired from the academic field in 1988, and at that time was selected to head the Latter Day Saints Missionary Program at the Genealogical Library in Salt Lake City, Utah. Through the years, Finchum had eight Ham call assignments, the last being W6HO (Extra)

FRANK A. GENOCHIO

Most persons entering the Navy Electronics Training Program did so to obtain the highly specialized service classification and then possibly receive assignments that were less combat-intensive. This, however, was not the objective of Frank Genochio when he enlisted in the Marines.

Born in 1922, Genochio was raised in Camanche, California, a small town founded during the 1849 Gold Rush and now inundated by the Camanche Reservoir. He had an early interest in radio and obtained Ham license W6RXU – one of the first in the county – when he was 16. After high school, he attended the College of the Pacific for two years, and then volunteered for the Marines in July 1942. Although aware of the electronics training, he initially did not even let it be known that he had radio skills and was sent to San Diego for Boot Training in a rifle company. Eventually realizing the Marines' critical need for electronic technicians, he applied for the training and, based on his Ham qualifications, was accepted and promoted to Sergeant. In October he went to Utah State for Primary School (prior to the initiation of Pre Radio), and from there was sent to Treasure Island for Secondary School.

Graduating from the program with high grades in June 1943, he was sent to Chicago to attend Eddy's Instructor School at 190 North State Street, and from there was posted to teach at Wright Junior College. Following a bout with rheumatic fever and a stay at the Great Lakes Naval Hospital, in early 1944 he joined the faculty of the newly opened Primary School at the College of the Ozarks in Clarksville, Arkansas. There he taught mathematics and electronic circuits for six months.

Feeling that he was not doing what he had intended when joining the Marines, Genochio requested and received a transfer to the Fleet Forces Pacific. His official request, dated June 7, 1944, included the following:

> For the 22 months that I have served since my enlistment, I have had no foreign duty and being in good physical condition and with no dependents, I feel that I should have the opportunity to work in the field and be replaced by someone not physically able to withstand combat duty.

The USS *Wasatch* (AGC-9), with Genochio and a shipload of other Marines aboard, left Portsmouth, Virginia, on June 26 and sailed by way of the Panama Canal to Papua, New Guinea. From there they were sent to Pavuvu in the Solomon Islands for assignment to the First Marine Division.

This Division had recently been severely mauled in capturing Peleliu Island, and many replacements were needed. Genochio was initially attached to the First Signal Company in the Signal Repair Section, but, in a short while, was transferred to the Radio Section. In recalling his activities, he noted, "My amateur radio experience was far more useful than my advanced training in electronics; all of my work was with field radio equipment and operations."

The Division remained in the Solomon Islands for several months, training and preparing for the invasion of Okinawa, the last Japanese stronghold before the mainland. Genochio was assigned a number of Native American radio operators who communicated using the Navajo language – a "secret code" that was never broken by the enemy. Although their bravery in battle was acknowledged, the special role of the Navajo Talkers was classified information. They were forbidden from telling about their work, and their unique contribution was forgotten until 1968 when the Navajo code was declassified.

The battle for Okinawa – the largest and last major battle of the war – took place from the first of April to the end of June, 1945. The First Signal Company supported the front-line fighting throughout the operations. They strung telephone lines and installed radio communication equipment as soon as an area was invaded; many of the men were killed. After the Company lost their Radio Officer, Genochio was made acting Platoon Leader (normally a Lieutenant).

Okinawa officially fell on June 21. The First Marine Division remained on the island, preparing for the final invasion of Japan. On August 14, Japan surrendered. Three days later, the Division Commanding Officer submitted a recommendation for Genochio to attend the Marine Officers School. The recommendation included the following:

> During this [Okinawa] campaign, Staff Sergeant Genochio was a member of the forward echelon [and] was acting Officer in Charge of the Radio Platoon. At all times he was cool and deliberate in his actions and orders to his men. Under fire at Shuri, Wana, and Kunishi, this man carried out his duties methodically and with the highest degree of leadership. . . .This man possesses, to a marked degree, initiative, intelligence, tact, leadership, and force.

Genochio was transported from Okinawa to Guam aboard the escort aircraft carrier USS *Casablanca* (CVE-55). There he joined more than 4,000 other servicemen on the troop transport USS *Weigle* (AP-119) sailing to Hawaii and finally to Los Angeles, arriving on September 24. He then crossed the country and reported to the Candidates Refresher School at Quantico, Virginia. There, however, he was offered and accepted an "early-out," and was discharged in October, 1945.

Returning to California, Genochio decided on a career in teaching; this had been recommended to him when he was an instructor at the College of the Ozarks. He entered the University of Santa Clara, earning the B.A. degree in economics in 1948

Genochio then enrolled in the Graduate School of Stanford University. The Dean of Education, who had been an official in Navy training programs during the war, encouraged him to research possible applications of the Naval electronics program in similar programs then being offered at California junior colleges. His subsequent thesis, "The Training of Navy Electronic Technicians and a Terminal Course in Radio Electronics," provided an excellent summary of the technical content of the Navy's Pre-Radio, Primary, and Secondary Schools. He was awarded the M.A. degree in Education in June 1949.

For the next several years, Genochio taught mathematics at the Placerville (California) High School. Moving back to Santa Clara in 1953, he took business courses at Santa Clara University and also worked at Quement Electronics, a large wholesale firm in San Jose. In 1954, he took a position at KAAR Engineering in Palo Alto, marketing in the two-way radio and marine electronics industry. Continuing part-time, he completed his MBA degree from Santa Clara University in 1963.

In 1968, KAAR was acquired by Marconi Canada, and, rather than relocating, Genochio joined California Television (CATEL) as their president. Under his guidance, this firm expanded from 6 to near 200 employees. CATEL became well known in developing and supplying radio-frequency hardware for the emerging cable-television market, with recognition for introducing frequency-modulated multiplexing for the transmission lines. Genochio eventually retired in 1988.

JACK M. GOUGAR

Some trainees in the electronics maintenance program were Regular Navy with prior ratings as Radiomen. Also, some graduates of the program had long careers in the Navy following the war. Jack Gougar is a representative of both of these categories. In addition, he had extensive combat experience and was eventually commissioned as an officer.

Born in 1920, Gougar joined the Navy in January 1941, taking Boot Training at Great Lakes. In March, he was selected to attend the four-month Aviation Radio School at NAS San Diego. Upon graduating, he had temporary duty at NAS Norfolk, and then was assigned as a radioman / gunner on a TDB Douglas Devastator aboard the USS *Yorktown* (CV-5), an aircraft carrier patrolling the Atlantic. Following the bombing of Pearl Harbor, the *Yorktown* passed through the Panama Canal to reinforce the badly damaged Pacific Fleet.

Gougar first saw combat in February 1942, as the *Yorktown* participated in the Marshalls-Gilberts raid. Several other raids followed; then, in May, the *Yorktown* engaged in the Battle of Coral Sea. In this, Gougar's squadron took part in attacks on two Japanese aircraft carriers, resulting in the sinking of one and damage to the other. Suffering damage herself, the *Yorktown* was repaired at Pearl Harbor. She then steamed out to participate in the Battle of Midway on June 4-6, 1942 – the turning point of the war. The *Yorktown* was sunk, but essentially all of the crew was saved. The Japanese fleet, however, was devastated, with four carriers sunk. This effectively ended Japan's Pacific-island expansion and allowed the U.S. to seize the strategic initiative.

Following a brief recovery at NAS Kaneohe Bay in Hawaii, Gougar joined the USS *Hornet* (CV-8) as a radioman / gunner on a TBF Avenger. The *Hornet* was joined by the USS *Enterprise* (CV-6) to intercept a large Japanese fleet that included four carriers. On October 24, 1942, the Battle of Santa Cruz Island took place without engagements between surface ships. U.S. planes inflicted severe damage on the Japanese ships, but the *Hornet* also suffered so much damage from kamikaze attacks and torpedoes that she was abandoned. Gougar's TBF and other planes from the *Hornet* landed on the *Enterprise*. Attending destroyers eventually sank the *Hornet*.

In early 1943, Gougar briefly returned to the States. In April, he was back in the South Pacific, assigned to an escort carrier, the USS *Sangamon* (CVE-26), as a radioman / turret gunner in a TBM Avenger. For the next several months, the *Sangamon* operated in the New Caledonia – New Hebrides – Solomons area. She returned to the States for overhaul, then in November joined a task force in the Gilberts to support the assault on

Tarawa. At the end of the year, she briefly returned to the States, but, by February 1944, was again in the South Pacific and supported the assaults on Kwajalein, Enewetak, Tarawa, Hollandia, and New Guinea. In all of Gougar's combat actions, he flew as crew member with the same pilot, Evan K. Williams.

Gougar was rotated back to the States in April 1944. As a Radioman 2/c, he applied for and was accepted in the Electronics Training Program. Starting in May 1944, he attended Herzl Junior College for Pre-Radio, Texas A&M for Primary School, then NATTC Ward Island for Secondary School. He finished in April 1945 as an Aviation Electronic Technician 2/c, and immediately applied for and was accepted for training as an Aviation Pilot. He attended Flight Training at St. Olaf College (Minnesota), Pre-Flight School at the University of Georgia, and Primary Training at NAS Mempis. This was followed by training at NAS Corpus Christi (SNJ aircraft) and NAS Pensacola (PBY aircraft); he received his wings on April 27, 1947, as an Aviation Pilot 1/c – a rare enlisted rating.

Assigned to the NAS Roosevelt Roads in Puerto Rico, Gougar served as a utility pilot in SNJ, SNB, 6F6, and RD4 aircraft, and received a promotion to Chief Aviation Electronic Technician / Aviation Pilot. Between November 1949 and December 1951, he first attended Instructor's Basic Training at NAS Pensacola, then served as an instructor in instruments, aerobatics, and night flying. He transferred to NAS San Diego to serve as an instructor in anti-submarine warfare, followed by a return to Corpus Christi to complete the All-Weather Flight School.

In September 1953, Gougar was attached to the Military Assistance Advisory Group, working with the French in Saigon. He also flew SNB-5 aircraft in Vietnam, Laos, and Cambodia, and, in May 1954, was aboard the last aircraft leaving Hanoi after the fall of Dien Bien Phu. Back in the States, Gougar was assigned as a maintenance test pilot at NAAS Barin Field, Foley, Alabama, and was commissioned as an Ensign in July 1955. Between September 1955 and October 1958, he was again at NAS Pensacola, where he qualified as a formation flight instructor and also instructed in communications and meteorology. In January 1957, he was promoted to Lieutenant (jg). Starting in late 1958, he attended the five-month Aviation Electronics School for officers at NAS Memphis.

In March 1959, Gougar began a six-year assignment with the Naval Air Transport Squadron at McGuire AFB, New Jersey. He was promoted to Lietuenant in August, and qualified as a Navigator, Second Pilot, First Pilot, then Plane Commander on the Douglas C-118 Liftmaster (Navy R6D) aircraft. He served as Liaison Officer for Air National Guard transport squadrons in all states east of the Mississippi. Returning to NAS Pensacola, he served during 1965 as a Plane Commander on R4D and C-54 aircraft, and was promoted to Lieutenant Commander in March.

In December 1965, Gougar was voted into the Blue Angels Flight Demonstration Team. For them, he served as Commander of their C-54 transport plane and Maintenance Officer during their 1966 and 1967 show seasons. In September 1967, flying from Tunisia to Turkey, his C-54 had an engine failure, resulting in an emergency landing on Rhodes Island. Gougan suffered a collasped lung and was returned to the NAS Pensacola Naval Hospital. His condition was diagnosed as beyond recovery, and he was retired with disability on March 1, 1968.

In his 27 years of Navy service, Gougan received three Air Medals, three Presidential Unit Citations, and ten Battle Stars. Following his retirement from the Navy, he had a long period of recuperation, and, defying the original diagnoses, eventually recovered. For some time after that, he taught flying at a private field near his retirement home in Pensacola.

CHARLES A. HOBSON

Following the end of the war with Germany and then Japan, the Navy's electronics maintenance training program began a phase-down and reorganization. Captain Eddy retired (again) in December 1945, and with him went the various operations of Radio Chicago. BuPers allowed reservists (V-6) in the program to continue, but only if they finished by mid-1946, or signed over to the regular Navy. Chuck Hobson is representative of such persons. He is also representative of the many men who, with no formal education beyond the Navy electronics program, attained the classification of engineer, as well as high-level positions in the field.

Hobson was born 1927 at McKeesport, Pennsylvania. He passed the Eddy Test and joined the Navy reserves in January 1945. Following Boot Camp at Great Lakes, he attended Herzl Junior College for Pre-Radio, and then went back to Great Lakes for Primary School. While he was there, the Japanese surrendered (August 14), but Hobson's company continued, finishing in early October, and was then allowed to go on to Secondary School. Hobson selected Treasure Island, and was in the last class of the program in its wartime configuration.

Hobson graduated with high grades in June 1946; there were only 22 remaining in the class. Concerned with losing their electronics maintenance capability, the Navy made a special appeal to the reservists to "sign over." Hobson did this, but was the only person in the class to do so. He was sent to Yokosuka, Japan, to join the USS *Piedmont* (AD-17), then the USS *Frontier* (AD-25), both Destroyer Tenders, as the only rated electronic technician aboard. After several months, he was transferred to the newly formed Yokosuka Naval Repair Facility. For some time, he was the sole electronic technician at the facility, and the destroyer fleet in the district essentially all had inoperative radars, fire-control systems, and sonars.

In late 1947, Hobson was discharged from the Navy under an "early-out" option. He was then employed by Philco TechRep Division as a field engineer and had various assignments at Air Force radar sites, primarily training military technicians in radar maintenance. With the Korean War underway, he was called back into the Navy in August 1951. For the next year, he served on the cruiser USS *Worcester* (CL-144) while she was primarily in the Mediterranean with the 6th Fleet.

Following this service, Hobson returned to Philco in November 1952, and for the next six years his assignments included serving as a maintenance advisor at Air Force radar sites; as a radar instructor in the Army Electronics School at Ft. Bliss, Texas; and in electronics maintenance activities for the Navy at Norfolk, Virginia. In 1958, Hobson was assigned to Philco's Western Development Laboratory in Palo Alto, California, as a senior electronics engineer. Over the next several years, he was involved in the development of boresighting systems for precision satellite ground-tracking equipment.

Hobson became a civil service employee of the Navy in 1965, joining the Air Station in Alameda, California, as a supervisory electronics engineer. Here he was involved in aircraft avionics and electrical systems, supporting their maintenance activities and developing various rework specifications and airframe changes. During much of the 1970s he was engaged in air-launched guided-missile maintenance programs; his original work relating to Sparrow missile modifications resulted in two patents. He noted that, "I owe the US Navy a huge debt of gratitude for my rather rich career in electronics – they did it right!"

In 1986, Hobson retired and moved to live in England at Wimborn, Dorset. There he continued "non-commercial" activities in electronics, building equipment and operating amateur station G0MDK (in the U.S., he had K4PIO, WA6FRM, and WA6TIR). He became interested in designing and fabricating Tesla coils (for high-voltage generation); his coils have been displayed in a number of shows and museums. He also turned to higher education, earning the B.A. degree in 1992 and B.Sc. (Honors) degree five years later, both through the Open University, "where ageism is a no-no." He established a number of websites, with instructional information on subjects including antennas, electromagnetic theory, Van de Graff machines, and spark-gap transmitters.

LEON M. JAROFF

Many graduates of the Electronic Maintenance Training Program ultimately became writers; Leon Jaroff is an excellent example. Born in Detroit in 1927, he briefly attended Wayne State University, then took the Eddy Test and volunteered for the Navy. Following Boot Training at Great Lakes in 1944, he completed Pre-Radio at Hugh Manley, went on to 190 North State Street for Primary School, then to Treasure Island for Secondary School. Upon finishing the program, he remained at Treasure Island, serving as a technician at the radar installation atop Yerba Buena Island.

Discharged in 1946, he continued his education at the University of Michigan, earning degrees in both electrical engineering and mathematics in 1950. While there, he was the editor of the school newspaper, *Michigan Daily*.

Following a year as a writer for *Materials & Methods* magazine, Jaroff joined Time, Inc., as an editor and reporter for *LIFE* magazine, and then moved to *TIME* magazine. In the 1960s, he served as chief science writer, responsible for the cover stories on space and the moon-landing program. He later served as a senior editor. His articles in *TIME* won many awards, including best science stories in 1978 from the American Association for the Advancement of Science and in 1988 from the American Medical Association. In the 1980s, Jaroff was the founding managing editor of *DISCOVER* magazine, winning awards from the American Institute of Physics. He authored *The New Genetics* (Whittle Communications, 1991). Jaroff is still active as a contributing editor for *TIME*, with recent articles on subjects ranging from asteroid dangers to debunking psychics.

Jaroff's writing often reflected his strong skepticism, rejecting false science. He noted that his training in the Navy's electronics program laid a firm foundation for his approach. "It taught me to think clearly and rationally – to assemble all of the known information and draw the most likely conclusion."

RONALD K. JURGEN

Since it began in 1964, *IEEE Spectrum* has been a major source of information and inspiration to thousands of professionals in the electronics field. Ron Jurgen, the founding Managing Editor, was a graduate of the Navy's electronic technician program.

Born in January 1927, Jurgen was raised in New Britain, Connecticut. Following high school, he entered Rensselaer Polytechnic Institute to study chemical engineering. Having earlier passed the Eddy Test, he enlisted in the Navy in January 1945 and took Boot Camp at Great Lakes. Pre-Radio at Hugh Manley followed, then he returned to Great Lakes for Primary School. After this, he was sent back to Chicago for Secondary School at Navy Pier, graduating in 1946.

Jurgen was first briefly assigned to the USS *Scribner* (APD-122), a high-speed transport being decommissioned at Norfolk, then to the light cruiser USS *Portsmouth* (CL-102), whose home port was Newport News. After a period at sea in gunnery practice, she was ordered to the Mediterranean for a good-will tour. Given a choice of the cruise or being released, Jurgen chose the latter and was discharged in July 1946.

Returning to Rensselaer, Jurgen changed his major to electrical engineering – "I was greatly influenced by my Navy training" – and graduated in January 1950 with the B.E.E. degree. Having a bent toward writing and editing, his first position was with *Electrical Engineering*, the monthly magazine of the AIEE. This was followed by positions as one of the editors of *Electronics* magazine (McGraw-Hill), and editor of *Electronic Equipment Engineering* and *Industrial Electronics* magazines (Sutton Publishing), then back to AIEE to the editorial staff of *Electrical Engineering*. In 1964, one year after the AIEE and IRE merged to form the IEEE, Jurgen was named the Managing Editor of *IEEE Spectrum*, helping to develop this highly respected magazine and remained on its staff until retiring in 1992.

In 1975, realizing that there had been no publications on the Navy's electronics training that had greatly influenced his own career, Donald Christiansen, the editor of *Spectrum*, asked Jurgen to interview Captain William C. Eddy at his home in Michigan City, Indiana. The resulting article, "Captain Eddy: the man who 'launched a thousand EEs'," was published in the December 1975 issue of *Spectrum*. Fourteen years later, Jurgen wrote Eddy's obituary for the November 1989 issue of *The Institute* (IEEE monthly newspaper).

After retiring in 1992, Jurgen moved the following year to Fort Lauderdale, Florida, where he continued his career with free-lance editing and writing. He co-edited the *Electronics Engineers' Handbook* (1996 and 2005 editions), and edited the *Digital Consumer Electronics Handbook* (1997) and the *Automotive Electronics Handbook* (1994 and 1999 editions), all published by McGraw-Hill. He is also editing an ongoing series of books on automotive electronics (14 volumes have already been published) by the Society of Automotive Engineers. These include *Electric and Hybrid-Electric Vehicles; Electronic Braking, Traction, and Stability Control; Automotive Microcontrollers; Electronic Instrument Panel Displays; Sensors and Transducers;* and *Object Detection, Collision Warning and Avoidance Systems.*

ROBERT B. KAMM

Although the professional career of Robert Kamm had no relationship to his training in the Navy's electronics program, he always noted this experience when describing his background – it was obviously very important to him. Kamm was born in 1919 and raised in West Union, Iowa. Following high school, he attended the University of Northern Iowa, receiving the B.A. degree in English in 1940. After teaching for two years at the Belle Plane (Iowa) High School, he spent the next year as a civilian employee of the Army Air Forces, instructing radio communications at Bradley Field in Sioux Falls, South Dakota.

After starting graduate studies at the University of Minnesota, Kamm took the Eddy Test and entered the Navy in June 1944. Following Boot Camp at Great Lakes and Pre-Radio in Chicago, he was sent to Oklahoma A&M for Primary School. His wife, Maxine, was with him in Stillwater, and as they left, he remarked to her, "Wouldn't it be wonderful to return to this delightful campus and community in the future?" This would come true, but not until many years later.

Following Oklahoma A&M, Kamm attended Secondary School at NATTC Ward Island, finishing in June 1945. From there, he was sent to the Los Alamitos Naval Air Station in California, joining a carrier aircraft unit preparing for launching planes to support the invasion of Japan. In early August, just as the unit was ready to sail, nuclear bombs were dropped on Hiroshima and Nagasaki, ending the war. In a few months, Kamm was discharged and returned to graduate studies at the University of Minnesota.

After completing M.A. and Ph.D. degrees in 1946 and 1948, respectively, majoring in educational psychology, Kamm joined Drake University and served as the Dean of Students for seven years. From 1955 to 1958, he was with Texas A&M, first as Dean of Student Services, then as Dean of the Basic Division. In 1958, he finally returned to Oklahoma State University (the name had changed from Oklahoma A&M the previous year), initially as Dean of the College of Arts and Science, then Vice President of Academic Affairs. He was appointed President in 1966 and served for 11 years in this position, leading the institution to become a comprehensive university. After leaving the presidency, he continued as a Professor, then Director of the Centennial History Project until retiring as President Emeritus in 1992. In his 44 years at Oklahoma State, Kamm was widely recognized as an educator, received two honorary doctorates, was a member of the Board of UNESCO, and authored several books and articles.

ROBERT M. KELLEY

The day after Pearl Harbor, Bob Kelley, a native of Fort Dodge, Iowa, joined the Navy. At 21 years old, holding a Ham license since 1938 (then W9QVC), a pilot's license, and a brand new Bachelor's Degree in economics from Regis University, he should have immediately qualified for Officer Candidate School, but the Navy was desperate for radio technicians. The recruiting office, therefore, convinced him to enlist as an Apprentice Seaman, with a promise of training in radio and rapid promotion.

Kelley was destined for the Radio Materiel School when it opened at Treasure Island. His first month, however, was at the Naval Reserve Armory in Los Angeles, attending an abbreviated Boot Camp – most of the time still in civilian clothes – with a group of other amateur radio operators. Finally receiving his promised Radioman 2/c rating, he went to Treasure Island and assisted in preparing the training facilities.

The Treasure Island RMS opened in early February 1942; Kelley was in the first class. The lack of equipment available for laboratory work limited this portion of the program. Based on placement tests taken earlier at the Armory, Kelley's classroom studies were accelerated by several months, and he graduated in May. He was then assigned to the USS *Mississippi* (BB-41) – the third battleship with this name – just before she left for Pearl Harbor. There the ship's initial air-search (SC) and fire-control (FC) radars were installed. At this time, Kelley attended the Fleet Radar School, obtaining hands-on experience with the radars. After 11 months on the *Mississippi* in South Pacific operations, Kelley was returned to Treasure Island as an instructor.

In May 1943, Kelley applied to become a Warrant Officer, but, instead, was sent to Officer Candidate School in New York. Following commissioning as an Ensign in the Naval Reserves, he attended the Pre-Radar School at Harvard University for five months, then went to the Radar School at MIT, graduating in April 1944. Kelley was again assigned to Treasure Island, this time in the Operational Training School. This was a refresher school for upper command officers to learn the use of the new Combat Information Center being introduced into the fleet.

Kelley was discharged from active duty as a Lieutenant (jg) in November 1945, and remained in the Naval Reserve until 1962. He returned to Fort Dodge, joining his family's insurance firm and continued with this activity until retiring in 1988. Although his civilian career did not involve electronics, he remained an active Ham, being assigned W0QVZ (Extra) in 1952, then W0BW in 1968.

BERNARD L. KLIONSKY

"In the Navy's electronic technician program, I learned to diagnose a problem in five minutes, then fix it in ten. I have continued to use this training throughout my professional career." Bernard Klionsky was born in 1925 and raised in Binghamton, New York. He started pre-med studies at Harvard, then took the Eddy Test and volunteered for the Navy in 1944. After Boot Camp at Great Lakes, he attended Pre-Radio at Manley, went to Gulfport for Primary School, and then to Navy Pier for Secondary School.

After graduation, Klionsky shipped out from San Francisco on August 1, 1945, on a troop transport bound for Okinawa and rumored to be part of the invasion of Japan. The A-bomb was dropped five days later, and the war ended before his arrival at Okinawa. He spent the remainder of his service time maintaining electronics on several ships while shuttling between Japan and China on a tanker, the USS *Kishwaukee* (AOG-9).

Klionsky was discharged from the Navy in June 1946. Returning to his studies at Harvard, he completed his undergraduate degree in biochemistry in 1947. Entering Hahnemann Medical College in Philadelphia, he received his M.D. degree in 1952. After an internship in Wisconsin and a residency in Kansas, in 1961 he became the director of Laboratories at Magee-Women's Hospital in Pittsburgh. Appointed an associate professor at the University of Pittsburgh School of Medicine, he became a full-time professor of pathology in 1970. He created a centralized system of clinical laboratories for all of the hospitals affiliated with the University of Pittsburgh, and for many years was Vice-Chairman of the Department of Pathology and President of the Department's practice plan. During a sabbatical, he set up a Department of Pathology for the new Ben Gurion Medical School at Beersheba in Israel.

In an interview, Klionsky noted that he not only taught his students pathology, but, more important, he showed them how to think – "to become highly trained problem solvers."

Through the years, Klionsky made significant contributions to his profession. He invented the open-top cryostat for rapid diagnosis of tissues from the operating room, forever changing surgical pathology. He identified the chemical structure of the lipid in tissues of patients with Fabry's disease, and contributed to understanding the pathogenesis of yellow hyaline membrane disease and the associated low bilirubin kernicterus, present at autopsy in many premature infants. His collaboration with his clinicians and contributions to the understanding of the biochemistry brought an end to these conditions.

In 2000, he officially retired from the University of Pittsburgh School of Medicine, but remained as Professor Emeritus, regularly teaching and interviewing candidates for medical school.

GEORGE W. MARTI

Born of humble roots on a Texas farm in 1920, and with his higher education from the Radio Materiel School, George Marti went on to become very successful and gain considerable wealth in the electronics field

After obtaining his amateur (W5GLJ) and FCC radiotelephone licenses at age 16, Marti started work at KTAT, then joined KFJZ, both in Forth Worth. For the next several years, he was with the Tarrant Broadcasting, maintaining equipment at several locations. In early 1942, he enlisted in the Marines and, based on his radio licenses, received the rank of Staff Sergeant. After basic training at San Diego, he was sent to the Radio Materiel School at the Naval Research Laboratory, graduating in the consecutive Primary and Secondary Schools in July 1943.

Having finished first in his class, Marti received an offer to remain at the RMS as an instructor, but he asked for overseas duty. He joined other Marines at San Diego, destined to join the First Marine Division in the Solomon Islands. They boarded the high-speed attack transport USS *Feland* (APA-11) bound for American Samoa. Three days out, their SA early-warning radar failed. The *Feland* had no radar technicians, so Marti was asked to attempt repairs. Having studied this particular radar at the RMS, he had it operating in a short while. Immediately, a surfaced submarine showed up on the remote display on the bridge. The Captain quickly took evasive action and escaped by outrunning the submarine. For the remainder of the voyage, Marti served on the Captain's staff, maintaining all of the radar and communications equipment.

Upon reaching American Samoa, and at the Captain's recommendation, Marti was given a special assignment as an adjunct to the Admiral's staff, responsible for maintenance of all critical electronic equipment. As islands in the Solomon Group were retaken, he would fly in and establish the primary communications. This continued over the next year, with work shifting to the Marshall Islands and his being promoted to Master Technical Sergeant.

In February 1945, Marti was rotated back to the States. After assignments at Cherry Point, North Carolina, and Washington, D.C., he was sent to Kingsville, Texas, where preparations were being made for the invasion of Japan. In August, just days after the atomic bombs were dropped and through special orders in reward for his services, he received his discharge.

Marti returned to Tarrant Broadcasting for a year, then started his own business. He opened KCLE in Cleburne, Texas, in early 1947, followed in two years by KCLE-FM, then KKJO, St. Joseph, Missouri, in 1953. Throughout the 1950s, Marti applied his knowledge of VHF – gained at the RMS as well as through subsequent Marine communications experience – in developing a system for providing radio-linked broadcasts from remote locations. In 1960, he sold the KCLE stations and started Marti Electronics, Inc., in Cleburne. Other radio product lines were added, including equipment that greatly improved audio quality. By 1994, when Marti sold his interest in the company, its products could be found in about 80 percent of broadcasting stations worldwide.

In 1992, Marti purchased the Cleburne State Bank, saving it just days before its failure. Five years later, this bank was acquired by First Financial of Abilene. Proceeds from this sale were applied to the Marti Foundation, established earlier to assist needy persons in college. Marti considers the Foundation his proudest achievement; with $10 million in existing investments, it assists over 100 students a year – most being the first in their family to ever attend college.

"When I was a struggling young man," Marti has noted, "The Radio Materiel School gave me an excellent education that led to my engineering career. Through the Marti Foundation, I want to similarly help as many needy young people of today as possible."

Over the years, Marti received many honors for his work. This included being named to the Texas Radio Hall of Fame, and earning the 1991 Engineering Achievement Award from the National Association of Broadcasters. He served as the Mayor of Cleburne from 1972 to 1984.

ARCH H. MCCLESKEY, JR.

On December 7, 1941, Arch McCleskey was operating his Ham radio when he heard that Pearl Harbor had been bombed. He was living with his family on Breezy Hills farm in Cobb County, Georgia, and teaching radio and electricity in the National Youth Administration program in nearby Marietta. Learning that the Navy was giving Second-Class Petty Officer ratings to licensed radio amateurs, he volunteered and entered the service in February 1942. He was sent to Charleston, South Carolina, for a brief Boot Camp, and then served a short time as an operator at the Charleston Navy Radio while waiting for the Houston University Primary School to open.

After the three months in Houston, McCleskey went to NATTC Ward Island for Secondary School, training in one of the first classes. He later noted that at Houston, "We were never told what we were training for; just that it was 'classified'. When we got to Ward Island, we learned that the great secret was radio detection and ranging – radar. We were not even allowed to mention this word outside Ward Island's secure Training Compound."

Upon graduating, McCleskey requested and received an assignment to the lighter-than-air service at Lakehurst, New Jersey. Here he received instruction in the operation of Navy blimps, as well as additional training on the radars, communication sets, and magnetic anomaly detector (MAD) carried by the large airships. In January 1943, he joined Airship Patrol Squadron Twenty-one (ZP-21) in NAS Richland, just below Miami, Florida. For the next year, he was almost continuously in the air, patrolling convoys and performing search-and-rescue missions up and down the U.S. and Cuban coasts. While the blimps carried depth charges as well as machine guns for defense, their primary duty as convoy escorts was in spotting enemy submarine and surface ships, providing this information for attacking aircraft and vessels.

In the early part of the war, the German submarine threat around the U.S. was great. Between January and July 1942, U-boats sank almost 400 merchant and naval ships in coastal waters, with a loss of near 5,000 seamen. The blimp patrols, however, were highly effective; from their start and to the end of hostilities, 134 of these airships made near 60,000 flights escorting more than 89,000 vessels – and never lost a ship to enemy action!

As the war expanded, so did the ZP-21 patrolled areas, extending over the Caribbean and the upper coast of South America. An operational base was set up on the Isle of Pines off the south coast of Cuba, and there were also operations out of Guantanamo Bay. McCleskey was promoted to Chief Electronic Technician Mate, and, in addition to flights on ZP-21, he set up electronic repair facilities on the bases. In September 1945, a hurricane destroyed the NAS Richland facilities, including 25 blimps, and the operation was transferred to NAS

Glynco, Georgia. McCleskey was the Leading Chief at Glynco until being discharged in December. While in the Navy, he had logged over 3,000 flight hours in blimps.

Entering Georgia Institute of Technology in the spring of 1946, McCleskey studied electrical engineering for three years, then became an instructor at Television Tech in Atlanta. He was employed at Lockheed Aircraft in Marietta in 1950, supervising their electronic laboratories, then in 1956 he moved to their Customer Field Service. Until retiring in 1981, he worked as a Tech Rep in 49 states and 54 countries.

CALVIN MOON

Electronic technicians found a wide variety of assignments following their training. Calvin Moon had both Navy and civilian experience that was different from most. Born in 1924 in New Jersey, he started at Rutgers College in 1942. To escape the draft, he volunteered into the Navy in January 1943 and was sent to the Bainbridge Naval Training Center for Boot Camp, taking the Eddy Test after arriving. His electronics training was Pre-Radio at Wright Junior College, Primary School at Texas A&M, and Secondary School at Treasure Island.

Upon graduating, he volunteered for submarine service and was sent to New London for special training on the SJ radar. After completing the eight-week course, Moon was assigned to the USS *Razorback* (SS-394) in July 1944, just before her departure for Pearl Harbor by way of the Panama Canal. The *Razorback's* first patrol started in August, supporting the Paula landings in the Philippines. Over the next year, the *Razorback* completed four other patrols, during which she sank a destroyer, damaged several freighters, and, in surface-gun actions, sank 12 sea-trucks. During a patrol in the Luzon Straits, the *Razorback* was heavily attacked with depth charges, and only escaped by 17 hours of silent running and diving to depths of 600 feet. On her fourth patrol, assigned to lifeguard duty in the Napo Shoto and Tokyo Bay areas, she rescued a downed P-51 fighter pilot, then picked up four B-29 Superfortress crewmen shot down following an air raid over Kobe. On August 31, 1945, the *Razorback* entered Tokyo Bay with 11 other submarines to take part in the formal surrender of Japan.

As the electronic technician aboard the *Razorback* – the only one on some patrols – Moon was responsible for the maintenance of all of the radar, sonar, and radio equipment. As with all diesel-powered submarines, it was necessary to surface periodically and recharge her batteries, primarily at night. At such times, the radar was vital to watch for enemy planes. The Airborne Pulse Receiver (APR) was another vital piece of equipment.

"What the APR did was sweep through frequencies to pick up those of the plane's radars. It was not directional, but as the signal got stronger, you knew he was homing on you, and you dive – pull the plug." Moon was awarded the Navy Commendation Ribbon and Medal for battle station proficiency.

Moon transferred from the *Razorback* in December 1945, and was discharged at Pearl Harbor the following February. He reentered Rutgers and graduated in 1948 with a B.S. degree in Agriculture Research. He entered the School of Veterinary Medicine at the University of Pennsylvania, receiving the V.M.D. degree in 1952.

For the next 47 years, Moon was a practicing veterinarian in Trenton, New Jersey, and was a partner in establishing the Central Veterinary Hospital in Columbus. He was President of the New Jersey Veterinary Medical Association, and, in 1984 and 2002, received the Alumni Award of Merit from the University of Pennsylvania. Although his Naval training and experience remained relatively unapplied during his professional career, he maintained a strong interest in electronics and, after retiring in 1999, obtained his Ham radio license (KC2CKI - Extra).

Moon was honored with the New Jersey Distinguished Service Medal, the State's highest military honor, for his wartime activities. In 2004, after service in the Viet Nam War and 21 years in the Turkish Navy, the USS *Razorback* was permanently berthed in the Arkansas River near Little Rock.

CALVIN H. MORTON

Operation Crossroads was a series of nuclear weapon tests conducted during the summer of 1946 at the Bikini Atoll in the South Pacific. Many Navy electronic technicians were involved in these tests, some volunteering to extend their service time to participate. Calvin Morton was one of these.

Morton was born in 1926 and raised in Bloomfield, New Jersey. Upon graduating from high school (he was the first person in his family to do so), he entered Rutgers College in the summer of 1944 as a student in electrical engineering. To ensure that he would be drafted into the Navy, he took the Eddy Test and entered the Navy in February 1945. Following Boot Camp at Great Lakes, he attended Pre-Radio at Wright Junior College, then Primary School at Gulfport, and Secondary School at Navy Pier.

When about a month from finishing the program, he accepted an offer of a graduation certificate and Electronic Technician Mate rating for supporting the Bikini tests. He shipped out from San Francisco in May 1946 aboard the USS *Avery Island* (AG-76), an electronics repair ship then carrying some 800 men, including many civilian technicians and instrumentation specialists.

At the Bikini Atoll, a fleet of about 90 vessels had been assembled as targets. These included older U.S. capital ships; a number of surplus cruisers, destroyers, and submarines; three captured Japanese and German ships; and many amphibious and auxiliary vessels. The tests were to study the effect of nuclear weapon detonations on the ships, equipment, and supplies. The *Avery Island* was among a support fleet of more than 150 ships that provided quarters, experimental stations, and workshops for most of the 42,000 men – including 37,000 Navy personnel – conducting the tests.

Morton assisted in setting up experimental equipment aboard the USS *Bladen* (APA-63), a 426-foot transport ship in the target fleet. This equipment was mainly radiation counters and recorders in various locations. During the tests, the ship also carried caged mice and small animals. The tests consisted of two detonations of 21-kiloton atomic devices. On July 1, Able was dropped from a B-29 Superfortress and detonated at 520 feet above the surface. Baker, suspended 90 feet beneath a landing craft, followed on July 25. Vessels closest to the detonations were severely damaged or sunk. Essentially all of the target fleet was bathed in radioactive water spray and radioactive debris from the lagoon bottom. For several weeks, the targets remained too contaminated for more than brief on-board visits. The *Avery Island* personnel eventually departed Bikini and returned to San Franscisco.

Morton was discharged at Brooklyn in October 1946. Returning to Rudgers, he completed another year in engineering, then switched to history and political science. He received his bachelor's degree in 1949, then went to New York University for graduate study, being awarded the M.A. degree in history in 1950. Morton's master's thesis concerned the Navy in the 1880-1890 period. He joined Prudential Insurance at their headquarters in Newark, New Jersey, and had a 36-year career in the claims and pension departments.

Residing in retirement in Ponte Vedra Beach. Florida, Morton became active in the Navy League, and, through this, joined the faculty of Program Afloat for College Education (PACE). From 1988 through 1998, he served as a professor of history under Central Texas College, teaching aboard naval vessels throughout the world.

ROBERT E. MUELLER

Some of the graduates of the Navy's Electronic Technician schools followed this with related work for earning a living, but gained personal satisfaction from a diverse profession. Such is the case for Bob Mueller. He was born above his family's bake shop in St. Louis, Missouri, in 1925. A Ham radio operator (W9AXB) since he was 13, he volunteered for the Navy in 1942 and received a rating as a Radio Technician 2/c. Following Boot Training at Great Lakes, he went directly to Bliss Electrical School for Primary School and the to Treasure Island for Secondary School.

Upon graduation, Mueller was sent to Gulfport, Mississippi, joining the skeleton crew of the destroyer escort USS *Forester* (DE-334) being built at nearby Orange, Texas. Beginning sea duty in early 1944, the *Forester* sailed from Norfolk in a convoy bound for Tunisa. Off the coast of Africa, the convoy came under heavy attack from German submarines and bombers, sustaining considerable damage to the electronic equipment and external wiring. In making quick repairs while under attack, Mueller so impressed the radio officer that he was recommended for V-12 training.

When the *Forester* returned to Norfolk, Mueller was transferred to Asbury Park for officer preparatory courses, then to MIT in the electrical engineering program. After the war, he was discharged from the Navy but remained at MIT as a co-op student engaged in radar development at Philco. He completed his B.S.E.E. degree in 1948.

Art had been Mueller's passion since early childhood. He had made many sketches while in the Navy and had taken art courses in the Architecture Department at MIT. Thus, after a period as a radio proctor at the U.S. Naval Academy, Mueller entered New York University under the G.I. Bill as an art major. He completed his B.A. degree in aesthetics in 1951, working as a technical writer and illustrator at McGraw-Hill after expiration of his G.I. Bill. Following this, he returned to engineering, with employment at Dumont Television, RCA Astro-Electronics, Bell Laboratories, and Belcore. He also continued art studies at the Brooklyn Art Museum School and the New School for Social Research and gained recognition in several visual-art styles.

Mueller has published a number of books, including *Science of Art: The Cybernetics of Creative Communications* (John Day, 1967), as well as many articles relating art and technology, especially computer-generated art. Eventually retiring from his "hobby" of engineering, Mueller continued in his "profession" of artist/writer. His wood-cut prints have been exhibited in museums worldwide, including the Metropolitan in

New York, Victoria and Albert in London, and state museums in Germany, Austria, Russia, and elsewhere.

NEIL PIKE

Captain Eddy said that he did not care what a man's education or background was; if they did well on the Eddy Test they were good candidates for his training program. Neil Pike is an example of such a person. He was born in 1922 at Meridian, Idaho, where his father worked in a lumber mill and his mother ran a small dairy farm. Pike attended a one-room school in Grand View, and rose early to help his sister deliver milk before school started. After the ninth grade, he left school to join his father in the lumber mill. For the next several years, he worked as a laborer and also studied radio through a correspondence school.

As the draft neared, Pike joined the Navy in August 1942 and was sent to Bremerton, Washington, for Boot Camp. There he passed the Eddy Test and, after only a few weeks of boot training, was sent to the Chicago Naval Reserve Armory for Pre-Radio. Following this, he went to the Houston University for Primary School, and finally to Treasure Island for Secondary School. Graduating near the top of his class, he was given his choice of assignments and selected the Submarine Service.

At New London in the fall of 1943, he completed the special six-week course on submarine electronics and was assigned to the USS *Spadefish* (SS-41), then being constructed at the Mare Island Navy Yards near San Francisco. After being commissioned in March 1944, the *Spadefish* went to Pearl Harbor and in July began her maiden patrol. In Philippine waters and the East China Sea, she sank ten Japanese ships before returning to port. The second patrol of the *Spadefish* centered on the Yellow Sea, resulting in four sinkings; the third and fourth patrols in the Yellow and East China Seas led to four sinkings each.

On the fifth and final patrol in June and July 1944, the *Spadefish* was equipped with a new mine-detecting sonar device and, with a pack of eight other submarines, made her way into the Sea of Japan between Japan and Korea. There she sank nine ships, some with gunfire attacks. Altogether, in 13 months the *Spadefish* sank 21 major ships and numerous smaller vessels.

During most of his stay on the *Spadefish*, Pike was the only Electronic Technician Mate aboard. The submarine was awarded two Presidential Unit Citations, and Pike personally received the Silver and Bronze Stars and was promoted to Chief Electronic Technician Mate. They returned to Mare Island in October 1945, and Pike was discharged the next month.

Pike settled in Washington, DC, and, while working full time at Remington Rand, returned to high school studies and also took courses from the Capital Radio Engineering Institute. In 1950, he received an excellent scholarship from Rutgers and completed his electrical engineering degree there, standing first in his class. This was followed by graduate studies in Cornell under a Ford Instrument Fellowship; he received the master's degree in 1955.

Pike's professional experience included research positions at MIT's Lincoln Laboratories, Bell Laboratories, IIT Communications, and Computer Sciences Corporation. He helped establish MCI in 1972, and followed this with employment at the Federal Communications Commission for seven years. In 1982, he joined the U.S. Army Materiel Command, eventually retiring in 1993. He served three times as the President of the Rutgers Engineering Society.

IRWIN A. ROSE

The Nobel Prize is the highest recognition of scientific achievement. Irwin "Ernie" Rose, a former Electronic Technician Mate, attained this honor. He was born in 1926 at Brooklyn, New York, and grew up in Spokane, Washington. After a year of study at Washington State College, he passed the Eddy Test and enlisted in the Navy in June 1944.

Following Boot Camp at San Diego, Rose attended Pre-Radio at Wright Junior College, Primary School at the Del Monte Hotel, and Secondary School at Treasure Island. Graduating just after the end of the war, he was assigned to the attack transport USS *Dade* (APA-99), as it came to San Francisco bringing troops home from the South Pacific. From San Francisco, Rose went with the *Dade* through the Panama Canal to the East Coast, where he participated in decommissioning the ship.

When he was discharged from the Navy in June 1946, Rose returned to Washington State, then transferred to the University of Chicago, where, in 1948, he received a B.S. degree in chemistry, followed in 1952 by a Ph.D. degree in biochemistry. After post-doctoral appointments at Western Reserve and New York Universities, he served on the faculty of the Biochemistry Department of the Yale Medical School from 1954 to 1963. Following that, Rose became a senior staff member at the Fox Chase Cancer Center in Philadelphia, serving until 1995 and being elected to the National Academy of Sciences. "My early training in the systematic tracing of cause and effect while in the Navy's electronics program has greatly benefited me throughout my career."

Upon retiring from Fox Chase, Rose accepted a special appointment as a distinguished professor-in-residence in the Physiology and Biophysics Department at the University of California in Irvine.

In 2004, Rose shared the Nobel Prize in Chemistry with Drs. Aaron Ciechanover and Avram Hershko, researchers at the Israel Institute of Technology. The focus of their prize-winning research, performed primarily in 1979 and 1980 when they were all at Fox Chase, was understanding and controlling ubiquitin, the protein that serves as the internal "garbage disposal" in cells of animals and plants. Their research led others to develop a class of cancer-fighting drugs, known as proteasome inhibitors, that capitalizes on cells' ability to destroy damaged or disease-causing proteins.

ROBERT M. SIMPSON.

During the latter half of the 20th century, the U.S. Space Program benefited greatly from engineers, scientists, and technologists who had gained their basic knowledge through Navy electronics training. Robert Simpson is typical of this large number of people.

Born in 1926, Simpson was raised in Cleveland, Mississippi, and finished high school when he was 16 years old. Believing that the war would be over shortly, he enrolled at Delta State Teachers' College, intending to transfer later and pursue a degree in engineering at MIT. After completing a year studying mathematics and science, the war continued and the draft loomed. With his college studies and a practical background from working in his father's garage, he easily passed the Eddy Test and enlisted in the Navy. Boot Camp at Great Lakes during the fall of 1944 was followed by Pre-Radio at Wright Junior College. He attended Oklahoma A&M for Primary School, then went to Secondary School at the Naval Research Laboratory.

When Simpson graduated in November 1945, the war was over and there was little opportunity for shipboard assignments. He was sent to the Naval Aviation Supply Depot in Norfolk, Virginia, handling spare-parts orders for carrier aircraft, then discharged from active duty in August 1946.

Simpson realized that it was pointless to join the masses attempting admission to MIT, so he returned to Delta State. Majoring in both mathematics and business administration, and with some advanced studies in physics, he earned his B.S. degree in 1949, graduating first in his class. With his academic record and Navy training in radar, he was exactly what a civil-service recruiter from the Air Force was seeking; thus, he was soon employed as a civilian radar instructor at Keesler AFB,

Biloxi, Mississippi. Still in the Naval Reserves, Simpson was recalled to active duty in 1950 as the Korean situation escalated. After a year at the Green Cove Springs Naval Station on the St. Johns River in Florida, he was discharged from the reserves and returned to Keesler.

The Ordnance Guided Missile Center, under Dr. Wernher von Braun and his German rocket team, was getting underway at Redstone Arsenal in Huntsville, Alabama. In July 1952, Simpson transferred there, joining the relatively small organization working on the Army's first ballistic missile, the Redstone. Missile activities greatly expanded, and in 1956, the Army Ballistic Missile Agency (ABMA) was formed, with development of the Jupiter missile as its primary mission. On January 31, 1958, ABMA used the Jupiter-C to place Explorer I – America's first satellite – into orbit. This led to even further expansion of ABMA, including the start of Saturn boosters with the expressed purpose of lifting space payloads.

On July 1, 1960, 4,000 personnel and their laboratory facilities were transferred from ABMA to form NASA's Marshall Space Flight Center (MSFC). The primary mission of MSFC was the development of Saturn I (small) and Saturn V (large) launch vehicles for the Apollo program. Simpson, classified as an Aerospace Engineer, was assigned to the Astrionics Laboratory, where the guidance, control, and instrumentation electronics were developed.

Concerning his classification and performance as an Aerospace Engineer, Simpson has noted, "My academic background in mathematics and science officially qualified me for the position, but the Navy training provided the background in electronics and related disciplines needed for my work."

In 1965, Simpson was sent to Downey, California, as the liaison engineer between North American Aviation and the MSFC Astrionics Laboratory during development of the S-II stage of Saturn V. After a four-year stay in California, he returned to MSFC. On July 20, 1969, Apollo 11 landed the first men on the moon.

The next major program at MSFC was Skylab, America's first space station, using a modified 3th stage of a Saturn V. A major feature on Skylab was the Apollo Telescope Mount (ATM), with instruments to study the sun in infrared and ultraviolet. After returning to MSFC, Simpson spent five years on this program and worked closely with the Naval Research Laboratory in developing the ultraviolet cameras. When Skylab was launched in May 1943, Simpson served on the ATM support team in mission control.

In 1974, Simpson was named Configuration Manager on the program that would eventually be called the Hubble Space Telescope.

He continued in this activity through 1986, interfacing with the mirror developer, Perkin-Elmer in Danbury, Connecticut, and the prime contractor, Lockheed in Sunnyvale, California.

In January 1987, after 40 years service with the Government, Simpson retired from NASA. In retirement, he and his wife traveled extensively throughout the U.S., and have been active in civic and church affairs. Simpson has also built an extensive array of computers and other electronic equipment in his study. Recently, he has greatly assisted the author in reviewing materials for this book.

ANDREW N. SMITH

Many Navy radio technicians eventually had careers with various government agencies. Andrew Smith is representative of those who were civil service employees with the U.S. Navy.

Born in 1925, Smith was adopted by a university family in Lincoln, Nebraska, and, being raised in an academic environment, was relatively unaffected by the Great Depression. In early 1942, he enrolled as a physics major at the University of Nebraska, partially in hope of obtaining a draft deferment. The deferment did not occur, and, after completing his junior year, he was called up for induction. By passing the Eddy Test, he was able to enter the Navy in late August 1944, and was sent to the Naval Training Center at Farragut, Idaho, for Boot Camp

Near the end of the year, Smith was at Hugh Manley for Pre-Radio, and then was sent to Gulfport for Primary School. This was followed by Secondary School at Navy Pier, where he graduated near the top of his class in early November 1945. Sent to the West Coast for assignment, he was shipped out to Guam on the USS *Dawson* (APA-79). From Guam, he went to NAB Enewetok and took over the maintenance of a radar beacon station. This was followed by an assignment at the Radio Range and ADF station on Parry Island, adjacent to the area being prepared for the Bikini Atoll nuclear tests. (The *Dawson* was one of the test ships.) A few days before the July 1 Able Test, with test preparations completed, Smith returned to the States aboard the USS *Randall* (APA-224). (Smith documented the dates and locations on the flyleaf of Sokolinkoff's *Advanced Calculus*, a previous textbook he had carried during his overseas assignments.)

Discharged from the Navy in mid-July 1946, Smith returned to the University of Nebraska and completed his B.A. degree in physics and mathematics in 1947. He then pursued graduate studies in physics and worked as a teaching assistant at the University of Minnesota, receiving the M.S. degree in 1950. After a year as a physics instructor at The

College of Wooster in Ohio, he served nine years as a civilian physicist with the Naval Electronics Laboratory (NEL) in San Diego, involved first in passive detection of submarine, then in very-low frequency research.

In 1961, Smith moved to Boulder, Colorado, and engaged in studies of VLF propagation, first at the National Bureau of Standards, then from 1962 to 1970 with the Westinghouse Georesearch Laboratory. While with Westinghouse, he designed VLF antennas and traveled throughout the world overseeing their construction. Returning to San Diego, Smith rejoined the NEL, which later became the Naval Ocean Systems Center. Remaining there in a supervisory position until 1978, he continued in designing VLF transmitters and antennas, and participated in developing and testing Omega, the first truly global, radio-based navigation system.

Recognized as a world-class authority on VLF systems, Smith became an independent consultant in 1978, operating from a farm near Eldridge, Missouri. For the next 20 years he provided design and testing services, primarily for the Navy, on VLF and LF systems around the world. At the same time, he and his wife also established a beef-cattle operation on their 700-acre farm. He has held a private pilot license for many years, and has a strong interest in geological and topographical mapping.

EDGAR C. SMITH

A descendant of '49ers, Ed Smith was born and raised in Los Angeles. He took the Eddy Test during his last year of high school and entered the Navy in September 1944. After Boot Camp at Great Lakes, he spent December in Pre-Radio School at Wright Junior College. As 1945 started, he went back across the country to Monterey, California, for Primary School at the Del Monte Hotel. Graduating April 12, 1945, the day President Franklin Roosevelt died, Smith was sent up the coast to Treasure Island for Secondary School. Upon completing his training in November, he remained at Treasure Island as an instructor, teaching electronic instrumentation until being discharged in July 1946.

Going directly to Stanford University, Smith was awarded his B.S. degree in 1949, with a major in mathematics. After completing a master's degree, he attended Brown University, earning the Ph.D. degree, also in mathematics, in 1955. He taught one year each at the University of Oregon and the University of Utah while completing his doctorate. Employed by IBM California in 1955, he initially served as a specialist in university computer problems, representing IBM in eight Western states. This included a period at the Western Data Processing Center at the

University of California at Los Angeles as liaison between IBM and UCLA. In 1965, he transferred to the international part of IBM in New York and served in a number of marketing and software development positions, including a four-year assignment in France. He is the author of several papers in the *IBM Systems Journal* and other publications.

Retiring in 1989, Smith took up a second residence in Monterey, California, where he pursued his life-long interest in California history and served as a volunteer at the Maritime Museum of Monterey. He authored a number of articles that appeared in the *Dogtown Territorial Quarterly*, including "The Hotel Del Monte Goes to War," Spring 2002.

GEORGE C. WEISERT

During the war, the Navy had many types of surface vessels. While the limelight was on the more glamorous carriers, battleships, and destroyers, some 1,000 Landing Craft Infantry (LCI) vessels of several types and the men who served on them did the dirty work of bringing invasion troops right up to the fighting, and providing close-in fire support with machine guns and rockets. In doing so, they suffered many casualties. The larger of these, the Landing Craft Infantry (Large), were ocean-going ships, with the crew including a radio technician to maintain the communications and other electronic equipment (but no radars).

The LCI(L)s were designed to transport up to 200 soldiers a relatively short distance, make quick beach landings, and disembark the troops in a few minutes from ramps lowered on each side of the bow. They carried a typical complement of 3 to 6 officers and 20 to 30 enlisted men. Their flat-bottom hulls were designed for beaching and not originally intended for cross-ocean travel. Given the urgency of wartime, however, they did just that, sailing from the United States to the European and Pacific Theatres. They had an endurance range of 8,000 miles at 12 knots. With a length of about 158 feet, they were the smallest Naval vessels that regularly crossed the ocean, with those aboard feeling every wave; it was said that every person who sailed the ocean on an LCI(L) suffered seasickness. George Weisert represents the radio technicians who served on these vessels.

Weisert was born in 1924 and raised in Darien, Connecticut, close by the veterans' facility that was taken over for the Noroton Heights Naval Radio School. After briefly attending New York University, he was drafted into the Navy in March 1943. During Boot Camp at Sampson Naval Training Center in New York, he took the Eddy Test and was subsequently sent to Pre-Radio in Michigan City, Indiana. Primary

School at Bliss Electrical School in Takoma Park, Maryland, followed; then he went a few miles south to Secondary School at the Naval Research Laboratory.

Upon graduating in March 1944, Weisert was assigned to the USS *LCI(L)-582* (these ships were not given a name). After training at the Little Creek Amphibious Training Base near Norfolk, Virginia, *LCI-582* was part of a flotilla that crossed the Atlantic to prepare for the invasion of southern France. After entering the Mediterranean, the LCIs went to Mers-El-Kebir on the coast of Algeria. With Weisert responsible for the electronic equipment, the ship sailed back and forth between North Africa and Italy, ferrying allied troops and prisoners of war. In addition, they participated in practice landings on beaches near Salerno.

The Allied invasion of southern France, an operation first code-named Anvil and later Dragoon, started in August 1944. LCI-582, being a group leader, took on a contingent of communications soldiers and hit the beach at Saint Tropez on August 15, initiating the invasion. Ferrying of troops to other southern French ports followed, with the campaign ending in September. LCI-582 convoyed back to Norfolk in October. Like many men who served on LCI(L)s, Weisert suffered from chronic seasickness and, after a brief hospital stay, was confined to shore duty only. He spent the next 15 months on assignment in Virginia, including an antiaircraft testing and training center at Virginia Beach and a mine craft shakedown group at Little Creek. (LCI-582 went on to San Diego, and then crossed the Pacific to participate in the assault on Iwo Jima and Okinawa.)

Upon being discharged in March 1946, Weisert applied by mail to Tufts College and was immediately accepted. Using his G.I. Bill, he graduated with a B.S.E.E. degree in 1950. Hired by IBM in Boston, he worked as a customer engineer on large electromechanical accounting machines, and then switched to digital equipment when the IBM 650 came out. In 1957, he transferred to IBM Endicott, New York, the original home of IBM. Initially a hardware engineer in the test laboratory, he migrated to software testing and remained in this activity until retiring in 1984.

For many years, Weisert was involved with Literacy Volunteers, primarily tutoring Chinese graduate students attending the State University of New York at Binghamton. He also played the euphonium in a town band.

ROBERT L. WEITZEL

Although in the Navy, Bob Weitzel served for over three years during the war on the USS *Dickman* (APA-13), a Coast Guard-manned amphibious attack transport that participated in critical amphibious landings throughout the world.

Born in 1921, Weitzel was raised in Shamokin, Pennsylvania, and developed a passion for radio equipment during his youth. After high school, he went to Washington, D.C., and worked as a deliveryman for a department store and also obtained his Ham license (W3JXV). As the war neared, he – like many other Hams – received a letter offering a rating of Radioman 2/c for enlisting in the V-6 Naval Reserve. (His letter, dated 4 December, 1941, is included in the Illustrations. It will be noted that it twice refers to "RADAR" – then a classified subject.)

Weitzel accepted this offer, and entered the Navy early in January 1942. He was sent to a special one-month combined Boot Camp and intensive pre-radio "filtering" activity at the Noroton Heights Radio School in Darien, Connecticut. From there he crossed the country to Treasure Island, entering the first class at this Radio Materiel School in February. Completing the combined Primary and Secondary School in August, he returned to the East Coast to join the attack transport USS *Dickman* (APA-13) in Hampton Roads, Virginia.

The *Dickman* was larger than most similar transports, having a complement of about 600 men and a capacity for transporting some 1,800 troops with their full battle equipment. Between leaving Norfolk in August 1942, until the end of hostilities, the *Dickman* earned six battle stars. She continuously criss-crossed both the Atlantic and Pacific oceans, typically in convoys but sometimes on her own, ferrying troops to and from the war. Weitzel and the *Dickman* participated in the invasion of French North Africa, Salerno, Italy, and Southern France. Briefly hospitalized for surgery, Weitzel missed being on the *Dickman* as she carried Rangers to the June 6, 1944, D-Day landing at Utah Beach. On the other side of the globe she carried troops to Australia, Guadalcanal, and other South Pacific areas, and participated in the initial landing at Okinawa. After the war, the *Dickman* returned Bataan Death March survivors from the Philippines. Operations were often in U-Boat infested waters, and enemy aircraft attacked her several times.

For most of the time, Weitzel was the only electronic technician and one of only a few Navy personnel aboard this ship. This sometimes led to inter-service conflicts. In early 1943, as the *Dickman* was returning unescorted from Australia, its SC radar went out and the Navigator was forced to use dead reckoning under overcast skies while approaching the

Panama Canal. The close-in waters were mined; thus, knowledge of exact position was vital. The Chief Radioman – a Coast Guardsman – usurped Weitzel's authority and took charge of the repairs, ultimately reporting to the Communications Officer that the radar could not be salvaged. Weitzel took over, and in short order had it operating. A navigational check using reflections from the Galapagos Islands showed that the ship was 50 miles off course. From that point on, there was no challenge to Weitzel's capabilities. He was promoted to Chief Radio Technician in 1944.

Weitzel was discharged from the Navy in November 1945, and enrolled at The Catholic University in Washington, D.C. After completing his B.E.E. degree in 1949, he was employed by Philco as a Technical Representative. His work over the next three years included assignments in Tokyo and Okinawa. Following this, he developed test equipment for Sylvania, then in 1954 joined the Applied Physics Laboratory of Johns Hopkins University. Here he was primarily involved in developing antenna components for the Terrier and Tartar Missile System, early forerunners of the Aegis System, then worked on the Transit Navigational Satellite.

In 1962, Weitzel joined NASA's Goddard Space Flight Center at Greenbelt, Maryland. He was initially involved in the Apollo Telescope Mount used on Skylab (America's first space station). He then devoted nine years to the Tirus and the GOES-series weather satellite programs, and was finally involved in developing electronics hardware for the Hubble Space Telescope. He retired from NASA in 1983.

After moving to Sarasota, Florida, Weitzel continued his interest in electronics, operating his Ham equipment, building electronic organs, and assembling personal computers. He also exercised his musical talents, playing trombone in the Sarasota Pops Symphony Orchestra, church orchestras, and jazz groups.

JOSEPH A. WEYGANDT

Recruiting advertisements for the Navy electronics maintenance program described this training as having a value of $12,000 (a very large amount for education in that time) and directly preparing men for post-Navy employment in communications. Following the war, many former electronic technicians took advantage of their Navy training in establishing careers in the broadcasting field.

Joe Weygandt was born in 1926 and raised in Toledo, Ohio. At age 14, he received the Ham license W8VDD. Later, a high-school co-op program allowed him to work at radio station WTOL, during which time he obtained his First-Class Radiotelephone license. In February 1944, he passed the Eddy Test and enlisted in the Navy. After only four weeks of Boot Camp at Great Lakes, the entire RT company was sent to the Navy Pier, not to enter radio training but to labor for a month in clearing out a former warehouse in Chicago near the 190 North State Street operation. (This four-floor facility became an important part of the enlarged Radio Chicago.)

Eventually, Weygandt was sent to Hugh Manley for Pre-Radio. Primary School at Texas A&M followed, and then back to the Navy Pier for Secondary School. Upon graduating at the end of March 1945, he crossed the country to join the crew of the USS *Breton* (CVE-23). This was an escort carrier with a complement of about 1,200 men and carrying 28 aircraft. She had just returned from the assault on Okinawa and had returned to the States to take on a new fleet of aircraft. For the remainder of the war, the *Breton* was on the "pineapple run," ferrying new aircraft to Guam and Hawaii, and returning with salvageable junk. After the Japanese surrender, the *Breton* was briefly in Japan, then returned to Tacoma, Washington, to join the mothball fleet in Puget Sound.

Weygandt was discharged in May 1946, and returned to Toledo and WTOL. His Navy training and additional studies under the G.I. Bill through Capital Radio Engineering Institute were very helpful as the station added FM and TV facilities. When the TV station was sold, the new owners took the WTOL call letters. Weygandt remained with the radio operations and became Chief Engineer of the AM and FM station, then known as WCWA. After several years, he returned to WTOL-TV where he participated in construction of new studio and transmitter facilities and the addition of color, stereo, satellite reception, and remote control.

"When I retired in 1991, I had completed a career of 49 years – with time-out for the Navy – without ever filling out a job application. As they had promised me in 1944, the Navy electronics training was perfect for a life-time of interesting and productive work."

Upon retiring, Joe and his wife immediately turned to travel, spending the next five years on the road in a 5th-wheel trailer. They then settled in LaRue, Ohio, selecting this village because it was at the center of their distributed family, but continued to take extended trips in a mini-motorhome. They keep busy in their church and maintain an interest in theater pipe organs. Ham radio helps keep up Joe's interest in electronics.

CHESTER P. WILKES

Passing the Eddy Test was a singular event in the lives of many young men, giving an often-unexpected entry to highly demanding intellectual pursuits and opening the door for careers that were previously little more than wishful thinking.

Chester Wilkes was born in 1921 and raised on a small farm near the town of Algoa, Texas. Although the area did not yet have power lines, by the time Wilkes was 14 years old he had built simple radios and had aspirations of being an electrical engineer. After completing high school, he became the sole support of his semi-invalid mother and raised chickens on the farm – there was no hope of attending college. Two years later, he found work as a laborer in nearby Texas City and eventually became a carpenter. When called up by the draft board, Wilkes initially received a hardship deferment, but in March 1944, he was inducted into the Navy.

At the Houston induction station, Wilkes scored exceptionally high on the entrance tests. This led to his taking and scoring well on the Eddy Test, and he was immediately elevated to Seaman 1/c, normally not given to draftees. Boot Camp at Great Lakes lasted only four weeks, and he was sent to Wright Junior College for Pre-Radio. In May, he went to Houston University for Primary School – an excellent assignment for him since it was only a few miles from his home. After Houston, it was back to Chicago and Navy Pier for Secondary School.

Upon graduating from Navy Pier in March 1945, Wilkes was sent to Pearl Harbor for assignment on the USS *Baltimore* (CA-6), a heavy attack cruiser. This ship had just supported the invasion of Iwo Jima, where it sustained damages from kamikaze attacks, and was destined for Okinawa, but was damaged in a typhoon and had returned to Pearl Harbor for repairs. Just as it was returning to service, Japan surrendered. For the remainder of 1945, the *Baltimore* served as part of the naval occupation forces in Japan. The ship then returned to the States, and Wilkes was discharged in March 1946.

Using the G. I. Bill, Wilkes enrolled at the University of Texas and completed his B.S.E.E. degree in 1950. He was recruited to join the civilian engineering staff of the U.S. Air Force Security Service (AFSS) at Brooks AFB in San Antonio, Texas. The AFSS had just recently been established and was responsible for equipping and operating monitoring stations throughout the world, gathering electronic intelligence. Often called the Air Force's intelligence branch, its motto was *Freedom through Vigilance*. The information collected in the field was normally sent to co-located analysts who would interpret the data, format reports, and send them on to the National Security Agency and other recipients.

Electronic engineers in the AFSS were involved a wide variety of activities, ranging from the assembling of state-of-the-art receivers and recorders for ground-based and airborne use, to the evaluation of intercepted signals to determine characteristics of the originating systems. Wilkes noted that "My training and experience in the Navy provided a very good background for AFSS activities."

In 1953, the central operations of the AFSS relocated to Kelly AFB, also in San Antonio. Through the years, the mission grew in importance until the AFSS eventually had some 12,000 personnel in about 70 locations throughout the world. Members of the AFSS were required to hold top levels of security clearance and were not allowed to discuss their jobs with outsiders – in fact, AFSS members usually could not even discuss their work among themselves unless they were in a secure location. In 1977, the AFSS was redesignated the Electronic Security Command, and is now the Air Intelligence Agency. Only in recent years have some details of these activities been made known to the public.

Wilkes worked his entire professional career in the secrecy of the AFSS and its successor, retiring in 1986. His final position was Deputy Director of the Maintenance and Plant Engineering Directorate (near 1,400 personnel). Remaining in San Antonio, he continued an active interest in electronics, primarily through operating and maintaining the television facilities of the First Baptist Church, a service he performed starting in 1971

The Author

Raymond C. Watson, Jr., brings a personal insight into the history covered in this book. He was first a student, then an instructor in the Navy's WWII Electronics Training Program. Additionally, with subsequent academic study that included undergraduate minors in history and literature (unusual for an engineer), he also has a passion for researching and writing about activities of this nature.

Watson was born in 1926 at Anniston, Alabama, and raised on a small dairy farm near Fort McClellan. He gained an early interest in radio from "hanging around" the communications station (WUR) at this U.S. Army installation, and built his first receiver before he entered his teens. After completing a special curriculum through the Engineering Defense Training Program, he obtained a commercial radio license in 1942 and began work in broadcasting.

Having earlier passed the Eddy Test, Watson volunteered for the U.S. Navy in 1944. After Boot Training at Great Lakes, he attended Pre-Radio at Wright Junior College, Primary School at Oklahoma A&M, and Secondary School at NATTC Ward Island (Corpus Christi), Texas. Upon completing the program, he remained at Ward Island, first as an instructor, then later writing instructional manuals for the Navy's post-war aviation electronics training school.

Upon being discharged from the Navy in 1946, Watson returned to the radio field, building and serving as the chief engineer for several communications and broadcast stations. In 1950, he was recognized by the FCC as a consulting engineer, and for the next 10 years provided services for radio and television operations. In parallel with post-Navy engineering work, he also returned to part-time educational pursuits, studying at a number of institutions and eventually earning five degrees, including a Ph.D. in engineering..

In 1953, Watson began a second career in academia while continuing with his consulting activities. His positions at several institutions through the following years included professorships in engineering and physics; department and division chairmanships; director of a research institute; and researcher as a National Science Foundation Faculty Fellow. In 1976, he was a founder and served as President of Southeastern Institute of Technology, a graduate professional school primarily for employees of government agencies and space and defense industries. He presently continues as a thesis and dissertation advisor, and also Chairs the Engineering Advisory Board at Alabama A&M University.

Watson continued his private consulting practice until 1960, then became Director of Research with Brown Engineering Company (later Teledyne Brown Engineering), a space and defense firm in Huntsville, Alabama. There he served for many years in positions including Vice President for the Science and Engineering Group, an organization of near 1,000 persons. For several years, he also served as Directing Manager for the American LLC of an Israeli medical electronics firm, and as the President of Vision Technologies Kinetics, a Singapore Technologies Engineering company.

In personal technical work, Watson was an early participant in U.S. ballistic missile defense and had extensive experience in the intelligence field, both activities drawing on his Navy-initiated radar knowledge. He led a team in developing the laser Doppler velocimeter, forerunner of laser radar. He contributed to America's greatest technological achievement, the Apollo lunar-landing, for which he received the NASA Public Service Award.

In 1991, he established R. C. Watson & Associates as a vehicle for his engineering and management consulting endeavors, and continues today full time in this activity. Throughout his careers, he has authored more than 430 papers, reports, and significant presentations..

Watson has lived in Huntsville, Alabama, since 1960, but his work has involved activities across America and in many other countries. He is married and has four children, eight grandchildren, and a growing number of great-grandchildren. He may be contacted as follows:

E-mail: RCW-Assoc@comcast.net

Mail: P.O. Box 1485; Huntsville, AL 35807

Oklahoma A&M, 1945

Huntsville, Ala., 2007

Information Sources, Acknowledgments, and Selected Bibliography

The collection of materials on this subject was a major endeavor. I first started after reading the obituary of Captain William C. Eddy in the November 1989 issue of the *IEEE Institute,* and realized that this outstanding training program had never been fully documented. It would have been far easier had I started earlier, while the principal developers of the program were still living and available for interviews. After a few unsuccessful attempts to find knowledgeable individuals, or official documents in places such as the Naval Historical Center, I put the project aside.

In late 2003, I read *The Invention That Changed the World* by Robert Buderi (Simon & Schuster, 1996), and *A Radar History of World War II* by Louis Brown (Institute of Physics Publishing, 1999). Finding that these two books, the most highly acclaimed on the subject, had no mention of the major undertaking of maintenance training for this new technology, I returned to my research with renewed fervor.

I must here acknowledge the patience of my wife, Charlotte (Bet), for putting up with the evenings, weekends, and holidays of the four years that I spent in researching and writing what she calls "the Eddy book."

Personal Contacts and Acknowledgements

A major difference in conducting research in 1989 and the present is availability of the Internet and search engines, such as Google and Yahoo's People Search, as well as the ubiquitous e-mail. Although still unsuccessful in obtaining significant official information, I have found many websites and associated links relating to persons who had trained in the program between 1942 and 1946. My subsequent e-mail correspondence with many former RTs/ETs and ARTs/AETs has been most enjoyable and informative.

I returned (via e-mail) to the IEEE concerning the 1989 obituary, and was referred to the writer, Ronald K. Jurgen. Himself a graduate of the Navy's program, Ron was formerly the Managing Editor of the *IEEE Spectrum,* but is now retired and lining in Florida. He cited an article that he had authored in the December 1975 issue: "Captain Eddy: the man who 'launched a thousand EEs'." Ron had interviewed Eddy at his home in Michigan City, Indiana, and had much to say about Eddy as a person. Neither the article nor Ron's recollections on the interview, however, provided factual information on the origin of the program.

A telephone call to *The News-Dispatch* in Michigan City resulted in three lengthy articles about Eddy; Maribeth Peterka, Archivist for the newspaper, was very helpful in finding this material. In one of his interviews, Eddy described the initial meeting in Washington and the opening of the school at the State-Lake television facility in Chicago. He did not, however, discuss the more extended aspects of the program, such as the Pre-Radio Schools that he started.

The unpublished autobiography of Richard Stock also contained information on the initial formation of the program. In addition, this document contained information on program activities at Utah State, Houston University, and NATTC Ward Island and its predecessor at the Naval Academy.

The unique characteristics of this program were mainly from the Primary Schools conducted by six engineering colleges. Contacts with libraries at Oklahoma State University (Jennnifer Paustenbaugh and David Peters), Utah State University (Dan Davis), Grove City College (Mary Sodergren), and the University of Houston (David Chapman) turned up a number of articles on participation at their institutions. A very complete document from Utah State was co-authored by W. Arnold Finchum, a student in the Navy's first class there who had returned later as a faculty member. The archives at Texas A&M had nothing on the program. The facilities of Bliss Electrical School – previously a private institution – were absorbed into Montgomery College in 1950, and essentially all archived materials concerning their participation in this program were lost.

Assembling information on the Navy-operated Primary Schools was a "mixed bag." There were many sources for Eddy's "prototype" operation at 190 North State Street. Former RT Walter M. Chambers loaned me his original copy of "Radio Chicago," a 78-page souvenir picture booklet with a treasure of information on all of the Chicago-based activities. Lesley Martin, a historian with the Chicago Historical Society Research Center, found a number of newspaper clippings. Stuart Stelzer provided documents on the Navy's takeover of the facilities at the College of the Ozarks. As a result of my inquiry, Larry Isch, Ozark's Director of Public Relations, researched and wrote an excellent article for their alumni magazine on this activity. A contact with the Naval Postgraduate School in Monterey, California – present owner of the former Del Monte Hotel – led me to Edgar C. Smith, a graduate of the program, who had extensively documented the history of the school at Del Monte. However, no archived information was found on the Navy-operated Primary Schools at Gulfport, Dearborn, and Great Lakes; all information concerning these schools came from personal recollections of students and instructors.

Internet and e-mail searches led to many people who provided much information. I communicated with former students and/or instructors from every Pre-Radio, Primary, and Secondary School. Although essentially all of these persons were in their 80s or 90s, they usually had excellent memories and amazingly good computer-communications skills. Many are still active Hams, and others maintain correspondence with fellow graduates and shipmates; consequently, an inquiry from me to one of them was often distributed to many others.

A number of descendants of program participants provided valuable information. These included children of two of the program founders: William C. Eddy, Jr., and Nancy Jane McClure, the living children of Bill Eddy, and Reed Stock, son of Sidney Stock, who, like his father, was a professor at Utah State. Unfortunately, no descendants of either Nelson Cooke or Wallace Miller were found.

Contributions from one individual must be given special recognition. Frank A. Genochio was a Marine student in the program, attended Eddy's State-Lake Teachers School, and then became an instructor at both Wright Junior College and College of the Ozarks. He went on to serve in field combat in the South Pacific and was instrumental in the famous "Navaho talker" secure communications. After the war, he completed a master's degree in 1949 at Stanford University by writing a thesis comparing the Navy's program to electronics programs then given in California junior colleges. This document includes very complete information on the program curricula at various levels. Frank personally provided me a copy of the thesis; other materials were later found and contributed by his daughters, Mary Jane Genochio and Benmi Genochio.

A small number of other former instructors in the program were also found and contributed interesting information. Dale Smock was on the electrical engineering faculty at Grove City College and participated in the development and subsequent instruction at this Primary School. Before his recent death, Dale provided historical materials on this activity. His wife, Ruth, described the hectic 10 days between being notified by the Navy and when the first students arrived.

Initial contacts with the Naval Historical Center (NHC) indicated that there were no archived documents on the program. However, with the outstanding assistance of Kathleen Lloyd, Head of the Operational Archives Branch, and Staff Archivist Timothy Pettit, two relatively comprehensive reports were unearthed: one on NATTC Ward Island and the other on electronics training at Treasure Island, each written by the respective Commanding Officer upon his return to retirement. Ms. Lloyd and her staff also found official biographies on all of the founding personnel of the program.

At the NRL, Records Manager Dean Bundy provided a number of class photographs that contributed information on the school at this facility. Bundy also called my attention to a special publication, *Pushing the Horizon,* released in 1998 on the 75th anniversary of the NRL, that included information on early electronic developments at this facility, as well as some data on the origin and operation of the RMS.

Contacts with students in the program also led to contributions of their letters home while in school -- very guarded, since much of the information about the training program, particularly the Secondary Schools, was classified at that time -- as well as a great variety of memorabilia. My personal collection included a picture booklet on NATTC Ward Island released by the Navy in 1943. Several people sent copies of graduating class booklets from Navy Pier and Treasure Island.

Memorabilia provided by former student and instructor Ralph M. Caldwell included a receipt from the Oklahoma A&M Bookstore showing that in 1942 he had purchased one of the issued books: *Radio Physics Course,* by Ghirardi. Recalling that I had a copy of this book, I opened it and, in my greatest surprise in collecting material for this book, found Caldwell's name written on the flyleaf! Years ago I had bought at an antique store the very book that Caldwell had originally studied in the program at Oklahoma A&M.

Internet searches turned up many websites with information on program graduates, as well as on electronics of the WWII era. Sites of antique bookshops resulted in a surprising number of contributing documents. These included copies of essentially every applicable book (those that were not already in my personal collection), as well as newsletters and instructional materials from many of the schools. One highly prized document obtained through this route was a 1934 study guide for the admission examination at the NRL Radio Materiel School. Another was *The Naval History of Treasure Island,* a hard-back book compiled in 1946 by the U.S. Naval Training and Distribution Center, San Francisco; this book includes a 16-page chapter on the Secondary School at Treasure Island, complete with pictures on every page.

Here should be mentioned that collected materials were often from sources that were either very parochial or second-person relayed, and there was sometimes conflicting information. Critical materials that were not corroborated by independent sources were omitted. An exception to this was information that I knew from personal participation as a student and/or instructor. I remember well my activities in the three schools, and still have my lecture notes from Pre-Radio and Primary Schools. Additionally, I have over 100 letters sent to my family while I was in the training. Even so, I found that six decades is a long time to retain accurate memories, and I revised some of my recollections based on inputs from others.

I must give special recognition to a present neighbor and friend, retired NASA engineer Robert M. Simpson. Robert went through the program at the same time as I did, even attending the same Pre-Radio and Primary schools, but we only came to know this in recent years. Like me, he has a large collection of letters written to his family while he was in the program and used these and his memory to make excellent contributions, particularly concerning the Secondary School that he attended at NRL/Bellevue. Robert read and commented on the book sections as they were developed, and also read in detail the final draft, finding many of my errors.

For the sections on the history of radar, there is a wealth of information available through the Internet from a multitude of history organizations, as well as personal postings. As in the material that I collected from individuals, however, there was sometimes conflicting information, primarily in dates and crediting people. The previously mentioned books, by Robert Buderi and Louis Brown, were believed to contain the most authentic, well-researched information and were often used in conflict resolution.

Dr. Brown's book, *A Radar History of World War II* (Institute of Physics Publishing, 1999), is especially recommended for readers who want a detailed, unbiased treatment of early radar history. My writing effort benefited greatly from Dr. Brown's review of a draft of this manuscript and calling my attention to items that needed correcting and/or clarifying. Unfortunately, Dr. Brown died a few days later. This final version contains some of his early comments in the Foreword.

Three additional sources of historical information must be mentioned. *History of Communications-Electronics in the United States Navy* by Captain Linwood S. Howeth (Government Printing Office, 1963, and at http://earlyradiohistory.us/1963hw.htm on the Internet) provided an excellent review of Naval electronics development. An outstanding summary of early radars is provided in a book-length document, "The Wizard War: WW2 & The Origins of Radar," by Greg Goebel, that may be found on the Internet at http://www.vectorsite.net/ttwiz.html. Similarly, a good summary of U.S. radars in the U.S. Navy is given in "Morgan McMahon and Radar," an extensive document posted at http://www.smecc.org, the Internet website of the Southwest Museum of Engineering, Communications and Computation.

The Publications Department of Teledyne Brown Engineering is acknowledged as the source of invaluable information on getting the initial fragmented notes and diverse pictures into publishable form. Individuals included Janet Bentley (Director), Lisa Jose and Darrell Osborn (Graphics), and Wanda Smith (Word Processing).

Last, but not least, thanks are given to Wallace E. Kirkpatrick, CEO of DESE Research, for providing facility support in the final phases of this effort.

Picture Sources

There are many pictures contained in this book; I consider them very important in communicating the information. It is customary to identify the sources of pictures; however, with the number of pictures provided the size of their captions, and ambiguity in many electronic documents, the source identification for each was not practical.

Many of the pictures came from public-domain Web sites; these are usually of low resolution, but suitable for the small sizes contained herein. A number of such pictures, particularly those from the early era of Navy electronics, are from the above-noted book by Howeth. Several of the pictures of specific equipment came from Goebel's document, and essentially all of the others of hardware came from Web sites devoted to particular units.

Essentially all of the pictures related to the Chicago-based activities, including the Pre-Radio Schools and 190 North State Street, came from the souvenir booklet "Radio Chicago." An article published in the November 1942 edition of *QST* magazine was the source of a number of pictures from college-based primary schools; these have been included with the permission of the American Radio Relay League, Inc. Most others related to these schools were provided by the associated colleges.

Many pictures related to Secondary School activities, especially of students in the laboratories, came from the Navy-published book, *The Naval History of Treasure Island*. Others came from photographs of RMS facilities provided by the Naval Research Laboratory. Some also came from a Navy-published souvenir booklet on NATTC Ward Island.

Two pictures deserve specific attribution. These are paintings of the JB radar included in the section on South Africa in Appendix II. They were done by South African artist Geoffrey Long (date unknown, c. 1942). The originals are in the National Museum of Military History, Johannesburg. Copies were provided by Joan Marsh, Treasurer of the South African Military History Society.

Selected Bibliography

Aitken, Hugh G. J.; *Syntony and Spark: the Origins of Radio,* Princeton University Press, 1979

Aitken, Hugh G. J.; *The Continuous Wave: Technology and American Radio, 1900-1932,* Princeton University Press, 1985

Alekseev, N. F., and D. D. Malairov, "Generation of High-Power Oscillations with a Magnetron in the Centimeter Band," *Journal of Technical Physics,* vol. 10, p. 1297, 1940 (in Russian). Translated by I. B. Benson, *Proc. of the IRE,* vol 32, p. 136, 1944

Alexanderson, E. F. W.; "New Fields for Radio Signaling, *GE Review,* vol. 27, p. 266, 1925

Allen, E. W. and H. Garlan; "Evolution of Regulatory Standards of Interference," *Proc. of the IRE,* vol. 50, p. 1306, 1962

Alvarez, Luis W.; *Adventures of a Physicist,* Basic Books, 1987

Amato, Ivan; *Pushing the Horizon: Seventy-Five Years of High Stakes Science and Technology at the Naval Research Laboratory,* Special Publication, Naval Research Laboratory, 1998

Atkinson, J. D.; *DSIR's First Fifty Years,* Department of Scientific and Industrial Research (New Zealand), 1976

Austin, Brian; *Schonland: Scientist and Soldier,* Institute of Physics Publishing, 2002

Avery, Donald H.; *The Science of War: Canadian Scientists and Allied Military Technology During the Second World War,* University of Toronto Press, 1999

Baker, William John; *A History of the Marconi Company,* Methuen, 1970

Baldwin, Ralph B.; *The Deadly Fuse,* Jane's Publishing, 1980

Barrow, William T.; "Transmission of electromagnetic waves in hollow tubes of metal," *Proc. of the IRE,* vol. 24, p. 1298, 1936

Baxter, James Phinney; *Scientist Against Time*, Little, Brown & Company, 1946

Beckley, Donald E.; "Teaching in a Naval School," *J. Higher Education*, vol. 14, pp. 122-126, March 1944

Beech, Bill; "MilList;" http://hereford.ampr.org (comprehensive listing of radio equipment)

Belevitch, Vitold; "Summary of the History of Circuit Theory," *Proc. of the IRE*, vol. 50, p. 848, 1962

Bowen, E. G.; *Radar Days*, Institute of Physics Publishing, 1987

Bowen, Edward George, editor; *A Textbook of Radar: A Collective Work by the Staff of the Radiophysics Laboratory, C.S.I.R.O., Australia*; Cambridge University Press, 1954

Brain, Peter, Sheilagh Lloyd, and Frank Hewitt; *South African Radar in World War II*, SSS Radar Book Group, 1993

Breckel, Harry F.; "History of Establishment and Operation, Naval Training School (Electronics Materiel), 1941-1945, Treasure Island, San Francisco, California," U. S. Navy Bureau of Naval Personnel, 1948

Brittain, James E.; "The Magnetron and the Beginning of the Microwave Age," *Physics Today*, vol. 38, p. 60, 1985

Brothers, J. T.; "Historical Development of Component Parts Field," *Proc. of the IRE*, vol. 50, p. 912, 1962

Brown, Louis; *A Radar History of World War II: Technical and Military Imperative*, Institute of Physics Publishing, 1999

Bryant, J. H.; "The First Century of Microwaves – 1886 to 1986," *IEEE Trans. on Microwave Theory and Tech.*, vol. 36, p. 830, 1988

Buderi, Robert; *The Invention That Changed the World*, Touchstone, 1997

Bugher, Elmer E.; *The Wireless Experimenter's Manual*, Wireless Press, 1920

Bukowski, Douglas; *Navy Pier: A Chicago Landmark,* Metropolitan Pier and Exposition Authority, 1996

Bullard, W. H. G.; "Some Facts Concerned with the Past and Present Radio Situation of the United States," *Proc. of the U. S. Naval Institute,* vol. 49, p. 1623, 1923

Bureau of Naval Personnel; "Introduction to Radio Equipment," Training Manual 10172, Department of the Navy, 1946

Bureau of Naval Personnel, "Radar Operator's Manual," Training Manual 01090, Department of the Navy, 1945

Bureau of Naval Personnel; "Submarine Sonar Operator's Manual," Training Manual 16167, Department of the Navy, 1944

Bureau of Naval Personnel; "Sonar," Training Manual 10884, Department of the Navy, 1953

Bureau of Naval Personnel; "The Radio Materiel School," Training Bulletin, Department of the Navy, 1948

Burns, Russell W.; *Television: an International History of the Formative Years,* IEE Press, 1998

Burns, Russell W. (editor); *Radar Development to 1945,* rev. ed., IEE Press, 1989

Carnearl, Georgetta; *A Conqueror of Space; An authorized Biography of the Life and Work of Lee de Forest,* H. Liveright, 1930

Cheney, Margarer; *Tesla: Man Out of Time,* Prentice-Hall, 1981

Clark, Ronald W.; *Sir Henry Tizard,* Meuthen Press, 1965

Cole, Larry S; *The History and Development of Electrical Engineering at Utah State University,* Utah State University Press, 1972

Colin, Robert I., and Otto Scheller; "The Radio Range Principle," *IEEE Trans. on Aerospace and Electronic Systems,* vol. AES-2, p. 481, 1966

Collins, George B.; *Microwave Magnetrons, MIT Radiation Laboratory Series*, vol. 6, McGraw-Hill, 1948

Colton, Roger B.; "Radar in the United States Army," *Proc. of the IRE*, vol. 33, p. 740, 1947

Conant, Jennet; *Tuxedo Park: A Wall Street Tycoon and the Secret Palace of Science That Changed the Course of World War II*, Simon & Schuster, 2002

Cooke, Nelson M.; "Mathematics for Electricity and Radio," Invited paper, Mathematics-Science Panel of the New York Society for the Experimental Study of Education, March 20, 1943. Published in *The Mathematics Teacher*, vol. 36, p. 329, Dec. 1943

Cooke, Nelson M.; *Mathematics for Electricians and Radiomen*, McGraw-Hill, 1942

Cooke, Nelson M.; "Fundamental Analysis of Radio and Electronic Theory: Effective Method of Teaching," *Radio News*, p. 46, June 1944

Cressman, Robert J.; *The Official Chronology of the U.S. Navy in World War II*, Contemporary History Branch, Naval Historical Center, 1999; Web version, http://www.ibiblio.org/hyperwar/USN/USN-Chron.html

Dayton, David M.; "The Naval Training School," *'Mid the Pines: A History of Grove City College*, Grove City Alumni Association, 1971

Dellinger, J. H., *et al*; "The Principles Underlying Radio Communications," National Bureau of Standards Pamphlet No. 40, Government Printing Office, 1918, 1921

Denfeld, Louis E.; "Research Activities of the Bureau of Naval Personnel," *J. Applied Physics*, vol. 15, pp. 289-290, March 1944

DeSoto, Clinton B.; "The Navy Trains Radio Technicians," *QST*, November 1942

"Directory of SCR Components;" U.S. Army Signal Corps document; http://www.gordon.army.mil/ocos/museum/equipment.asp

Eddy, Captain William C.; *Television: The Eyes of Tomorrow*, Prentice-Hall, 1945

Eddy, Lieut. Comdr. W. C., *et al*; *Wartime Refresher in Fundamental Mathematics*, Prentice-Hall, 1945

Edwards, Lee; "The War Effort [at Grove City College]," *Freedom's College*, Regnery Publishing, 2000

Ewell, Cyril Frank; *The Poulsen Arc Generator*, Van Nostrand, 1923

"Facilities of the Naval Research Laboratory," Internal Document, Naval Research Laboratory, Washington D.C., June 1944

Fagen, M. D. (editor); *A History of Engineering and Science in the Bell System; National Service in War and Peace (1925-1975)*, Bell Telephone Laboratories, 1978

Farnsworth, Philo T.; "Television by Electron Image Scanning," *J. Franklin Inst.*, vol. 218, p. 411, 1934

Fessenden, Helen M.; *Fessenden, Builder of Tomorrow*, Coward-McCann, 1940

Fessenden, R. A.; "Wireless Telegraphy," *The Electrical Review*, vol. 60, p. 252, 1907

Fessenden, Reginald Aubrey; "Autobiography," *Radio News*, May through Nov. 1925

Finchum, W. Arnold, Doran Baker and Darwin L. Salisbury; "The Navy Training Station at Utah State Agricultural College During World War II," Archive Document, Utah State University Library, 1995

Fort Monmouth Historical Office; *A Concise History of the U. S. Army Communications-Electronics Life Cycle Management Command and Fort Monmouth, New Jersey*, CE LCMC Historical Office, 2005; electronic (Web) version, http://www.monmouth.army.mil/historian/pub.php

Friedman, Norman; *Naval Radar*, Naval Institute Press, 1981

Gallegher, Pat; "Where did the TV engineers come from?" History and Archives, Web site of the Order of the Iron Test Pattern; http://www.oitp.org/archive.htm

Garnett, E. B.; "A 'gadget man' [Bill Eddy] develops electronics [technicians] for the Navy," *The Kansas City Star*, May 2, 1945

Gebhard, Louis A.; *Evolution of Naval Radio-Electronics and Contributions of the Naval Research Laboratory*, Report 8300, Naval Research Laboratory, 1979

Genochio, Frank Albert; "The Training of Navy Electronics Technicians and a Terminal Course in Radio Electronics," Thesis, submitted to the School of Education, Stanford University, 1949

Goebel, Gregory V.; *The Wizard War: WW2 & The Origins of Radar*, a book-length document on the Web; http://www.vectorsite.net/ttwiz.html

Guierre, Maurice; "The History and the Creation and Progress of the Radar [in France]," *Waves and Men, History of the Radio*, Julliard, 1951

Habann, Erich; "A New Generator Tube," *Zeitschrift f. Hochfrequenz*, vol. 24, p. 115, 1924 (in German)

Hall, John S.; *Radar Aides to Navigation, MIT Radiation Laboratory Series, vol. 2*, McGraw-Hill, 1947

Hanbury Brown, Robert; *Boffin: A Personal Story of the Early Days of Radar, Radio Astronomy and Quantum Optics*, Taylor & Francis, 1991

Hayes, Harvey C.; "Detection of Submarines," *Proc. Am. Phil. Soc.*, vol 59, no. 1, p. 1, 1920

Heil, Oskar E., and Agnesa Arsenjewa-Heil; "On a New Method for Producing Short, Undamped Electromagnetic Waves of High Intensity," *Zeitschrift fur Physik*, vol. 95, p. 752, 1935 (in German)

Hepcke, Gerhard (translated by Hannah Liebmann); *The Radar War, 1930-1945*, M. Holliman; http://www.radarworld.org/radarwar.pdf

"Here's How Navy Grinds Out Radio Ace in 3 Months," *Chicago Daily Tribune*, March 8, 1942

Hewett, F. J.; "South Africa's Role in the Development and Use of Radar in World War II," *Military History Journal*, vol 3, no 3, June 1975

Hewitt, Frank J.; "50 Years of Radar in South Africa," *Radioscientist*, vol 1, p. 71, 1990

Hollmann, Hans E., *Physik und Technik der ultrakurzen Wellen, Band 1 und 2, [Physics and Technique of Ultrashort Waves, Book 1 and 2]* Springer, 1938 (in German)

Hollmann, Martin; *Hans Erich Hollmann, Pioneer and Father of Microwave Technology*, Aircraft Designs, Inc., 2003

Hornung, J. Lawrence; *Radar Primer*, McGraw-Hill, 1948

Howeth, Linwood S.; *History of Communications-Electronics in the United States Navy*, Government Printing Office, 1963; electronic (Web) version, http://earlyradiohistory.us/1963hw.htm

Hugill, Peter J.; *Global Communications Since 1844: Geopolitics and Technology*, Johns Hopkins University Press, 1999

Hull, Albert W.; "The Effect of a Uniform Magnetic Field on the Motion of Electrons Between Concentric Cylinder," *Phys. Rev.*, vol. 18, p. 31,1921

Hülsmeyer, Christian; "Hertzian-Wave projecting and receiving apparatus adapted to indicate or give warning of the presence of a metallic body, such as ship or train, in the line of projection of such a wave," British Patent issued Sept. 2, 1904

Hunt, Frederick V.; *Electroacoustics*, Harvard University Press, 1954

IEEE Center for the History of Electrical Engineering; *RAD Lab: Oral Histories Documenting World War II Activities at the MIT Radiation Laboratory*, IEEE Press, 1993

Isch, Larry; "The Navy Invasion of 1944," *Today – A Magazine for Ozarks' Alumni and Friends,* Fall/Winter 2004, University of the Ozarks, Fall/Winter 2004

Jenkins, Charles F.; "Radio Vision," *Proc. of the IRE*, vol. 15, p. 958, 1927

Jones, R. V.; *The Wizard War: British Scientific Intelligence*, Coward, McCann and Geoghegan, 1978

Jurgen, Ronald K.; "Captain Eddy: the man who launched a thousand EEs," *IEEE Spectrum*, Nov. 1975

Jurgen, Ronald K.; "William Eddy, devised the Eddy Test," Obituaries, *The Institute* (IEEE), Nov. 1989

Kamm, Robert B., *et al*; *First Hundred Years: Oklahoma State University: People, Programs, Places,* Oklahoma State University Press, 1990

Kennedy, John F.; "Speech, Aug. 1, 1963, U.S. Naval Academy," *Public Papers of the Presidents of the United States: John F. Kennedy, January 1 to November 22, 1963,* U.S. Government Printing Office, p.620, 1964

Kilbon, Kenyon; *A Short History of the Origins and Growth of RCA Laboratories, 1919 to 1964,* David Sarnoff Research Laboratory, RCA, 1964; electronic (Web) version, http://www.davidsaranoff.org/kil.htm

Killer, Peter A.; *The Cathode-Ray Tube: Technology, History, and Applications,* Palisades Press, 1992

Kingsley, F. A. (editor); *The Development of Radar Equipment for the Royal Navy, 1935-45,* McMillian Press, Ltd., 1995

Kroge, Harry von (translated by Louis Brown); *GEMA: Birthplace of German Radar and Sonar,* Institute of Physics Publishing, 2000

Lange, Henry; "Genius Bill Eddy never stopped dreaming," *The News-Dispatch* [Michigan City, Indiana], Sept. 18, 1989

Labonav, M. M.; *Iz Proshlovo Radiolokatzii [Out of the Past of Radar],* Military Publisher of the [Soviet] Ministry of Defense, 1969 (in Russian)

Lange, Henry; "The Eddy Genius," Four-part series on William Crawford Eddy. *The News-Dispatch* [Michigan City, Indiana], March 25-28, 1985

Latham, Colin, and Anne Stobbs; *Radar: A Wartime Miracle,* Sutton Publishing Ltd., 1996

Le Pair, C. (Kees); "Radar in the Dutch Knowledge Network," Telecommunication and Radar Conference, EUMW98, Amsterdam, 1998; electronic (Web) version, http://www.clepair.nl/radar-web.htm

Lewis, David L.; "Henry Ford and His Magic Beanstalk," *Michigan History Magazine,* May / June 1995

Lockwood, Charles A.; "Electronics in Submarine Warfare," *Proc. of the IRE*, vol. 35, p. 712, 1947

"Loop Sailors," *Time Magazine,* March 23, 1942

Lovell, Sir Bernard; *Echoes of War: the Story of H2S Radar,* Adam Hilger, 1991

MacKenzie, John A.; "World War II Ground Radar," *The History of Canadian Military Communications and Electronics,* Canadian Forces Communications and Electronics Museum, 2003

MacLauran, W. R.; *Invention and Innovation in the Radio Industry,* Macmillan Company, 1949

MacLeod, Roy M. (editor); "The Boffins of Botany Bay: Radar at the University of Sydney, 1939-1945," *Historical Records of Australian Science,* vol.12, p. 411, 1999

"Marconi Calling," an extensive Web site dedicated to the life, science, and achievements of Guglielmo Marconi, Marconi Corporation, 2001; http://www.marconicalling.com

Marconi, Degna; *My Father, Marconi,* Guernica Editions (translation from Italian), 1996

Martinez, Deborah; "Ward Island was hush-hush radar school," *Corpus Christi Caller-Times,* March 7, 2000

McDevitt, E. C. (editor); *The Naval History of Treasure Island (1941-1946);* U. S. Naval Training and Distribution Center, San Francisco, 1946

McKellar, Ian C.; *History and Memories of 14 Radar Station, Wilsons Promontory.* [Australia], McKellar Publisher, 2005

Megaw, Eric C. S.; "The High-Power Magnetron: A Review of Early Developments," *J. of the IEE,* vol. 93, p. 928, 1946

Middleton, D. E. K.; *Radar Development in Canada: The Radio Branch of the National Research Council of Canada 1939-1946*, Wilfrid Laurier University Press, 1981

Muller, Werner; *Ground Radar Systems of the Luftwaffe, 1939-1945*, Schiffer Publishing Ltd., 1998

"Mystery Ray Locates 'Enemy': U.S. Army Tests Detector for Hostile Ships and Planes," *Popular Science*, Oct. 1935

Naeter, Albrecht; "Naval Training School: Elementary Electricity and Radio Materiel," *The Oklahoma State Engineer*, Dec. 1943

Nakagawa, Yasudo; *Japanese Radar and Related Weapons of World War II*, translated and edited by Louis Brown, John Bryant, and Naohiko Koizumi, Aegean Park Press, 1997

"Naval Air Technical Training Center, Corpus Christi, Texas," souvenir picture brochure distributed by the Navy, date unknown, c. 1944

Navy Department; "Admiral Louis Emil Denfeld, U.S. Navy, Retired." Navy Office of Information, Dec. 1967

Navy Department; "Commander Sidney Richard Stock, U.S. Navy Reserve," Navy Office of Information, Nov. 1946

Navy Department; "Lieutenant Commander Nelson Magor Cooke, U.S. Navy, Deceased," Navy Office of Information, Aug. 1967

Navy Department; "Rear Admiral Wallace J. Miller, U.S. Navy, Deceased," Navy Bibliographies Section, Nov. 1951

Navy Department; "Transcript Service Record of Captain William Crawford Eddy, U.S. Navy, Retired," Bureau of Naval Personnel, May, 1946

"Navy Establishes New Radio School," *Chicago Daily Tribune*, Jan. 13, 1942

"Navy Sonar," NAVPERS 10884, Training Division, Bureau of Naval Personnal, 1953

Nicholson, Patrick J.; *In Time: An Anecdotal History of the First Fifty Years of the University of Houston*, Pacesetter Press, 1977

Norberg, Arthur L. "The Origins of the Electronics Industry on the West Coast," *Proc. of the IEEE*, vol 64, p. 1314, 1976

Okabe, Kinjiro; "On the Applications of Various Electronic Phenomena and the Thermionic Tubes of New Types," *Japanese IEEE*, vol. 473 (Suppl. Issue), p. 13, 1927 (in Japanese)

"Operational Characteristics of Radar Classifiecd by Operational Characteristics," Radar Research and Development Subcommittee of the Joint Committee on New Weapons and Equipment, FTP 217, August 1943; Web, http://www.history.navy.mil/library/online/radar.htm

"Outstanding opportunity for amateurs to serve their country in Class V6 of the Naval Reserve," *QST*, Feb. 1942

Page, Robert Morris, "The Early History of Radar," *Proc. of the IRE*, vol. 50, p. 1231, 1962

Page, Robert Morris; *The Origin of Radar*, Doubleday, 1962

Parcher, James V.; *A History of the Oklahoma State University College of Engineering, Architecture and Technology*, Oklahoma State University Press, 1988

Pardini, Albert L.; *The Legendary Norden Bombsigh*t (Schiffer Publishing, 1999)

Parrott, D'Arcy Grant; "LORAN," *The Coast Guard At War, IV*, Public Information Division, U.S. Coast Guard Headquarters, 1944

Pierce, John R.; "History of the Microwave-Tube Art," *Proc. of the IRE*, vol. 50, p. 978, 1962

Pierce, J, A., A. A. McKenzie, and R. H. Woodward (editors); *Loran: Radiation Laboratory Series, Vol. 4*, McGraw-Hill, 1947

Pierrepoint, J. J., J. S. Weigand, and N. M. Cooke; "Preparation for Candidates, Radio Materiel School," 3rd ed., Naval Research Laboratory, Oct. 1934

Price, Alfred; *Instruments of Darkness: the History of Electronic Warfare*, Macdonald and Jane's Publishers, 1977

"Radar Countermeasures – Electronics War Report," *Electronics Magazne*, p. 92, Jan. 1946

"Radio Chicago," souvenir picture brochure prepared for the Navy by Albert Love Enterprises, undated, c. 1945

Randal, J. T. and H .A. H. Root, "Historical Notes on the Cavity Magnetron," *IEEE Trans. on Electron Devices*, vol. 39, p. 363, 1976

Redhead, Paul A.; "The Invention of the Cavity Magnetron and Its Introduction into Canada and the U.S.A.," *Physics in Canada*, Nov./Dec. 2001

Reich, Herbert J. and Radio Research Laboratory Staff; *Very High-Frequency Techniques, Vols. I and II*, McGraw-Hill, 1947

Reintjes, J. Francis, and Godfrey T. Coate (editors); Members of the Staff of the Radar School, MIT; *Principles of Radar*, McGraw-Hill, 1944 (Secret), 1946, 1952

Ridenour, Louis N.; *Radar System Engineering, MIT Radiation Laboratory Series*, vol. 1, McGraw-Hill, 1947

Roberts, Arthur; *Radar Beacons, MIT Radiation Laboratory Series*, vol. 3, McGraw-Hill, 1947

Romano, Salvatore; "History of the Development of Radar Technology in Italy," Regia Marina Italiana [Royal Italian Navy], 2004; Web, http://www.regiamarina.net/others/radar/radar_one_us.htm

Rosco, Theodore; *United States Submarine Operations in World War II*, United States Naval Institute, 1949

Rowe, Albert P.; *One Story of Radar*, Cambridge University Press, 1948

Sarnoff, David; *Looking Ahead*, McGraw-Hill, 1968

Schatzkin, Paul; *The Boy Who Invented Television*, TeamCom Books, 2002

Schedvin, C. B.; *Shaping Science and Industry: A History of Australia's Council for Scientific and Industrial Research 1926-1949,* Allen and Unwin, 1987

Schmidt, Edmund F.; *Noroton Heights -- A Neighborhood for Generations,* Darien Historical Society, 1992

Schooley, Allen H. "Pulse Radar History," *Proc. of the IRE,* vol. 37, p. 404, 1949

Schottky, Walter; "On the Origin of the Super-heterodyne Method," *Proc. of the IRE,* vol. 14, p. 695, 1926

Shembel, B. K., "U istokoc radiolokatzii v SSSR [The origin of radar in the USSR]," *Soviet Radio,* 1977 (in Russian)

Sherman, Don; "The Secret Weapon," *Smithsonian Air & Space Magazine,* Feb./Mar. 1995

Sieche, Erwin F., "German Naval Radar to 1945," translated by Heinz-Gerhard Schoeck from *Warship,* vol. 21 and 22, 1982; electronic (Web) version, http://www.navweaps.com/Weapons/WRGER_01.htm

Skolnik, Merrill I.; "Fifty Years of Radar," *Proc. of the IEEE, Special Issue on Radar*, vol. 73, p. 182, 1985

Smith, Edgar C.; "The Hotel Del Monte Goes to War," *Dogtown Territorial Quarterly* (California History), Spring 2002

Smith, Philip H.; "Transmission line calculator," *Electronics* Magazine, p. 29, Jan. 1939

Solby, Earl; "Simplicity's Godchild: The Navy's Bill Eddy," *Esquire* Magazine, April 1943

Soller, Theodore Merle A. Star, and George E. Valley, Jr.; *Cathode Ray Tube Displays, MIT Radiation Laboratory Series,* vol. 22, McGraw-Hill, 1948

Soulounac, A; "The story of IFF (Identification Friend or Foe)," *IEE Proc,,* vol.132, pt. A, p. 435, Oct. 1985

Southworth, George C.; "Hyperfrequency waveguides – general considerations and experimental results," *Bell Sys. Tech. Journal*, vol. 15, p. 284, 1936

Southworth, George C.; "Survey and History of the Progress of the Microwave Arts," *Proc. of the IRE*, vol. 50, p. 1199, 1962

Spires, Al; "It's Like This," Seven-part series on Captain William C. Eddy, *The News-Dispatch* [Michigan City, Indiana], Nov. 7-15, 1956

Stalnaker, John M.; "Construction and Application of Psychological Tests in the Armed Services," *Review of Educational Research*, vol. 14, p. 102, 1944

Stewart, Irvin; *Organizing Scientific Research for War*, a volume in the series *Science in World War II*, Office of Scientific Research and Development, Little, Brown & Company, 1948

Stock, Sidney R. and Josephine L. Stock; "The Life History of Sidney Richard Stock," unpublished, 1969

Stoddard, George K; "Naval Air Technical Training Center, Ward Island, Corpus Christi, Texas, 1942-1944" Commander's Summary Report to the U. S. Navy Bureau of Naval Personnel, 1944

Strong, K. W.; "He did the impossible," [Eddy at RCA Television], *The New York Sun*, Jan 27, 1940

"Submarine Sonar Operator's Manual," NAVPERS 16167, Training Division, Bureau of Naval Personnal, June 1944

Swords, Sean S.; *Technical History of the Beginnings of Radar*, Peter Peregrinus, Ltd, for the IEE Press, 1986

Taylor, A. H. and Hulburt, E. O; "The Propagation of Radio Waves Over the Earth," *Physical Review*, vol. 27, February 1926

Taylor, A. Hoyt; *Radio Reminiscences: A Half Century*, Naval Research Laboratory, 1948, 1960

Terrett, Dulany; *The Signal Corps: The Emergency (to December 1941)*, 4th edition, Government Printing Office, 2002

"The Navy Trains Radio Technicians: Radio Hams Prepare to Learn a Fascinating New Art," *QST*, Nov. 1942

"The 'VT' or Radio Proximity Fuze," Applied Physics Laboratory, Johns Hopkins University. 1945

Tomlinson, John D.; *International Control of Radiocommunications*, J. W. Edwards, 1945

Truxal, John G.; "Review of Control Developments," *Proc. of the IRE*, vol. 50, p. 781, 1962

Tuev, Merle A.; "Early Days of Pulse Radio at the Carnegie Institution," *J. of Atmospheric and Terrestrial Phys.*, vol. 36, p. 2079, 1974

Tyne, Gerald F.; *Saga of the Vacuum Tube*, Howard W. Sams, 1977

"U.S. Radar – Operational Characteristics of Radar Classified by Tactical Operations," Radar Research and Development Subcommittee of the Joint Committee on New Weapons and Equipment, Joint Chiefs of Staff, FTP 217 [formerly Secret], 1 August 1943

Vaeth, J. Gordon; *Blimps and U-Boats: U.S. Navy Airships in the Battle of the Atlantic*, Naval Institute Press, 1992

Varian, R. H., and S. F. Varian, "A High Frequency Oscillator and Amplifier", *J. Appl. Phys.*, vol. 10, p. 321, 1939

Walls, H. J.; "The Civil Airways and their Radio Facilities," *Proc. of the IRE*, vol. 17, p. 2141, 1929

Warner, John W.; "A Former SecNav Reminisces," in "Captain Eddy: the man who launched a thousand EEs," Ronald K. Jurgen, *IEEE Spectrum*, Nov. 1975

Watson-Watt, Sir Robert A.; *The Pulse of Radar: the Autobiography of Robert Watson-Watt*, Dial Press, 1959

Watson-Watt, Sir. Robert A.; *Three Steps to Victory*, Odhams Press, 1957

Weihe, Vernon I.; "Fifty Years in Aeronautical Navigational Electronics," *Proc. of the IRE*, vol. 50, p. 658, 1962

White, W. C.; "Early History of Industrial Electronics," *Proc. of the IRE*, vol. 50, p. 1129, 1962

Whiteside, George W.; "Naval Training School – EE&RM," *Oklahoma A&M College Magazine*, July 1943

Whitman, Howard; "Radar Man," *This Week* Magazine, August 25, 1945

Wilkins, Arnold; Colin Latham and Anne Wilkins (editors); *The Birth of British Radar*, Speedwell, 2006

"William Eddy, devised Navy's Eddy Test," Obituary; *The Institute*, IEEE, November 1989

Willoughy, Malcom Francis; *The Story of LORAN in the U.S. Coast Guard in World War II*, Arno Pro, 1980

"Wright College Launches Naval Radio Course, *Chicago Daily Tribune*, July 11, 1943

Yagi, H.; "Beam Transmission of Ultra-Short Waves", *Proc. of the IRE* vol. 16, p. 715, 1928

Zaloga, Steven J.; "Soviet Air Defense Radar in the Second World War," *J. of Soviet Military Studies*, vol. 2, p. 104, 1989 (in Russian)

Zázek, Napsal August; "A New Method for Generation of Undamped Oscillations," *Casopis pro Pest. Matem. a Fys.*, vol. 53, p. 378, 1924 (in Czech)

Zimmerman, David; *Top Secret Exchange: the Tizard Mission and the Scientific War*, McGill-Queen's University Press, 1996

Zworykin, Valdimir K.; "Television with cathode ray tube for receivers," *Radio Engineering*, Dec. 1929

Zworykin, Valdimir K.; "Description of an Experimental Television System and the Kinescope," *Proc. of the IRE*, vol. 21, p. 1655, 1933

Index